T0354449

The Carabidae (Coleoptera) of Fennoscandia and Denmark

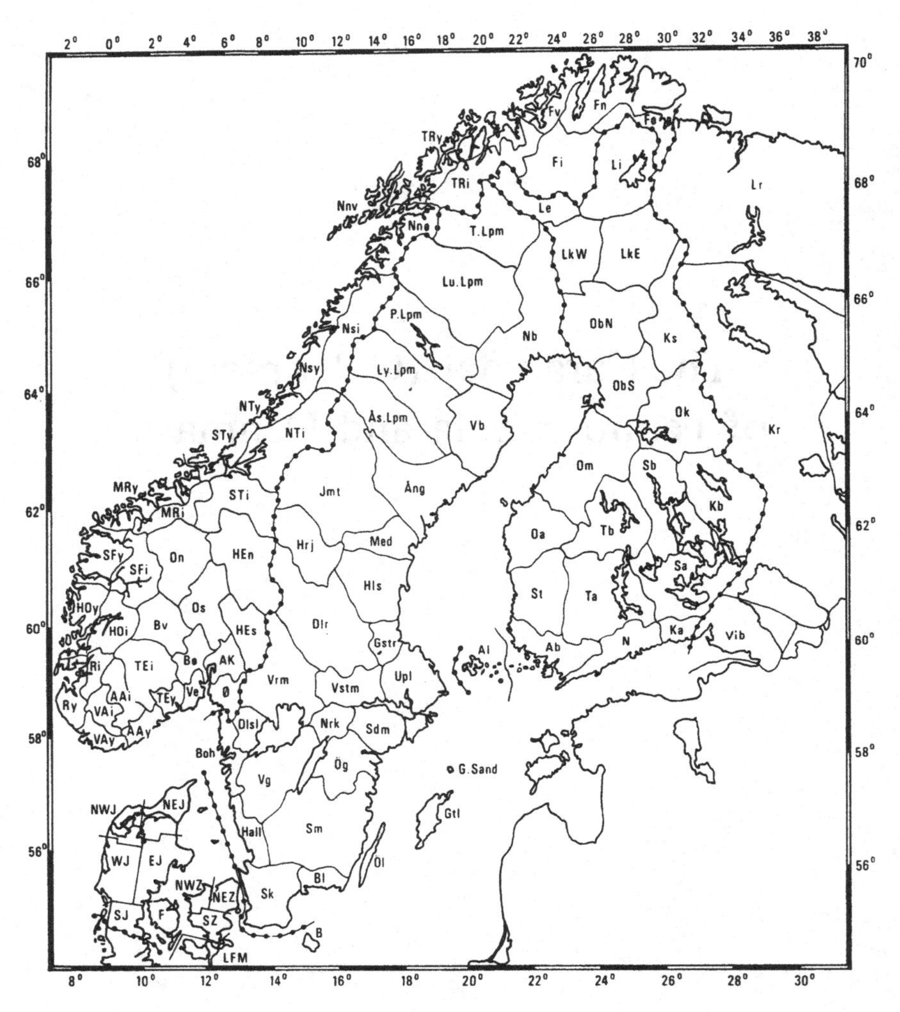

List of abbreviations for the provinces used throughout the text, on the map and in the following tables.

DENMARK

SJ	South Jutland	LFM	Lolland, Falster, Møn
EJ	East Jutland	SZ	South Zealand
WJ	West Jutland	NWZ	North West Zealand
NWJ	North West Jutland	NEZ	North East Zealand
NEJ	North East Jutland	B	Bornholm
F	Funen		

SWEDEN

Sk.	Skåne	Vrm.	Värmland
Bl.	Blekinge	Dlr.	Dalarna
Hall.	Halland	Gstr.	Gästrikland
Sm.	Småland	Hls.	Hälsingland
Öl.	Öland	Med.	Medelpad
Gtl.	Gotland	Hrj.	Härjedalen
G. Sand.	Gotska Sandön	Jmt.	Jämtland
Ög.	Östergötland	Ång.	Ångermanland
Vg.	Västergötland	Vb.	Västerbotten
Boh.	Bohuslän	Nb.	Norrbotten
Dlsl.	Dalsland	Ås. Lpm.	Åsele Lappmark
Nrk.	Närke	Ly. Lpm.	Lycksele Lappmark
Sdm.	Södermanland	P. Lpm.	Pite Lappmark
Upl.	Uppland	Lu. Lpm.	Lule Lappmark
Vstm.	Västmanland	T. Lpm.	Torne Lappmark

NORWAY

Ø	Østfold	HO	Hordaland
AK	Akershus	SF	Sogn og Fjordane
HE	Hedmark	MR	Møre og Romsdal
O	Opland	ST	Sør-Trøndelag
B	Buskerud	NT	Nord-Trøndelag
VE	Vestfold	Ns	southern Nordland
TE	Telemark	Nn	northern Nordland
AA	Aust-Agder	TR	Troms
VA	Vest-Agder	F	Finnmark
R	Rogaland		

n northern s southern ø eastern v western y outer i inner

FINLAND

Al	Alandia	Kb	Karelia borealis
Ab	Regio aboensis	Om	Ostrobottnia media
N	Nylandia	Ok	Ostrobottnia kajanensis
Ka	Karelia australis	ObS	Ostrobottnia borealis, S part
St	Satakunta	ObN	Ostrobottnia borealis, N part
Ta	Tavastia australis	Ks	Kuusamo
Sa	Savonia australis	LkW	Lapponia kemensis, W part
Oa	Ostrobottnia australis	LkE	Lapponia kemensis, E part
Tb	Tavastia borealis	Li	Lapponia inarensis
Sb	Savonia borealis	Le	Lapponia enontekiensis

USSR

Vib Regio Viburgensis Kr Karelia rossica Lr Lapponia rossica

FAUNA ENTOMOLOGICA SCANDINAVICA
Volume 15, part 2 1986

The Carabidae (Coleoptera) of Fennoscandia and Denmark

by

Carl H. Lindroth (†)

with the assistance of

F. Bangsholt, R. Baranowski, Terry L. Erwin, P. Jørum, B.-O. Landin, D. Refseth and H. Silfverberg

Also including an appendix on the family Rhysodidae

E. J. Brill/Scandinavian Science Press Ltd.
Leiden · Copenhagen

Fauna entomologica scandinavica
is edited by “Societas entomologica scandinavica”

World list abbreviation
Fauna ent. scand.

Text composed and printed by
Vinderup Bogtrykkeri A/S
7830 Vinderup, Denmark

ISBN 90 04 08182.8
ISBN 87-87491-33-8
ISSN 0106-8377

Contents

Tribe Pterostichini

This is one of the largest Carabid groups and cosmopolitan in distribution. Many species are large or at least middle-sized. It is a rather heterogeneous assemblage, divided into several subtribes, of which *Agonum* and related genera often are regarded as forming a tribe of their own, the *Agonini*. The head has 2 supra-orbital setae, and the pronotum one seta at the hind-angle (in our species). The dorsal punctures of elytra vary in number from one to several; they are lacking in *Stomis*. Male almost constantly with 3 dilated pro-tarsal segments; parameres much variable in form.

Genus *Stomis* Clairville, 1806

Stomis Clairville, 1806, Ent. Helv. 2: 46.
Type-species: *Carabus pumicatus* Panzer, 1796.

Our single species is similar to *Pterostichus minor* or a *Patrobus* but it is at once separated by the straightly protruding mandibles and the long first antennal segment (Fig. 247). Elytra without abbreviated scutellar stria as well as dorsal punctures. Wings rudimentary.

171. ***Stomis pumicatus*** (Panzer, 1796)
Fig. 247; pl. 6: 2.

Carabus pumicatus Panzer, 1796, Fauna Ins. Germ. 30: 16.

6.8-8.3 mm. Piceous to brown, antennae, palpi and legs dark rufous. Pronotum cordate. Elytra with deep, punctate striae.

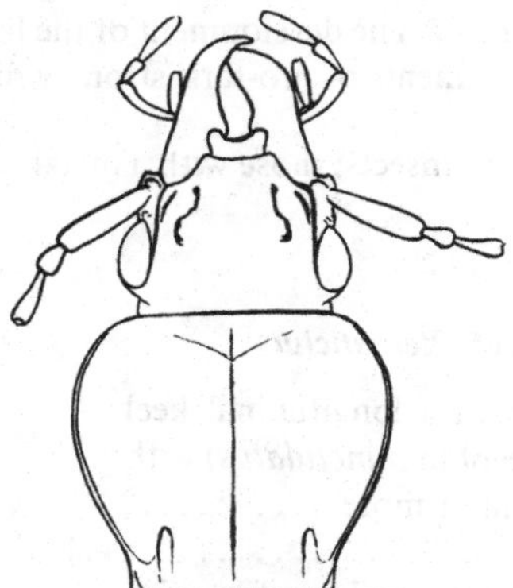

Fig. 247. Head and pronotum of *Stomis pumicatus* (Pz.).

Distribution. Denmark: rather common in EJ and in the six eastern districts. Only few finds in SJ, WJ, NWJ and NEJ; not found north of the Limfjord. - Sweden: from Sk. to Dlr. (not recorded from G. Sand., Dlsl. and Vrm.). Rather common and widespread in the south of the area, rare and local in the north of the area. - In Norway restricted to AK. - Finland: only known from southern Finland, but locally abundant near Helsinki, earliest find is from 1945. - Entire Europe north to 60°N, south to Portugal, S.Italy and Asia Minor, east to the Urals.

Biology. Notably on clayey and mull-rich, moderately humid soil in meadows and fields; also in gardens, parks and open deciduous forest. Regularly found in the nests of moles and rodents. It is clearly favoured by human influence. *S. pumicatus* is a nocturnal species and occurs mainly in spring and early summer when the eggs are laid; appears singly in the autumn.

Genus *Pterostichus* Bonelli, 1810

Pterostichus Bonelli, 1810, Obs. Ent. 1 (Tab. Syn.).
Type-species: *Carabus fasciatopunctatus* Creutzer, 1799.
Platysma Bonelli, 1810, Obs. Ent. 1 (Tab. Syn.).
Type-species: *Carabus niger* Schaller, 1783.
Feronia Latreille, 1817, *in* Cuvier, Reg. Anim. 3: 191.
Type-species: *Carabus cupreus* Linnaeus, 1758.
Other names are listed as subgenera below.

As here treated, this is a very large genus; by some authors the species are placed in several genera. The species vary in size from 5 to 21 mm, though are usually above average. They are normally of a rather uniform appearance: pronotum stout, strongly sclerotized and not much narrower than elytra; antennae thick; legs rather long and with heavy tibiae (notably the anterior pair). Claws simple. Mandibles long and sharp (Fig. 248). Pronotum with a single or double latero-basal fovea. Elytral epipleura "crossed" (as in Fig. 36), except in *burmeisteri*. The development of the hind wings is much varying in this genus. Male with 3 segments of pro-tarsi strongly dilated. Also last sternite sometimes modified.

Most species are carnivorous, night-active insects; those with a metallic coloration are diurnal.

Key to species of *Pterostichus*

1 Three basal antennal segments with a longitudinal keel above (Fig. 249). Entire body (except in *punctulatus*) with a brilliant metallic reflection. 9 mm or more 2
– All antennal segments cylindrical 6
2(1) Uniformly dull black. Elytral striae exceedingly fine with flat intervals .... 172. *punctulatus* (Schaller)

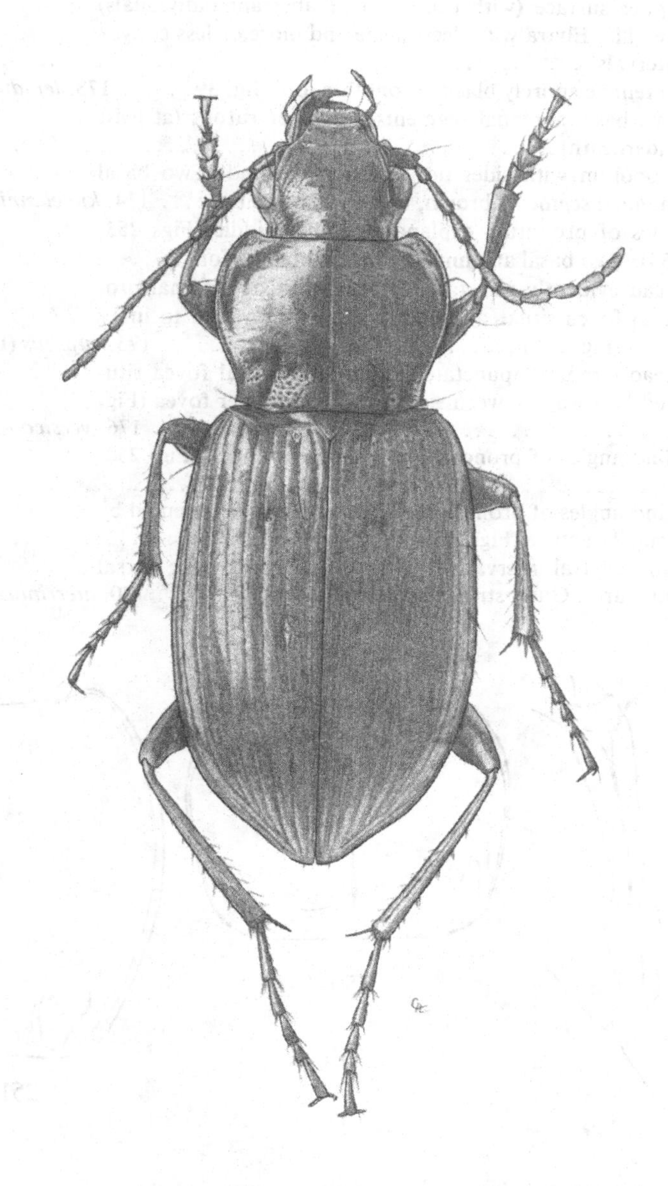

Fig. 248. *Pterostichus quadrifoveolatus* Letzn., length 9.5-11 mm.

–	Upper surface (with exception of aberrant individuals) metallic. Elytra with deep striae and more or less convex intervals	3
3(2)	Antennae entirely black. Pronotum as in fig. 39	173. *lepidus* (Leske)
–	Two basal antennal segments brown or rufous (at least underneath)	4
4(3)	Pronotum with sides not explanate behind. Two basal antennal segments brown, usually darker above	174. *kugelanni* (Panzer)
–	Sides of pronotum explanate behind middle (Figs 253, 254). Two basal antennal segments bright rufous	5
5(4)	Head evidently punctate. Deepest part of external pronotal fovea situated closer to side-margin than to inner fovea (Fig. 253)	175. *cupreus* (Linnaeus)
–	Head almost impunctate. External pronotal fovea situated half-way between side-margin and inner fovea (Fig. 254)	176. *versicolor* (Sturm)
6(1)	Hind-angles of pronotum completely rounded (Figs 250, 251)	7
–	Hind-angles of pronotum evident, at least represented by a small denticle (Figs 255-258, 260-266)	9
7(6)	Third elytral interval with 3 or 4 strongly foveate dorsal punctures. Outer striae obsolete anteriorly	180. *aterrimus* (Herbst)

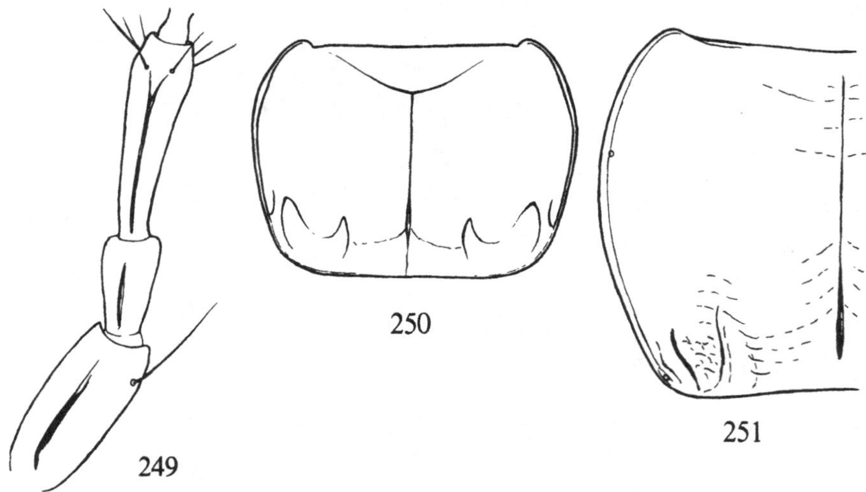

Fig. 249. Antennal base of *Pterostichus versicolor* (Sturm).
Figs 250, 251. Pronotum of *Pterostichus*. - 250: *aterrimus* (Hbst.) and 251: *madidus* (F.).

– Elytra with 1 to 3 fine, not foveate, dorsal punctures. All striae well impressed 8

8(7) Third elytral interval with 3 dorsal punctures. Basal fovea of pronotum obscurely delimited externally .... 181. *aethiops* (Panzer)

– Third elytral interval normally with a single dorsal puncture (rarely 2, exceptionally 3). Basal fovea of pronotum delimited externally by a blunt carina (Fig. 251) .. 182. *madidus* (Fabricius)

9(6) Dorsal punctures of third elytral interval foveate, 4 or more in number (except 3 in some individuals of *quadrifoveolatus*) 10

– Dorsal punctures not foveate, 1-3 in number (except 4-5 in subgenus *Cryobius*, with short met-episterna, Fig. 268) 12

10(9) Base of pronotum oblique laterally (Fig. 248). Dorsal punctures of elytra 3 or 4. First antennal segment much shorter than third .... 185. *quadrifoveolatus* Letzner

– Base of pronotum almost straight (Figs 261, 262). Dorsal punctures almost constantly 5 or more. First antennal segment barely shorter than third 11

11(10) Sides of pronotum clearly sinuate in front of the prominent hind-angles (Fig. 261). Tibiae pale . 183. *oblongopunctatus* (Fabricius)

– Sides of pronotum not or faintly sinuate (Fig. 262). Tibiae black or piceous .... 184. *adstrictus* Eschscholtz

12(9) Elytral epipleura not crossed. Upper side strongly metallic .... 197. *burmeisteri* Heer

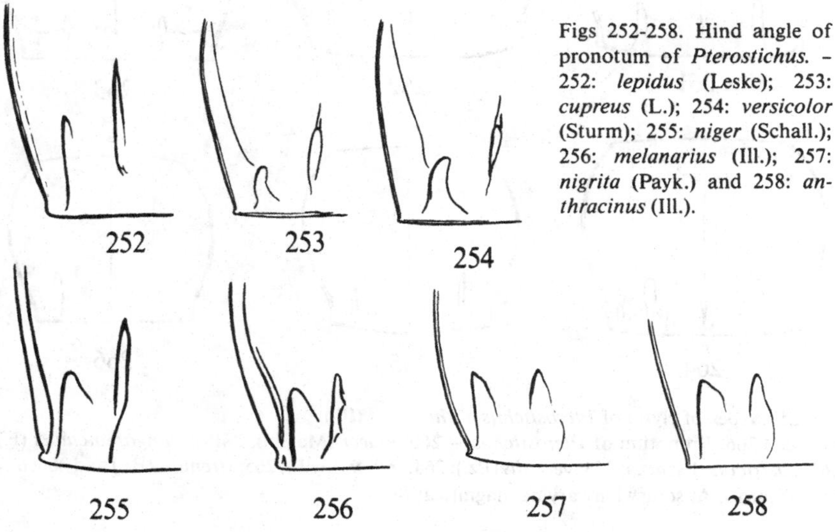

Figs 252-258. Hind angle of pronotum of *Pterostichus*. – 252: *lepidus* (Leske); 253: *cupreus* (L.); 254: *versicolor* (Sturm); 255: *niger* (Schall.); 256: *melanarius* (Ill.); 257: *nigrita* (Payk.) and 258: *anthracinus* (Ill.).

- Elytral epipleura crossed (as in Fig. 36). Body black or piceous, sometimes with a blue hue 13

13(12) Elytra with a single dorsal puncture. 4th-6th abdominal segments transversely impressed at base ..... 177. *longicollis* (Duftschmid)

- Elytra with at least 2 dorsal punctures. Abdominal sternites not impressed 14

14(13) Tarsal segments longitudinally furrowed above. Elytra without abbreviated scutellar stria .... 178. *vernalis* (Panzer)

- Tarsi n ot furrowed on dorsum. Abbreviate scutellar stria

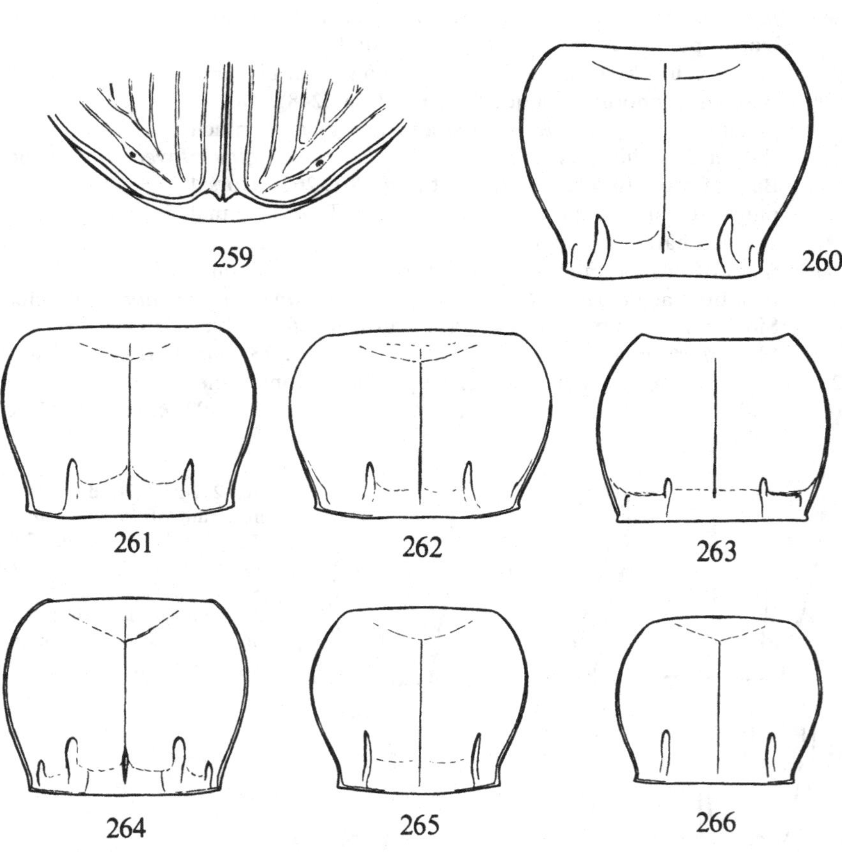

Fig. 259. Apex of elytra of *Pterostichus anthracinus* (Ill.).
Figs 260-266. Pronotum of *Pterostichus.* - 260: *macer* (Marsh.); 261: *oblongopunctatus* (F.); 262: *adstrictus* (Eschtz.); 263: *vernalis* (Pz.); 264: *minor* (Gyll.); 265: *strenuus* (Pz.) and 266: *diligens* (Sturm). At somewhat varying magnification.

present (rarely rudimentary or even virtually absent in some individuals of *brevicornis*)........................ 15

15(14) Pronotum (Fig. 260) strongly constricted at base. The posterior of the 3 dorsal punctures very fine, situated close to apex .... 179. *macer* (Marsham)

– Pronotum less constricted. Posterior dorsal puncture well removed from apex 16

16(15) Pronotum on each side with 2 latero-basal foveae, the outer seperated from lateral bead by a keel (Figs 255-258)............ 17

– Basal fovea of pronotum simple (Figs 269, 270). If a rudiment of an external fovea it is present, then it is not delimited by a convexity. Less than 7.5 mm 24

17(16) Last tarsal segment setose underneath. Elytra with 2 dorsal punctures... 18

– Last tarsal segment glabrous underneath. Elytra with 3 or 4 dorsal punctures.. 19

18(17) Lateral bead of pronotum strongly widened near base (Fig. 256) .... 187. *melanarius* (Illiger)

– Lateral bead of pronotum thin, not widening at base 196. *melas* (Creutzer)

19(17) Large species, at least 15 mm. Inner basal fovea of pronotum prolonged forwards (Fig. 255).............. 186. *niger* (Schaller)

– Under 13 mm. Inner pronotal fovea not or slightly prolonged (Figs 257, 258, 269).................................. 20

20(19) Hind-angles of pronotum denticulate (Fig. 257); sides in front of hind-angles evenly rounded .... 188. *nigrita* (Paykull)

– Pronotum with sides straight or sinuate posteriorly and hind-angles not denticulate (Figs 258, 264, 269)..................... 21

21(20) Pronotum with long sinuation in basal part of sides; hind-angles very acute, prominent (Fig. 269) ... 195. *middendorffi* (J. Sahlberg)

– Sides of pronotum straight or with shallow sinuation; hind-angles right-angled or faintly prominent (Figs 258, 264) 22

22(21) Abdominal sternites with dense, fine, more or less conflu-

Figs 267, 268. Met-episternum with epimeron (epm) of *Pterostichus*. – 267: *strenuus* (Pz.) and 268: *brevicornis* Kirby.

ent punctuation. Last sternite of ♂ with a longitudinal fovea. Elytra of ♀ with a sutural tooth (Fig. 259) 189. *anthracinus* (Illiger)

– Abdominal sternites not or obsoletely punctate. Last sternite of ♂ keeled or unarmed. Elytra of ♀ without sutural tooth 23

23(22) Inner pronotal fovea hardly prolonged. Elytra clearly iridescent, microsculpture very dense, transverse. Last sternite of ♂ smooth .... 190. *gracilis* (Dejean)

– Inner pronotal fovea prolonged forwards (Fig. 264). Elytra at most faintly iridescent, microsculpture weak, more irregular. Last sternite of ♂ with a longitudinal keel .... 191. *minor* (Gyllenhal)

24(16) Wings strongly rudimentary, not reaching first visible abdominal tergite. Elytra firmly coalescent. Met-episternum (Fig. 268) very short. On the tundra .... 194. *brevicornis* Kirby

– Wings full or, if reduced, the rudiment reaches at least beyond first visible tergite. Met-episterna longer (Fig. 267), also in brachypterous specimens. Only below timber limit 25

25(24) Pronotum with sides clearly sinuate before hind-angles (Fig. 265); its disc shiny, without microsculpture. Legs entirely brown .... 192. *strenuus* (Panzer)

– Pronotum with an indistinct posterior sinuation (Fig. 266); with reticulate microsculpture also on disc (more faintly in the ♂). Legs darker, at least the femora ... 193. *diligens* (Sturm)

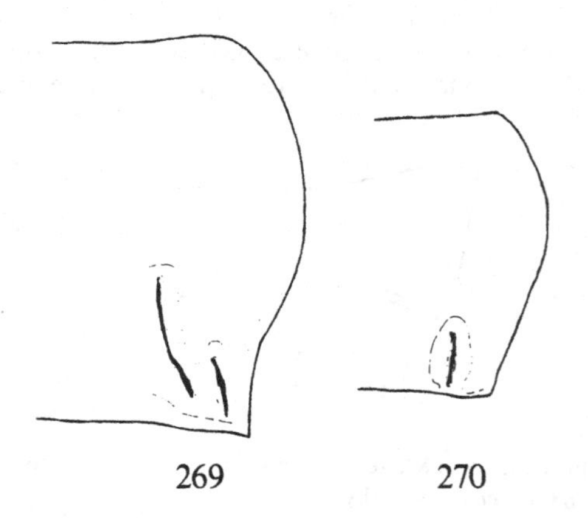

Figs 269, 270. Pronotum of *Pterostichus*. – 269: *middendorffi* (J. Sahlbg.) and 270: *brevicornis* Kirby.

Subgenus *Poecilus* Bonelli, 1810

Poecilus Bonelli, 1810, Obs.Ent. 1 (Tab. Syn.).
Type-species: *Carabus cupreus* Linnaeus, 1758.

172. ***Pterostichus punctulatus*** (Schaller, 1783)

Carabus punctulatus Schaller, 1783, Abh. Naturf. Ges. Halle 1: 318.

11-14 mm. Unmetallic coal black in colour. Upper surface dull due to very strong microsculpture. Basal foveae of pronotum obliterate. Elytral striae extremely fine, almost transformed into rows of punctures; intervals entirely flat.

Distribution. Denmark: very rare and in strongly decreasing numbers. Since 1900 only the following finds, usually of single specimens: SJ (Bevtoft, Gram), WJ (Fanø, Esbjerg, Estrup skov), EJ (Boes, Løvenholt, Salten, Gødvad, Mols bj., Pinds mølle), NWJ (Mønsted), F (Ristinge), NWZ (Svebølle), NEZ (Rude skov, Jægerspris), B (Sose, Arnager). - Sweden: very rare, earlier in Sk., Hall., Öl. and Gtl., but during the last 25 years only found in Sk. (Harlösa and Vomb 1960, Sandhammaren 1972 on the seashore, Revinge 1978 in numbers but later not rediscovered). - In Europe north to 57°N, but not in the west and south of the continent, south to N.Spain and N.Italy, east to W.Siberia.

Biology. A xerophilous species, occurring on sun-exposed, dry ground, notably in sandy grass areas and on heather-ground; in C.Europe also on clayey, sometimes cultivated soil. The species is a pronounced spring breeder, and is only occasionally found after the end of May.

173. ***Pterostichus lepidus*** (Leske, 1785)
Fig. 252; pl. 3: 6.

Carabus lepidus Leske, 1785, Reise Sachsen 1: 17.

11-15.2 mm. Largest species of the subgenus. Extremely varying in colour, normally strongly metallic: coppery, green, sometimes blue, violaceous, rarely almost unmetallic black. Appendages pure black, except that the spines and spurs of tibiae and tarsi are dark reddish brown. Female with elytra more strongly microsculptured by transversely arranged meshes, and therefore dull. Pronotum (Fig. 252) with narrow, parallel, deep basal foveae, the outer delimited externally by a strong convexity. Elytral striae almost impunctate. Wings either full or, usually, strongly reduced.

Distribution. Denmark: rather distributed but uncommon; apparently absent from the areas along the northwestern coast of NWJ and NEJ. Only two finds in LFM. - Sweden: distributed throughout the country from Sk. to Lu.Lpm., rather local and uncommon, and somewhat decreasing during the last 20 years. - Norway: fairly common and distributed north to NT. - Finland: rather common north to Lk; also in the adjacent districts of the USSR. - Entire Europe with the exception of the northern and extreme southern parts; also in Siberia east to the Amur region.

Biology. A xerophilous species, occurring on open, sun-exposed ground among sparse vegetation, notably on sandy or gravelly soil in *Calluna*-heath and grassland. Also in cleared areas of forests. It is predominantly carnivorous, the prey being ants, lepidopterous larvae and other arthropods. According to Schjøtz-Christensen (1965) the propagation period extends from May to August. The majority of the population reproduces in July-August, giving rise to winter larvae. A rather great proportion of the imagines may hibernate and enter upon a second breeding period in the following spring. Their offspring develops during the summer.

174. ***Pterostichus kugelanni*** (Panzer, 1797)

Carabus kugelanni Panzer, 1797, Fauna Ins. Germ. 39: 8.
Carabus dimidiatus Olivier, 1795, Ent. 3: 72; *nec* Rossi, 1790.

12-14 mm. Similar to *lepidus* in the pronotum but with paler antennal base and upper surface normally bicoloured: coppery forebody and green elytra, very rarely entirely black. Antennae with the 2 brown basal segments normally darker above. Pronotal foveae as in *lepidus,* but shallower. Elytral striae evidently punctate, intervals flattened also apically. Wings full.

Distribution. Denmark: very rare, only 6 specimens from NEZ (Lundtofte) about 1820). - Not in Sweden, Norway or East Fennoscandia.- A West and Central European species.

Biology. On sun-exposed, dry grassland and heath-ground, occurring on sandy or gravelly soil with sparse vegetation.

175. ***Pterostichus cupreus*** (Linnaeus, 1758)
Fig. 253.

Carabus cupreus Linnaeus, 1758, Syst. Nat. ed. 10: 416.
Feronia puncticeps Thomson, 1867, Skand. Col. 9: 35.

11-13.4 mm. Shorter than *lepidus,* with much broader elytra. Black, underside with faint, upper surface with more pronounced though somewhat dull, metallic colour varying from bluish green to coppery; quite black individuals are rare. The 2 basal antennal segments clear yellowish red, rarely also femora rufous (ab. *"affinis"* Sturm). Head evidently punctate. Pronotum not wider than elytra over shoulders; deepest part of external fovea situated closer to side-margin than to inner fovea (Fig. 253). Inside of meta-tibia with a row of 8-10 rather long and thin spines. Wings full and probably functionary.

Distribution. Denmark: rather distributed in eastern Jutland north to the Limfjord, and also in LFM, NEZ and B. Only a few finds in WJ (Esbjerg, Nissum), NWJ (Harboør, Ø.Assels), NEJ, F, SZ and NWZ. - Sweden: in all districts from Sk. to Gstr. and Dlr. (except G. Sand.). Rather common but local. - Norway: mainly restricted to the southern coastal areas from Ø to HOy. - Finland: common in the southern districts, found to Ob in the north; also Vib and Kr in the USSR. - Entire Europe, south to S.Spain, S.Italy and Greece; Asia Minor; Siberia east to River Lena.

Biology. In open, not too dry meadows and fields, preferably on clayey soil with rather dense vegetation of grasses and sedges. The species often occurs on agricultural land, mostly among cereal crops. *P. cupreus* is a diurnal species, feeding on both animal and vegetable matters. It is most numerous in May-June - the time of propagation -, and in the autumn when the new generation of beetles appears. Krehan (1970) found that very few individuals reproduced more than once.

176. ***Pterostichus versicolor*** (Sturm, 1824)
Figs 249, 254.

Platysma versicolor Sturm, 1824, Deutschl. Fauna Ins. 5 (5): 99.
Pterostichus coerulescens auctt.; *nec* (Linnaeus, 1758).
Feronia pauciseta Thomson, 1867, Skand. Col. 9: 36.

9-12.2 mm. Shorter than *cupreus,* with pronotum wider than elytra over shoulders. More shiny and more variable in the metallic lustre, from bluish to golden, often mottled. Two (sometimes 3) basal antennal segments rufous. Head almost impunctate. External fovea of pronotum situated halfway between inner fovea and side-margin, inside of which the surface is more broadly flattened (Fig. 254). Inside of meta-tibia with a row of 5-7, exceptionally 8, spines; these shorter and stouter than in *cupreus.* Wings full but, due to the short apex, probably not functioning.

Distribution. In Denmark generally distributed and rather common, known from all districts. - Sweden: found in all districts except P.Lpm. Common in the south, less so in the north, and no records from the high altitude areas in the west. - Norway: north to NT. - East Fennoscandia: common in southern and central Finland, north to Lk; also Vib and Kr of the USSR. - Entire Europe except the extreme north and south, through Siberia east to the Pacific Ocean.

Biology. On different kinds of open ground, e.g. meadows, fields and heaths; often also on cultivated soil. It is more xerophilous than *P. cupreus,* preferring rather dry, sandy soil with scattered vegetation. Occasionally found together with *cupreus* on clay-mixed sandy soil. It is a diurnal species, and is often seen running about in the sunshine. The reproduction takes place in spring, and newly emerged adults occur in the autumn. Only a few beetles seem to enter upon a second breeding period (Krehan, 1970).

Subgenus *Pedius* Motschulsky, 1850

Pedius Motschulsky, 1850, Käfer Russl., p. 50, pl. IX.
Type-species: *Carabus inaequalis* Marsham, 1802.

177. ***Pterostichus longicollis*** (Duftschmid, 1812)

Carabus longicollis Duftschmid, 1812, Fauna Austriae 2: 180. Primary homonym of *Carabus longicollis* Lichtenstein, 1796.
Carabus inaequalis Marsham, 1802, Ent. Brit.: 456; *nec* Panzer, 1796.

5-6 mm. Separated from all other *Pterostichus* by the presence of only a single dorsal puncture behind middle of elytra in third interval. Brown or piceous, appendages pale. Similar to *vernalis* and like this without, or with a very rudimentary, scutellar stria on elytra; striae strongly punctate. Pronotum with a single lateral fovea. Lower surface with a coarse punctuation laterally; sternites 4-6 each with a deep transverse impression.

Distribution. In our area only found in Denmark: EJ, Hyby strand north of Fredericia, 1 specimen 1942; NWZ, Kongstrup klint (Røsnæs) a few specimens 1977-80, Overbjerg (Røsnæs) in number 1979. - W. and C.Europe, south-west to the Pyrenees, south-east to the Carpathians.

Biology. The habitats on Røsnæs (NWZ, Denmark) are moist to very wet clayish soil. In Britain and in C. Europe the species also favours clayey, often chalky soil, occurring for instance on limestone.

Subgenus *Argutor* Dejean, 1821

Argutor Dejean, 1821, Catal. Coleopt., p. 11.
Type-species: *Carabus vernalis* Panzer, 1796.
Lagarus Chaudoir, 1838, Bull. Soc. Nat. Moscou 11: 10.
Type-species: *Carabus vernalis* Panzer, 1796.

178. ***Pterostichus vernalis*** (Panzer, 1796)
Fig. 263.

Carabus vernalis Panzer, 1796, Fauna Ins. Germ. 30: 17. Primary homonym of *Carabus vernalis* Müller, 1776.
Carabus crenatus Duftschmid, 1812, Fauna Austriae 2: 92.

6-7.5 mm. A small species with the broad, flat pronotum denticulate at hind-angle; side-margin non-sinuate (Fig. 263). Piceous to black, elytra faintly iridescent; at least first antennal segment, tibiae and tarsi reddish. Pronotum with outer basal fovea obsolete or evanescent; base strongly punctate laterally. Elytra with short apex, without scutellar stria, the outer striae deep and strongly punctate. Tarsal segments longitudinally furrowed above. Wings full or moderately reduced.

Distribution. Denmark: rather distributed but not common, apparently absent from district NWJ and most of district WJ. - Sweden: distributed in the southern half of the country (Sk.-Med.) and usually not rare. Also found in the coastal areas of Vb. and Nb. - Norway: fairly common in the south-eastern districts and along the coast to MR. - Finland: fairly common in the southern and central parts, north to Ob. In the adjacent parts of the USSR in Vib and southern Kr. - Most of Europe except the north and extreme south; N.Africa; Siberia east to River Lena.

Biology. A very hygrophilous species, preferring eutrophic fens and moist meadows with grasses and sedges, usually near water. Occasionally found in water-meadow forest and more rarely in oligotrophic bogs. It is a typical spring breeder.

Subgenus *Adelosia* Stephens, 1835

Adelosia Stephens, 1835, Ill. Brit. Ent. Mand. 5: 378.
Type-species: *Carabus macer* Marsham, 1802.

179. ***Pterostichus macer*** (Marsham, 1802)
Fig. 260; pl. 6: 6.

Carabus macer Marsham, 1802, Ent. Brit.: 466.
Carabus picimanus Duftschmid, 1812, Fauna Austriae 2: 159.

11-15 mm. A large species with an extremely characteristic basal contriction of the pronotum (Fig. 260). Very flat, with long parallel-sided elytra. Piceous to brown, legs rufous. Outer basal fovea of pronotum small or obsolete. Elytral striae smooth with almost flat intervals; the posterior of the 3 dorsal punctures very fine, situated close to apex.

Distribution. In our area only in Denmark, and here very rare. SJ: Ydre Bjerrum near Ribe 1 specimen 1857, Ribe marsk 1 specimen 1940, Ballum sluse 1 specimen 1961; WJ: Esbjerg a few specimens 1901-19 under sea-weed; F: Fårevejle skov on Langeland 1 specimen about 1840; LFM: Handermelle 1 specimen 1898, Strognæs 1 specimen 1904 and 1920; NWZ: Kongstrup klint on Røsnæs repeatedly captured since 1964. - In Europe from the British Isles east to the Baltic States and south to Italy and Greece.

Biology. In open country, in Denmark exclusively on clay-soil. At Røsnæs (NWZ, Denmark) the species occurs in gently sloping country with a meadow-like vegetation. It has further been encountered on clayey salt-marsh ground. In Britain and C.Europe *P. macer* occurs on rather moist soil rich in humus, often in arable fields and parks. It is a nocturnal species; during daytime often subterraneous or under bark and stones. Most specimens from Denmark have been found in spring.

Subgenus *Omaseus* Dejean, 1821

Omaseus Dejean, 1821, Catal. Coleopt., p. 12.
Type-species: *Carabus aterrimus* Herbst, 1784.
Lyperosomus Motschulsky, 1850, Käfer Russl., p. 47, pl. IX.
Type-species: *Carabus aterrimus* Herbst, 1784.

180. ***Pterostichus aterrimus*** (Herbst, 1784)
Fig. 250; pl. 3: 7.

Carabus aterrimus Herbst, 1784, Arch. Insectengesch. 5: 140.

13-15 mm. Coal black, very shiny, easily recognized on the rounded pronotum (Fig. 250) and the foveate elytra. Pronotum with a deep anterior transverse impression and a single basal fovea. Outer striae of elytra obsolete anteriorly, third interval with 3 or 4 strongly foveate dorsal punctures.

Distribution. Denmark: only very few captures in Jutland: EJ (Sattrup mose near Horsens, Marselisborg) and NEJ (Læsø). No records from F, but more finds in the eastern districts including B. Breeding populations are probably only present in periods and is dependent on repeated immigration from the south. - Sweden: very rare and local in the south-east, north to Vstm. - Norway: only VAy and Ry. - East Fennoscandia: in Finland recorded to Kb in the north, but mostly unfrequent. Also records from the borders of the lakes Onega and Ladoga in the neighbouring part of the USSR. - In Europe north to 60-62°N and south to N.Spain and N.Italy; N.Africa; W.Siberia.

Biology. An extremely hygrophilous species, living at the border of standing waters, always near the water edge. Its habitats is usually oligotrophic bogs with soft, muddy or peaty soil and a luxuriant vegetation of *Carex, Eriophorum,* etc.; less often in eutrophic fens. The species sometimes occurs together with *Carabus clathratus* and *Chlaenius tristis.* It is a spring breeder, and is most numerous in May. Newly emerged beetles occur in the autumn.

Subgenus *Eosteropus* Tschitschérine, 1902

Eosteropus Tschitschérine, 1902, Hor. Soc. Ent. Ross. 35: 500.
Type-species: *Platysma creperum* Tschitschérine, 1902.

181. ***Pterostichus aethiops*** (Panzer, 1797)

Carabus Aethiops Panzer, 1797, Fauna Ins. Germ. 37: 22. Primary homonym of *Carabus aethiops* Herbst, 1784.

12-14 mm. Agreeing with the preceding and the following species through the broadly rounded hind-angles of pronotum. Black, appendages piceous. Basal fovea of pronotum obscurely delimited externally. Elytra with 3 dorsal punctures, intervals convex, striae impunctate. Basal segments of meta-tarsi with a deep external furrow. Penultimate abdominal sternite of ♂ with a transverse carina, the last one with an impression.

Distribution. In our area only found in the southern parts of Finland and adjacent parts of the USSR (Vib and Kr). - W., C. and E.Europe, south to SE.France, S.Austria and Serbia. In C. Europe mainly montane.

Biology. A woodland species living among moss and leaves or under bark of tree-stumps in deciduous as well as coniferous forests. In C.Europe predominantly in mountain forests.

Subgenus *Steropus* Dejean, 1821

Steropus Dejean, 1821, Catal.Coleopt., p. 13.
Type-species: *Carabus madidus* Fabricius, 1775.
Corax Putzeys, 1846, Mem.Soc.Sci. Liège 2: 406.
Type-species: *Corax ghilianii* Putzeys, 1846.

182. ***Pterostichus madidus*** (Fabricius, 1775)
Fig. 251.

Carabus madidus Fabricius, 1775, Syst.Ent.: 241; *nec* Paykull, 1790.

13-17 mm. Third elytral interval normally with a single (rarely 2, exceptionally 3) dorsal puncture. Black, legs either entirely dark or, usually, with rufous femora ("*concinnus*" Sturm). Basal fovea of pronotum delimited externally by a blunt carina (Fig. 251). Elytral intervals almost flat, striae punctulate. Tarsal furrows obsolete. Last abdominal sternite of ♂ with a transverse carina.

Distribution. Very common in W.Europe including the British Isles. In our area apparently only accidental. – Denmark: two records from EJ before 1900, both probably representing stray individuals. Finland: one record from N: Ekenäs, 15.7.1937, P. Brinck (?introduced).

Biology. A eurytopic species, in C.Europe mainly occurring in woodland; in Britain notably in open country, often on cultivated soil.

Subgenus *Bothriopterus* Chaudoir, 1838

Bothriopterus Chaudoir, 1838, Bull. Soc. Nat. Moscou 11: 9.
Type-species: *Carabus oblongopunctatus* Fabricius, 1787.

183. ***Pterostichus oblongopunctatus*** (Fabricius, 1787)
Fig. 261; pl. 6: 7.

Carabus oblongopunctatus Fabricius, 1787, Mant. Ins. 1: 202.

9.5-12.6 mm. The members of this subgenus are characterized by 3 or more foveate dorsal punctures in third elytral interval. This species has normally 5 (from 4 to 7) such strongly deepened punctures. Black to dark piceous, upper surface with a brassy lustre, at least in the ♂, rarely bluish or greenish. Tibiae and tarsi evidently pale. Pronotum (Fig. 261) with sides more sinuate posteriorly, lateral bead evident and reaching almost to front-angles. Elytra with rounded sides and greatest width behind middle; intervals somewhat convex. Female with elytra more dull due to stronger microsculpture.

Distribution. Denmark: generally distributed and common. - Sweden: found all over the country; no records from Hrj. and T.Lpm. Common in the southern half. - Norway: generally distributed except in the extreme northeast. - East Fennoscandia: distributed all over Finland and common except in the very north. Also Vib, Kr and Lr in the USSR. - A common Euro-Siberian species and generally distributed in its area.

Biology. A eurytopic woodland species, occurring in both deciduous and coniferous forests, usually in light stands on moderately dry, mainly sour humus soil. In the atlantic climate of W.Norway also in open country. It is predominantly nocturnal, but also shows some diurnal activity. The species propagates in spring; newly emerged individuals occur in the autumn .

184. ***Pterostichus adstrictus*** Eschscholtz, 1823
Fig. 262.

Pterostichus adstrictus Eschscholtz, 1823, Mem. Soc. Nat. Moscou 6: 103.
Feronia vitrea Dejean, 1828, Spec. Gén. Col. 3: 320.
Harpalus borealis Zetterstedt, 1828, Fauna Ins. Lapp. 1: 32.

10.4-13 mm. Closely allied to the preceding species, but separated in the following respects: elytra longer and narrower, legs somewhat stouter. Colour darker, upper surface pure black or with a quite faint brassy hue; palpi and tibiae, usually also tarsi, dark piceous to black. Pronotum (Fig. 262) broader, sides not or hardly sinuate posteriorly, lateral bead evident only in posterior half; sides more depressed laterally. Elytral striae fine, intervals flatter.

Distribution. Not in Denmark. - Sweden: common in the northern districts, south to Vstm.. Isolated occurrence in the highlands of Sm. and Vg. - Norway: fairly common in the north, scattered in the southern areas. - East Fennoscandia: common in the north, rare in the southern districts of Finland but recorded even from N and southern Kr. - An inhabitant of the northern coniferous region; not in C.Europe but common on Iceland, the Faroes and in N. Britain. Also in North America.

Biology. Predominantly in open grassland, living on moderately humid, notably gravelly soil mixed with clay and humus. In the north frequently on cultivated land; also in open forests and forest edges. It is mainly associated with the coniferous region, but occurs also in the subarctic region.

185. ***Pterostichus quadrifoveolatus*** Letzner, 1852
Fig. 248.

Pterostichus angustatus var. *quadrifoveolatus* Letzner, 1852: 209.
Carabus angustatus Duftschmid, 1812, Fauna Austriae 2: 162; *nec* Fabricius, 1787.

9.5-11 mm. Elytra with only 3 or 4, less foveate, dorsal punctures (none at apex of first striae). Base of pronotum oblique laterally. Elytral striae evidently punctate. First antennal segment shorter (as compared with third). Upper surface black or with an extremely faint bronzy hue. Appendages darker than in *oblongopunctatus.* The microsculpture forms more distinct transverse rows of meshes than in the two preceding species.

Distribution. Denmark: rare, but an increasing number of records after 1950. Now found scattered over the southern half of Jutland and on the islands; no finds in northern Jutland. - Sweden: very rare and scattered, in Sk., Bl., Hall., Sm., Ög., Vg., Sdm. and Upl. - Norway: only AAy. - Finland: rare in the south; also in Kr of the USSR. - W., C. and E.Europe; the Caucasus.

Biology. More xerophilous and heat-preferent, and also more resistent to dryness, than *P. oblongopunctatus,* occurring mainly in warm, dry forest-clearings, notably in open coniferous and mixed stands. In Scandinavia apparently always on recently burned areas, sometimes in company with *Agonum quadripunctatum.* In C.Europe also on localities without any evidence of earlier forest fires. Occasionally in cities (e.g. Göteborg), probably attracted to sites of fire. Paarmann (1966) demonstrated that the species is well adapted to this unstable habitat: it lays twice as many eggs as *oblongopunctatus* at a time, and more females of *quadrifoveolatus* reproduce in a second breeding period. Under favourable climatic conditions the species is therefore able to build up large populations within a short time. *P. quadrifoveolatus* has also a high power of dispersal. Shortly after emerging in autumn, the young beetles pass through a swarming period, after which they never fly again. It is a typical spring breeder.

Subgenus *Platysma* Bonelli, 1810

186. ***Pterostichus niger*** (Schaller, 1783)
Fig. 255; pl. 3: 8.

Carabus niger Schaller, 1783, Abh. Naturf. Ges. Halle 1: 315.

15-20.5 mm. Largest species of the genus. Dull black and rather flat. Inner basal fovea of pronotum (Fig. 255) prolonged forwards; lateral bead only slightly dilated posteriorly. Elytral striae very deep, intervals very convex, the 9th as wide as the 10th; 3 dorsal punctures. Meta-tarsi with 3 basal segments sharply keeled laterally; last tarsal segment glabrous underneath. Wings of full size but probably not functionary. Male with a sharp, angulate, longitudinal keel on last sternite.

Distribution. Denmark: very distributed and very common. - Sweden: very common and distributed in the southern half from Sk. to Gstr., Hls. and Med. In the northern half rare or completely lacking over large areas. A few finds in Nb. is apparently associated with the Finnish distribution-area. - Norway: north to TR but scattered in the central and eastern areas. - Finland: very common in the south, recorded as far north as in Ob. - Main part of Europe, Siberia, W.Asia.

Biology. A eurytopic species, especially characteristic of woodland, occurring in almost every type of forest community, predominantly in deciduous and mixed stands on humus-rich, rather moist soil. Also in hedges, parks and gardens; less frequent in meadows and on arable land. It is a night-active carnivorous species, the food of which consists of insect larvae as well as of dead animals. *P. niger* normally propagates in August-September, hibernates in the third instar, and young adults emerge in spring and summer. A number of old individuals, and young individuals emerging late, hibernate and reproduce in the following spring or summer (Witzke, 1976).

Subgenus *Morphnosoma* Lutshnik, 1915

Morphnosoma Lutshnik, 1915, Ent. Obozr. 14: 424.
Type-species: *Carabus vulgaris* Linnaeus, 1758 s. Lutshnik.
Euferonia Casey, 1918, Mem. Coleopt. 8: 322.
Type-species: *Feronia stygica* Say, 1823.
Omaseidus Jeannel, 1942, Faune de France 40: 784.
Type-species: *Carabus vulgaris* Linnaeus, 1758 s. Jeannel.
Omaseus auct. nec Dejean, 1821.

187. ***Pterostichus melanarius*** (Illiger, 1798)
Fig. 256.

Carabus melanarius Illiger, 1798, Verz. Käf. Preuss.: 163.
Pterostichus vulgaris auctt.; *nec* (Linnaeus, 1758).
Carabus leucophthalmus Rossi, 1790, Fauna Etrusca 1: 207; *nec* Linnaeus, 1758.

12-18 mm. Superficially similar to *niger*, but more convex and shiny. Entirely black. Pronotum (Fig. 256) with lateral bead much widened posteriorly; inner basal fovea reaching only slightly before the outer one. Elytra with 9th interval much wider than 10th; 2 dorsal punctures. Legs and antennae shorter. Meta-tarsi with a lateral keel, last segment setose underneath. Wings usually quite rudimentary, but single long-winged individuals, no doubt capable of flying, occur. Male with a suggested impression on last sternite.

Distribution. In Denmark very distributed and common. - Sweden: distributed in about the same way as the preceding species and very common in southern Sweden; north to Nb. - Norway: north to NTy. - Finland: common in the south, found north to Lk; also Vib and Kr in the USSR. - In most parts of Europe, south to N.Spain, S.Italy and Bulgaria; the Caucasus; Siberia east to the Amur region.

Biology. Very eurytopic, usually occurring on open and not too dry and sandy ground, e.g. meadows and grassland. It has clearly been favoured by human cultivation, being a common inhabitant of arable land, parks, gardens, etc. Also in forest edges and light woods, living in company with eurytopic forest-dwelling Carabidae, e.g. *P. niger*. The species is nocturnal and predominantly carnivorous, feeding on a broad spectrum of prey and probably playing an important role in the control of various insect pests. It may also feed on vegetable matter and has repeatedly been reported to damage strawberry fruits (Briggs, 1965). The propagation period of *P. melanarius* is chiefly August-September. Hibernation takes place in the third instar, and young beetles emerge in spring and summer. In addition, a great number of old adults, and also young beetles emerging late, hibernate and reproduce in the following spring and summer (Krehan, 1970).

Subgenus *Melanius* Bonelli, 1810

Melanius Bonelli, 1810, Obs. Ent. 1 (Tab. Syn.).
Type-species: *Carabus nigrita* Paykull, 1790.

188. ***Pterostichus nigrita*** (Paykull, 1790)
Fig. 257; pl. 6: 8.

Carabus nigrita Paykull, 1790, Mon. Car. Suec.: 129.

8.8-12.8 mm. A diminutive imitation of *melanarius* but with small denticulate hind-angles of pronotum (Fig. 257). Black, shiny; appendages sometimes in part piceous. Pronotum with a narrow even lateral bead. Microsculpture of elytra (strongest in the ♀) reticulate, with in part almost isodiametric meshes. Last abdominal sternite of ♂ with a small tubercle or keel; ♀ without sutural tooth. Wings full and normally functionary. Small stout specimens, especially occurring in the north, have been named "var. *rhaeticus* Heer" (see note below).

Distribution. Denmark: very distributed and common. - Sweden: common all over the country, especially in the south. Recorded from all districts. - Norway: north to TR. - Common all over Finland and the Soviet part of E. Fennoscandia, especially in the south. - Entire Europe, N.Asia, Morocco, W.Asia.

Biology. The adult beetles are very sensitive to low humidities (Thiele, 1967), and the species is accordingly confined to wet habitats, usually occurring in the shore zone of ponds, lakes and rivers, preferably on clayey, humus-rich soil with rich vegetation (e.g. of *Carex*). It is most abundant in eutrophic fens; less often in oligotrophic bogs; also in damp places far from open water, for instance in forest-swamps. *P. nigrita* is a night-active species propagating in spring. Ferenz (1975), comparing a *nigrita*-population from Swedish Lapland (Messaure) with one from C.Europe, discovered a number of physiological adaptations to subarctic temperature and light conditions: the larvae of the Scandinavian population developed faster and had a reduced mortality at low temperatures compared to larvae from C.Europe. The egg-development was induced at far longer photoperiods in beetles from Messaure than in C.European beetles.

Note. *Pterostichus nigrita* has been shown to consist of two sibling species (Koch & Thiele, 1980, Entomologia Generalis 6: 135-150), both of which occur in Fennoscandia. In the 1980 paper they were referred to as α and β, but later Koch (1984, Entomol. Blätt. 79: 141-154) has shown that the α-species is the true *P. nigrita*, and the β-species should be known as *P. rhaeticus* Heer, 1837. Their distribution is not yet investigated in detail, but in the northern parts only *P. rhaeticus* has been found; the northernmost provinces for *P. nigrita* are in Sweden Ång., in Finland Ta and Sa. (H. Silfverberg).

189. ***Pterostichus anthracinus*** (Illiger, 1798)
Figs 258, 259.

Carabus anthracinus Illiger, 1798, Verz. Käf. Preuss.: 181.

10.5-12.5 mm. Closely allied to *nigrita* but with sides of pronotum almost straight posteriorly (Fig. 258) and hind-angles not denticulate. Flatter and somewhat narrower. Elytral intervals less convex, their sides more sinuate subapically. Microsculpture of elytra denser and only in spots forming strongly marked transverse meshes. Abdominal sternites with a dense, fine, confluent punctuation. Wings dimorphic, either full or highly reduced. Last sternite of ♂ with a longitudinal fovea. Elytra of ♀ with a sutural tooth (Fig. 259).

Distribution. Denmark: in Jutland a few scattered captures along the eastern coast north to Randers; also Endelave and Samsø. Only other finds i Jutland are WJ (Fanø, before 1900) and NWJ (Nors sø 1955 & 1981). Rather distributed on the islands, but no finds on Falster and only two finds in district F. - Sweden: rather distributed from Sk. to Gstr. (not recorded from Boh.) but usually rare. Has decreased in number recently. Rather common only on Öl. and Gtl. - Not in Norway or Finland, but one record from Kr in the USSR at the river Svir. - Europe north to 60°N, south to S.France, C.Italy and Bulgaria; Iran and the Caucasus.

Biology. A hygrophilous and rather warm-preferent species, living in damp sites on clayey or muddy soil rich in humus. It is frequently found in dark forest with a sparse field layer, often at the margin of ponds and temporary pools. Also in more or less shaded sites in open country, e.g. on the shores of eutrophic lakes. *P. anthracinus* is a spring breeder exhibiting brood care (Lindroth, 1946); the female digs out a nesting cavity in rotten wood and guards her eggs until the larvae have appeared.

190. ***Pterostichus gracilis*** (Dejean, 1828)

Feronia gracilis Dejean, 1828, Spec. Gén. Col. 3: 287.

8.5-10 mm. Smaller and more convex than *anthracinus* which it imitates, especially in the form of the pronotum. Pronotal foveae indistinctly delimited, the inner hardly prolonged forwards. Elytra clearly iridescent, microsculpture very dense, transverse, especially in the ♂. Elytral subapical sinuation faint. Abdominal sternites laterally virtually smooth but dull from microsculpture. Wings always full. Male with last sternite unarmed.

Distribution. Denmark: very rare and usually only found singly or in few specimens. No records from NWJ and NEJ. - Sweden: rather distributed in the lowlands from Sk. to Gstr.; no records from Hall. and Boh.; usually rare and in small numbers. - Not recorded from Norway. - In East Fennoscandia rare in southern Finland (north to Kb) and in southern Kr of the USSR. - Europe north to 60-61°N, south to Portugal, S.Italy and Albania, east to E.Siberia.

Biology. A hygrophilous species, occurring in open country, predominantly in the shore zone of eutrophic lakes and slowly running rivers, usually on clayey soil with dense vegetation of *Carex, Glyceria,* etc. Also in wet meadows at some distance from open water. It is occasionally found in forest, probably during hibernation. *P. gracilis* is most numerous in spring when reproduction takes place; newly emerged beetles have been found in abundance in late autumn.

191. ***Pterostichus minor*** (Gyllenhal, 1827)
(Fig. 264.

Harpalus minor Gyllenhal, 1827, Ins. Suec. 1(4): 426.

6.8-8.7 mm. Smaller than *gracilis,* more piceous than black, base of antennae paler. Elytra with only suggested iridescence, their microsculpture obsolete, without regular transverse lines. The basal foveae of pronotum (Fig. 264) divided by a faint convexity, the inner one clearly reaching more forwards. Abdomen laterally dulled from microsculpture. Wings dimorphic. ♂ with a longitudinal keel on last sternite.

Distribution. In Denmark very distributed and common. - Sweden: recorded from nearly all districts and very common and distributed in the south up to 60°N; uncommon in the north and almost completely lacking in the highland of the west. - Norway: mainly in the southern coastal areas north to HOy; also further north in ST and NTi. - Finland: common, especially in the south, distributed to Ks in the north; also in Vib and Kr of the USSR. - Over most parts of Europe, south to N.Spain, C.Italy and Greece, east to W.Siberia.

Biology. A hygrophilous species, occurring in almost every type of wet habitats with rich vegetation, both in open country and in forest. It is usually very abundant in the shore zone of eutrophic lakes, but occurs also in *Sphagnum*-bogs, alder-swamps, etc. It is frequently encountered in shore-drift. *P. minor* is a spring breeder, most numerous in May-June; the young beetles are often active in the autumn.

Subgenus *Phonias* des Gozis, 1886

Phonias des Gozis, 1886, Rech. espèc. typ., p. 8.
Type-species: *Platysma interstincta* Sturm, 1824 (= *Pterostichus ovoideus* (Sturm, 1824)).
Argutor auct. *nec* Dejean, 1821.

192. ***Pterostichus strenuus*** (Panzer, 1797)
Figs 265, 267; pl. 6: 5.

Carabus strenuus Panzer, 1797, Fauna Ins. Germ. 38: 6.
Carabus erythropus Marsham, 1802, Ent. Brit.: 461, *nec* Gmelin, 1790.

Feronia Wasastjernae J. Sahlberg, 1875, Notis. Sällsk. Faun. Fl. Fenn. Förh. 14: 99.

6-7.2 mm. This and the two following species are small and have a single latero-basal fovea on pronotum. Piceous, appendages reddish brown. Pronotum (Fig. 265) narrower than in *minor*, with short but well developed marginal sinuation. Elytral striae evidently punctate. Prosternum with coarse punctures. Microsculpture missing on the disc of pronotum, on the elytra consisting of slightly transverse meshes, in the ♂ suggested only. Wings dimorphic, either full or strongly reduced.

Distribution. Denmark: very distributed and very common. - Sweden: common and generally distributed in the south, north to Gstr., Dlr. and Vrm. Rare in the northern half of the country and here with a scattered distribution. Almost completely lacking at high altitudes. - Norway: north to TR, but scattered in the central eastern areas. - East Fennoscandia: common in southern and central Finland and in Vib and Kr of the USSR. - Most parts of Europe, south to N.Spain, C.Italy and Greece; through Siberia to the Amur region.

Biology. A characteristic species of the litter layer of damp deciduous forest on clayey, mull-rich soil, e.g. in beech forests and in the drier parts of *Alnus glutinosa*-swamps. Also in shaded sites in open country, for instance at the margin of eutrophic lakes, in meadows with tall vegetation, etc. It is a night-active spring breeder, mainly occurring in April-June; newly emerged beetles become active in late autumn.

193. ***Pterostichus diligens*** (Sturm, 1824)
Fig. 266.

Platysma diligens Sturm, 1824, Deutschl. Fauna Ins. 5(5): 81.
Pterostichus strenuus auctt.; *nec* (Panzer, 1796).
Feronia boreella J. Sahlberg, 1870, Notis. Sällsk. Faun. Fl. Fennica Förh. 11: 403.

5.3-6.7 mm. Easily confused with *strenuus* but darker and with well developed microsculpture. Pure black, on legs at least femora infuscated. Pronotum (Fig. 266) smaller and its latero-basal sinuation less developed. Entire upper surface dull, with reticulate microsculpture. Elytral striae almost impunctate. Prosternum smooth. Wings nearly always strongly rudimentary, but single full-winged specimens are known.

Distribution. Very distributed and common over the entire area. - Europe to the extreme north, including Iceland; south to N.Spain, C.Italy and Bosnia; Siberia east to river Lena.

Biology. Confined to wet habitats like the shore zone of lakes and rivers, meadows and swamps, both in forest and in open country, mainly on soil poor in nutrients. It is especially typical of oligotrophic and dystrophic lakes and *Sphagnum*-bogs, occurring among moss and litter, often in somewhat shaded sites, e.g. under alder. Less frequent in eutrophic fens. It is rather widespread in the Scandinavian mountains, reaching the lower arctic zone. *P. diligens* is a nocturnal and predominantly carnivorous species, preying upon small arthropods, e.g. Collembola and mites (Dawson, 1965). Reproduction takes place in spring; the young beetles are active in late autumn.

Subgenus *Cryobius* Chaudoir, 1838

Cryobius Chaudoir, 1838, Bull. Soc. Nat. Moscou 11: 11.
Type-species: *Poecilus ventricosus* Eschscholtz, 1823.

A large group with main distribution on the circumpolar arctic tundra. Only two species penetrate from the east to north-eastern Fennoscandia. They are very different, but many intermediate species occur in the Arctic. All have highly reduced hind-wings.

194. ***Pterostichus brevicornis*** Kirby, 1837
Figs 268, 270.

Pterostichus brevicornis Kirby, 1837, Fauna Bor.-Amer. 4: 31.
Cryobius fastidiosus Mannerheim, 1853, Bull. Soc. Nat. Moscou 17: 131.
Feronia arctica J. Sahlberg, 1880, K. Svenska VetenskAkad. Handl. 17: 31.

4.5-6.4 mm. Piceous to black, elytra usually faintly iridescent; appendages either entirely rufous or infuscated, but first antennal segment, femora and tibiae at least piceous. Head dull due to strong isodiametric microsculpture. Antennae stout. Pronotum as in *diligens,* with weak basal sinuation of sides and a single latero-basal fovea. Dorsal punctures of elytra 3-5; elytral microsculpture varying, forming transverse meshes. Met-episternum (Fig. 268) much shorter (broader) than in preceding subgenus. The rudimentary wing is very small, not reaching first visible abdominal tergite (in short-winged members of subgenus *Argutor* at least reaching beyond first tergite). The elytra seem to be constantly firmly coalescent along the suture. Male with 3 sclerites in the internal sac of penis (lacking in *Argutor*).

Distribution. Circumpolar. In Europe only in N.Russia, from the Kola Peninsula through Siberia and northern N.America to New Foundland.

Biology. In our area restricted to the tundra. In Siberia and N.America also in the northern parts of the taiga. Under leaves and moss.

195. ***Pterostichus middendorffi*** (J. Sahlberg, 1875)
Fig. 269.

Feronia Middendorffi J. Sahlberg, 1875, Notis. Sällsk. Faun. Fl. Fenn. Förh. 14: 102.
Pseudocryobius deplanatus Motschulsky, 1850, Käf. Russl.: 54; *nec* (Ménétries, 1832); *nec* (Chaudoir, 1842).

7.4-9.4 mm. Much longer and flatter than *brevicornis,* easily seperated on the pronotum. The including in the subgenus *Cryobius* is clear from the highly reduced hind-wing (as in *brevicornis*) and the presence of a large sclerite in the internal sac of penis. Shining black, elytra with a bronzy lustre. Appendages varying in colour, either entirely dark or with antennal base rufous. Pronotum (Fig. 269) with a long lateral sinuation and acute hind-angles; 2 well-delimited latero-basal foveae. Elytra with 3 or 4 dorsal punctures. Entire upper surface with reticulate, from isodiametric to (on the elytra) weakly transverse microsculpture. Elytra apparently never coalescent along suture.

Distribution. Siberia and NE.Europe; the westernmost finds are from Ponoj in the eastern Kola Peninsula.

Biology. On the tundra of the Kola Peninsula on dry, sandy river banks with sparse vegetation. In Siberia mainly in the taiga on sandy places along the rivers.

Subgenus *Feronidius* Jeannel, 1942

Feronidius Jeannel, 1942, Faune de France 40: 808.
Type-species: *Carabus melas* Creutzer, 1799.

196. ***Pterostichus melas*** (Creutzer, 1799)

Carabus melas Creutzer, 1799, Ent. Vers., Wien: 114.
Molops maurus Sturm, 1818, Deutschl. Fauna Ins. 5 (4): 169.

13-17 mm. Somewhat similar to *melanarius*, but more related to subgenus *Steropus*. Convex, with short elytra. Shiny black, of the appendages only palps pale. Pronotum short with sides sinuate only close to hind-angles, which are sharp, about right-angled. Marginal bead thin (cf. *melanarius*, Fig. 256), several times narrower than the convexity along the outer of the two well-defined latero-basal foveae. Elytra with 2 dorsal punctures, striae barely punctulate. Male with last sternite longitudinally impressed.

Distribution. S. and C.Europe. - One record from Denmark, EJ, probably representing an introduction.

Biology. Both in forest and on open ground.

Subgenus *Cheporus* Latreille, 1829

Cheporus Latreille, 1829, Règne Anim., 2e ed., IV: 396.
Type-species: *Carabus metallicus* Fabricius, 1792.

197. ***Pterostichus burmeisteri*** Heer, 1841
Pl. 3: 5.

Pterostichus metallicus var. *Burmeisteri* Heer, 1841, Fauna Col. Helv. 1: 79.
Carabus metallicus Fabricius, 1792, Ent. Syst. 1: 146; *nec* Scopoli, 1763.

12-15 mm. This is the only *Pterostichus* in our fauna with "non-crossed" elytral epipleura. Very broad and therefore similar to *Abax* in general habitus. Black, upper surface bright coppery, often with a greenish hue; palps and tibiae rufous brown. Pronotum on each side with 2 very deep linear latero-basal foveae; lateral bead wide;

sides with a shallow sinuation before the protruding hind-angles. Elytra short and broad with very fine and faintly punctuate striae, intervals flat; normally only 2 dorsal punctures. Last sternite of ♂ with a protruding longitudinal keel.

Distribution. In mountains of C.Europe. In our area a few specimens are recorded from Denmark, and one specimen from Sweden (Bl., Rödeby, viii.1957). All are certainly introductions.

Biology. Mainly in forests.

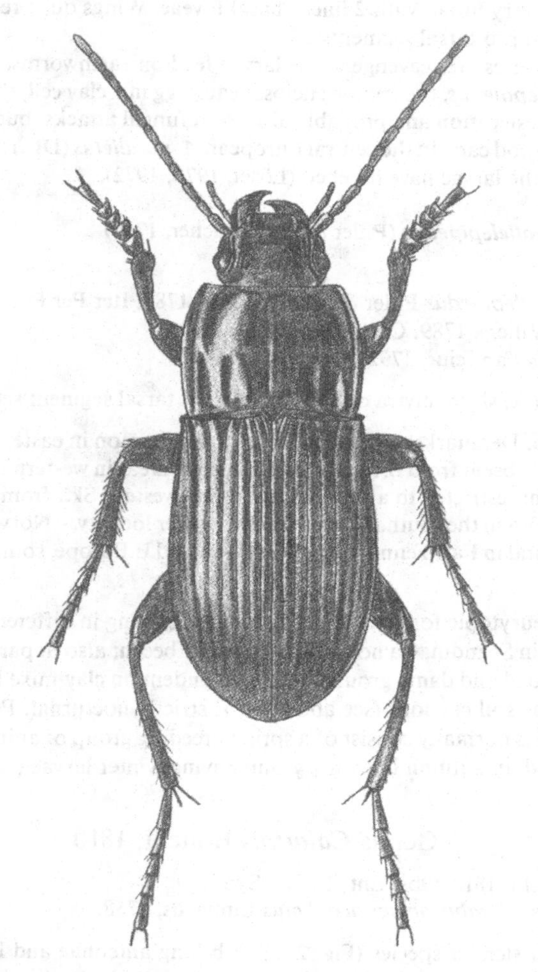

Fig. 271. *Abax parallelepipedus* (Pill. & Mitt.), length 18-22 mm. (After Victor Hansen).

Genus *Abax* Bonelli, 1810

Abax Bonelli, 1810, Obs. Ent. 1 (Tab. Syn.).
Type-species: *Carabus striola* Fabricius, 1792 (= *Carabus parallelepipedus* Piller & Mitterpacher, 1783).

Our single species (Fig. 271) is separated from *Pterostichus* by the presence of 2 deep extra striae along the side margin apically on the elytra and the carinate 7th interval behind the shoulder. Elytra without dorsal puncture, shoulder tooth protruding. Pronotum (Fig. 271) very broad with 2 linear basal foveae. Wings quite reduced. ♂ with 3 strongly dilated pro-tarsal segments.

The adult beetles are scavengers; the larvae feed on earthworms. In some species, e.g. *A. parallelepipedus*, the female encloses each egg in a clay cell, thereby protecting the egg from desiccation and probably also from fungal attacks, but she does not exhibit further brood care. In the central European *A. parallelus* (Dft.) the female guards her eggs until the larvae have hatched (Löser, 1970, 1972).

198. ***Abax parallelepipedus*** (Piller & Mitterpacher, 1783)
Fig. 271.

Carabus parallelepipedus Piller & Mitterpacher, 1783, Iter Per Pos: 105.
Carabus ater Villers, 1789, Car. Linn. Ent. 1: 364.
Carabus striola Fabricius, 1792, Ent. Syst. 1: 146.

18-22 mm. Black, shiny, elytra dull in the ♂. Last tarsal segment setose underneath.

Distribution. Denmark: rather distributed and common in eastern Jutland and on the islands, but absent from B. Also missing in large areas in western and northern Jutland. - Sweden: restricted to a small area in northwestern Sk., from Kullaberg in the north to Ramlösa in the south. Not rare in the former locality. - Norway: only AK and TEy. - Not found in East Fennoscandia. - W., C. and E.Europe, south to N.Spain and S.Italy.

Biology. A eurytopic forest-dwelling species, occurring in different kinds of forest communities, in Scandinavia notably in stands of beech; also in parks, etc. It prefers somewhat shaded and damp ground and is dependent on clay-mixed soil for the construction of the soil cocoons (see above). It is strictly nocturnal. Populations of *A. parallelepipedus* normally consist of a spring breeding group of animals having summer larvae, and an autumn breeding group having winter larvae (Drift, 1951; Löser, 1972).

Genus *Calathus* Bonelli, 1810

Calathus Bonelli, 1810, Obs. Ent. 1 (Tab. Syn.).
Type-species: *Carabus melanocephalus* Linnaeus, 1758.

Medium-sized, slender species (Fig. 272) with long antennae and legs (adapted for running with great speed). They are characterized by the ventral side being more convex than the upper surface. Elytra of female more or less dull. Pronotum with sides straight or very little rounded, parallel or convergent in basal half; latero-basal foveae flat or only suggested. Elytra with at least 2 dorsal punctures on third interval. All claws with a comb on inside. Male with 3 dilated pro-tarsal segments (except in *rotun-*

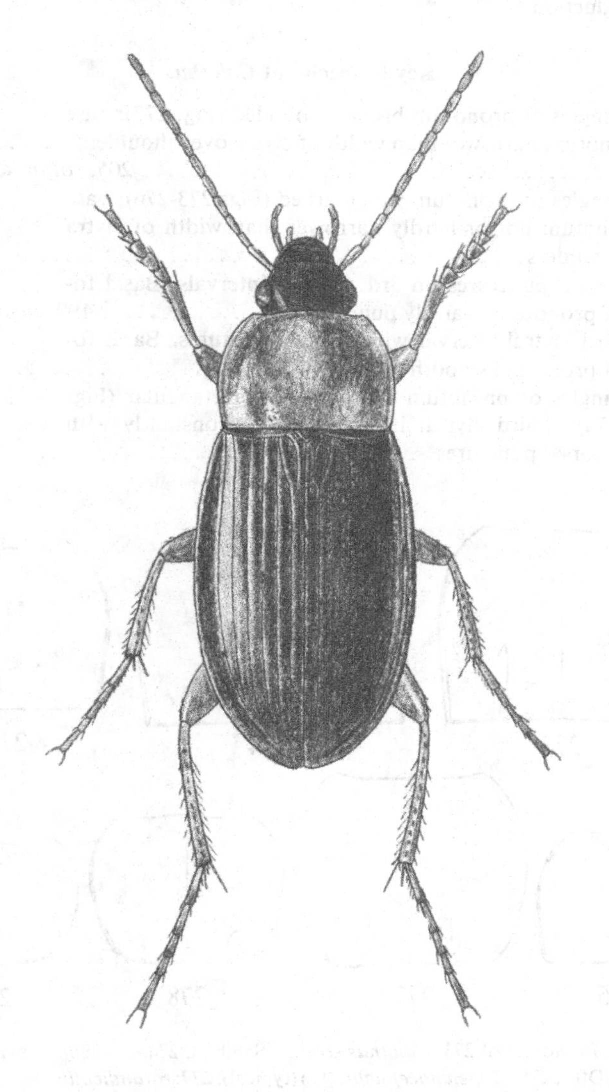

Fig. 272. *Calathus melanocephalus* (L.), length 6-8.8 mm. (After Victor Hansen).

dicollis). Right paramere very long and slender, usually hooked at apex. Wings highly variable.

All species are more or less xerophilous, and most of them are found in open country with sparse vegetation. They are nocturnal, chiefly carnivorous, and exhibit autumn reproduction .

Key to species of *Calathus*

1. Hind-angles of pronotum broadly rounded (Fig. 277); base of pronotum narrower than width of elytra over shoulders .... 205. *rotundicollis* Dejean
- Hind-angles of pronotum well marked (Figs 273-276); base of pronotum not or hardly narrower than width of elytra over shoulders 2

2(1) Elytra with punctures on 3rd and 5th intervals. Basal foveae of pronotum coarsely punctate .... 199. *fuscipes* (Goeze)
- Only 3rd elytral interval with dorsal punctures. Basal foveae of pronotum smooth and finely punctate 3

3(2) Hind-angles of pronotum sharp, almost rectangular (Figs 273, 274). Third elytral inverval almost constantly with only 2 dorsal punctures 4

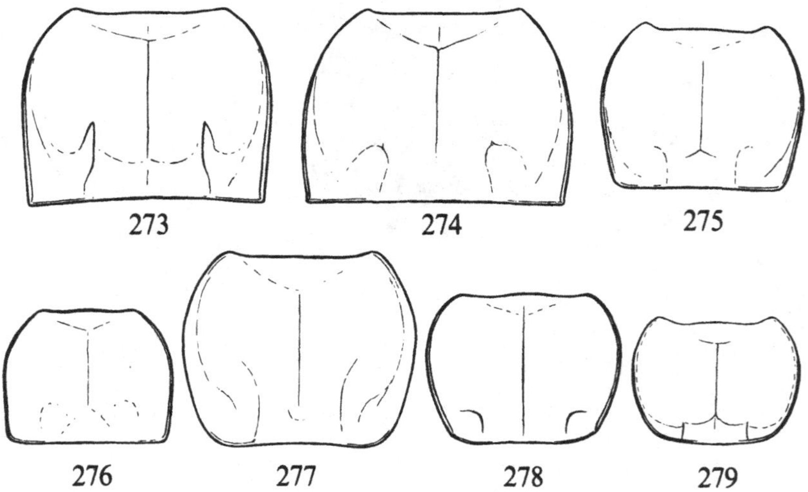

Figs 273-279. Pronotum of 273: *Calathus erratus* (Sahlbg.); 274: *C. ambiguus* (Payk.); 275: *C. micropterus* (Dft.); 276: *C. melanocephalus* (L.) (typical); 277: *rotundicollis* Dej.; 278: *Synuchus vivalis* (Ill.) and 279: *Olisthopus rotundatus* (Payk.).

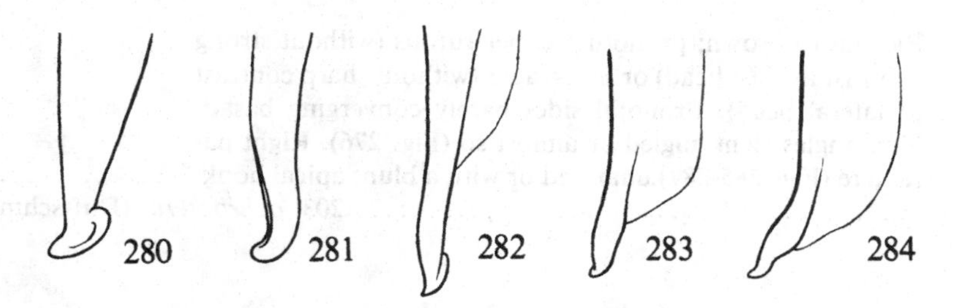

Figs 280-284. Apex of penis of *Calathus*. – 280: *erratus* (Sahlbg.); 281: *ambiguus* (Payk.); 282: *micropterus* (Dft.); 283: *ochropterus* (Dft.) and 284: *melanocephalus* (L.).

– Hind-angles of pronotum rounded at tip (Figs 275, 276). Third elytral interval with 3 or more punctures 5

4(3) Greatest width of pronotum before middle (Fig. 273). First metatarsal segment without internal furrow though keeled externally .... 200. *erratus* (Sahlberg)

– Greatest width of pronotum behind middle (Fig. 274), often close to base. First meta-tarsal segment with a shallow internal furrow .... 201. *ambiguus* (Paykull)

5(3) Pronotum bright rufous, contrasting against the black head; if infuscated, then also appendages or at least tarsi more or less darkened. Right paramere (Figs 288-290) widened and hooked at tip .... 202. *melanocephalus* (Linnaeus)

– Head and disc of pronotum concolorous (or head very slightly darker): black, piceous or brown; all appendages entirely pale. Right paramere (Figs 285-287, 291) not widened at apex 6

6(5) Piceous black, only extreme margins of pronotum paler. Sides of pronotum converging basad, hind-angles clearly obtuse (Fig. 275). Right paramere with arcuate apex (Fig. 291) .... 204. *micropterus* (Duftschmid)

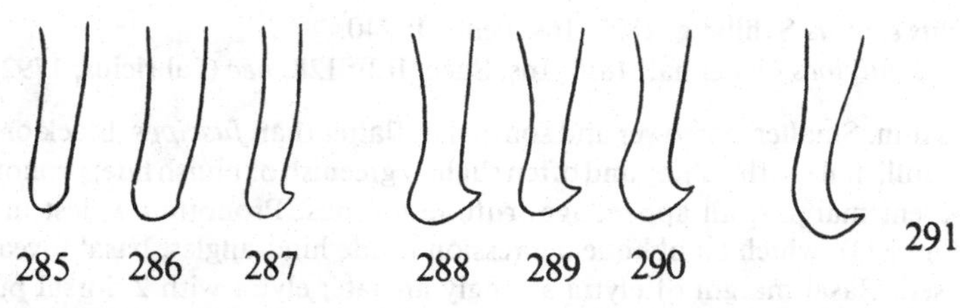

Figs 285-291. Apex of right paramere of *Calathus*. – 285-287: *ochropterus* (Dft.); 288-290: *melanocephalus* (L.) and 291: *micropterus* (Dft.).

- Piceous to brown; pronotum either rufous (without strong contrast against head) or infuscated (without sharp contrast to lateral bead). Pronotal sides barely converging basad, hind-angles right-angled or almost so (Fig. 276). Right paramere (Figs 285-287) unarmed or with a blunt apical hook .... 203. *ochropterus* (Duftschmid)

Subgenus *Calathus* s. str.

199. ***Calathus fuscipes*** (Goeze, 1777)
Pl. 6: 9.

Carabus fuscipes Goeze, 1777, Ent. Beytr. 1: 666.
Carabus cisteloides Panzer, 1793, Fauna Ins. Germ. 11: 12.

10-14.4 mm. Easily recognized on the presence of elytral dorsal punctures both on 3rd and 5th intervals. Black, female with dull elytra, sides of pronotum often pale transparent, antennae (with paler first segments) and mandibles rufo-piceous; legs either piceous or rufous, but apices of tarsal segments always dark. Basal foveae of pronotum coarsely punctate. Wings constantly rudimentary.

Distribution. Denmark: very distributed and very common. - Sweden: generally distributed from Sk. to Gstr., common. - Norway: mainly in the southeast and in coastal areas north to Nsy. - Finland: locally common in the southwest, recorded north to Kb; also Vib and southern Kr of the USSR. - Most of Europe, N.Africa, W.Asia.

Biology. A eurytopic species, predominantly occurring in open country on rather dry, notably sandy or clayey soil more or less rich in humus, e.g. in meadows and grassland; often on cultivated soil and also in light forests. The food consists mainly of aphids, caterpillars and ants. Oviposition takes place in late summer and autumn. Larvae and some old adults hibernate; young imagines emerge in early summer.

200. ***Calathus erratus*** (Sahlberg, 1827)
Figs 273, 280.

Harpalus erratus Sahlberg, 1827, Ins. Fenn. 1: 240.
Harpalus fulvipes Gyllenhal, 1810, Ins. Suec. 1(2): 128, *nec* (Fabricius, 1792).

8.5-11.8 mm. Smaller, narrower and somewhat flatter than *fuscipes*. Black or piceous, female dull, male with a shiny and often slightly greenish or bluish hue; pronotum with transparent margins; all appendages rufo-testaceous. Pronotum widest in anterior half (Fig. 273), which an oblique depression inside hind-angles, basal foveae clearly impressed. Basal margin of elytra strongly arcuate; elytra with 2 dorsal punctures. Meta-tarsi: see the key. Wings either full or strongly reduced. Penis (Fig. 280) with apical disc.

Distribution. Denmark: very distributed and common. - Sweden: generally distributed and common to very common in the south, north to Hls., Med., eastern Jmt. and Ång. A few records from Vb. and Nb., and one doubtful record from Ly.Lpm. - Norway: north to TRi. - East Fennoscandia: common and distributed to Lk in Finland; also in Vib, Kr and Lr of the USSR. - Entire Europe except the extreme south.

Biology. A xerophilous species, usually occurring on dry, sandy or gravelly soil poor in humus and with sparse vegetation. Predominantly in open country, for instance on *Calluna*-heaths, in dunes, and in dry meadows and grassland; also in thin forests. It is frequently encountered on agricultural land on light soil, notably in root crop fields. The species preys on aphids, ants, etc. The main period of propagation is August-September. The larvae and a few old adults hibernate; young imagines appear in spring.

201. ***Calathus ambiguus*** (Paykull, 1790)
Figs 274, 281.

Carabus ambiguus Paykull, 1790, Mon. Car. Suec.: 130.
Carabus fuscus Fabricius, 1792, Ent. Syst. 1: 158.

8.4-11.6 mm. Broader than *erratus.* Dull piceous, all pale parts more pale testaceous and without metallic hue. Greatest width of pronotum behind middle (Fig. 274), its margins more widely translucent; basal foveae obsolete. Antennae thinner. Basal margin of elytra less arcuate. First meta-tarsal segment with a shallow internal furrow. Wings always full. Penis (Fig. 281) with a spoon-like apex but without disc.

Distribution. Denmark: scattered and uncommon, but recorded from all districts. - Sweden: from Sk. to Upl., usually rather rare but quite generally distributed. Common in the sandy areas of Sk., Öl. and Gtl. - Norway: rare, only Ø and VAy. - Finland: rare, recorded from Al to N, Sa and Om; and Vib in the USSR. - Entire Europe to 60°N.

Biology. A rather stenotopic species, living in open, dry country on sandy or gravelly, sometimes clay-mixed soil with sparse vegetation, notably on southern slopes. Also in agricultural land on sandy fields. The species is often found together with *C. erratus.* The food consists chiefly of ants and aphids (Smit, 1957). The main breeding period is September. Larvae and a small percentage of old imagines hibernate; young beetles emerge in early summer.

202. ***Calathus melanocephalus*** (Linnaeus, 1758)
Figs 272, 276, 284, 288-290; pl. 6: 10.

Carabus melanocephalus Linnaeus, 1758, Syst. Nat. ed. 10: 415.

6-8.8 mm. Pronotum either rufous (contrasting against the black head), or more or less infuscated; if so, then also appendages, or at least tarsi, more or less darkened. The palest form has all appendages pale; elytra always black. Pronotum (Fig. 276) almost parallel-sided in basal half. Normally with 4 dorsal punctures. Wings normally highly reduced but long-winged individuals are known. Penis (Fig. 284) with apex slightly bent ventrad; right paramere (Figs 288-290) with strong apical tooth.

Different melanistic forms have been described.

Distribution. Distributed and common throughout the entire area. - Entire Europe including Iceland.

Biology. This commonly distributed species usually lives in open country on different kinds of moderately dry ground with sparse vegetation, achieving its greatest abundance on sandy soil. It is a common inhabitant of dry meadows, grassland, dunes and heaths; also on agricultural land and in thin forests, mainly of *Pinus*. It is frequent in the fjelds up to the lower alpine region. It is probably chiefly carnivorous. The main period of reproduction is August-September. Larvae hibernate, young adults emerge in spring. A varying proportion of old beetles hibernate and reproduce for a second time. In northernmost Scandinavia development apparently takes two years (Forsskåhl, 1972).

203. ***Calathus ochropterus*** (Duftschmid, 1812)
Figs 283, 285-287.

Carabus ochropterus Duftschmid, 1812, Fauna Austriae 2: 124.
Carabus mollis Marsham, 1802, Ent. Brit.: 456; *nec* Strøm, 1768.
Calathus erythroderus Gemminger & Harold, 1868, Cat. Col. 1: 362.

6.6-9.2 mm. Closely allied to *melanocephalus* and often confused with this species. More slender, especially the pronotum, and the elytra more stretched; body also less shiny (also in the male), and the legs longer. The colour contrasts are less pronounced: head and elytra brown to piceous, the pronotum entirely pale in Scandinavian populations, and rarely with a faint infuscation of the disc (*erythroderus*). In western populations (SW.Norway and Jutland; also in Great Britain) the pronotum is more or less infuscated and even as dark as in *micropterus*. Appendages always entirely pale. The basal margin is more arcuate than in *melanocephalus*. The species is wing-dimorphic, with the long-winged form predominating or alone present in the northern and western parts of the area. Penis (Fig. 283) almost straight at apex; right paramere (Figs 285-287) without or with a blunt apical tooth.

Distribution. Denmark: spp. *mollis* is very distributed in dune-areas along the western coast of Jutland; also inland captures in central Jutland. Ssp. *erythroderus* is known from 22 localities representing all districts except NEJ and F. - Sweden: restricted to the south and generally distributed to 58°N; also G. Sand.; not rare. Only ssp. *erythroderus* is known from the area. - Norway: rare, only VAy and Ry in the southwest. - Not in East Fennoscandia. - Europe north to 58-59°N.

Biology. Strictly confined to dry, sandy habitats in open country, occurring in sparse xerophilous vegetation, both inland, e.g. in dry meadows and grassland, and on dune-sand at the coast. Often in tufts of *Elymus* and *Ammophila* in company with *Demetrias monostigma* and *Dromius linearis.* Reproduction takes place in the autumn, young adults appearing in early summer.

204. ***Calathus micropterus*** (Duftschmid, 1812)
Figs 275, 282, 291.

Carabus micropterus Duftschmid, 1912, Fauna Austriae 2: 123.

6.5-8.8 mm. Piceous black, only extreme margin of pronotum paler; appendages pale. Sides of pronotum (Fig. 275) converging basad, hind-angles distinctly obtuse. Elytra with narrower and more rounded base than in the two preceding species; their basal margin arcuate as in *ochropterus.* Wings constantly reduced. Penis (Fig. 282) with long, straight apex; right paramere as in fig. 291.

Distribution. Denmark: very distributed in Jutland, in north and central Zealand, and on Bornholm. Absent from districts F, LFM, SZ and large parts of NWZ (see map 256 in Bangsholt 1983). - Sweden: very common and distributed throughout the country, known from all areas. - Norway: common throughout the country. - In East Fennoscandia very common over the entire area. - Europe except the southwestern and southern parts.

Biology. Predominantly a woodland species, living among litter and moss on moderately dry ground in light deciduous or coniferous forests, notably in warm, dry forest on sandy soil, often in company with *Pterostichus oblongopunctatus.* Also in somewhat shaded places in open country. It is often a dominant Carabid in the birch region of the fjelds. In C.Europe reproduction takes place in summer and autumn, young adults emerging in early summer following larval hibernation (Drift, 1959; Kůrka, 1972). In Scandinavia spring propagation occasionally seems to occur (Lindroth, 1945).

Subgenus *Amphyginus* Haliday, 1841

Amphyginus Haliday, 1841, Newman's Entomologist, p. 175.
Type-species: *Carabus piceus* Marsham, 1802.
Amphigynus auct.

205. ***Calathus rotundicollis*** Dejean, 1828
Figs 277, 296.

Calathus rotundicollis Dejean, 1828, Spec. Gén. Col. 3: 75.
Carabus piceus Marsham, 1802, Ent. Brit.: 444; *nec* Linnaeus, 1758.

8.5-10.5 mm. At once recognized on the pronotum (Fig. 277) which has completely rounded hind-angles, and is much narrower than elytra over shoulders. Dark piceous, all margins and usually elytral suture somewhat translucent. Appendages rufous but femora sometimes darker. Elytra with 4-5 dorsal punctures; elytral microsculpture remarkably scale-like. Wings full in specimens from our area, dimorphic in specimens in other areas. Male without dilated pro-tarsal segments. Right paramere hooked at apex.

Distribution. Denmark: rather distributed and rather common, but scattered occurrence in western and northern parts of Jutland. - Sweden: a southern species, only known from Sk. In that landscape not rare except in the north. Somewhat increasing in number in the last 30 years. - Not in Norway or East Fennoscandia. - A West European species.

Biology. A forest-dwelling species, occurring in the litter layer of deciduous, rarely coniferous, forests, notably in light stands of beech; also in parks and gardens. Preferably on mull-soil (Drift, 1959). The main period of reproduction is August-September; newly emerged beetles occur in spring.

Genus *Sphodrus* Clairville, 1806

Sphodrus Clairville, 1806, Ent. Helv. 2: 86.

Type-species: *Carabus planus* Fabricius, 1792 (= *Carabus leucophthalmus* Linnaeus, 1758).

To this genus belongs one of our largest carabids. Uniformly dark, unmetallic, some-

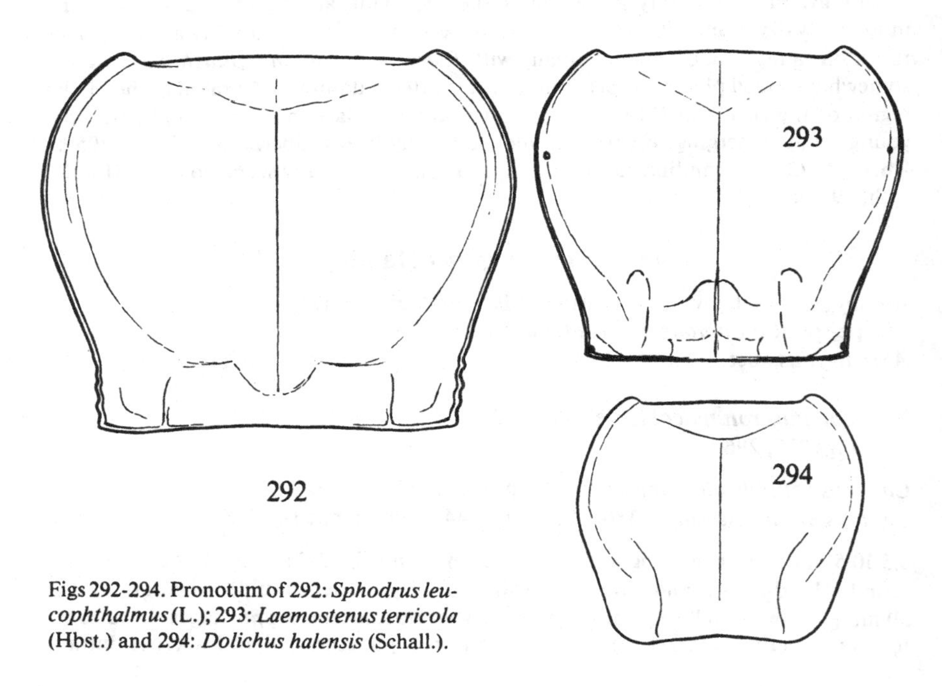

Figs 292-294. Pronotum of 292: *Sphodrus leucophthalmus* (L.); 293: *Laemostenus terricola* (Hbst.) and 294: *Dolichus halensis* (Schall.).

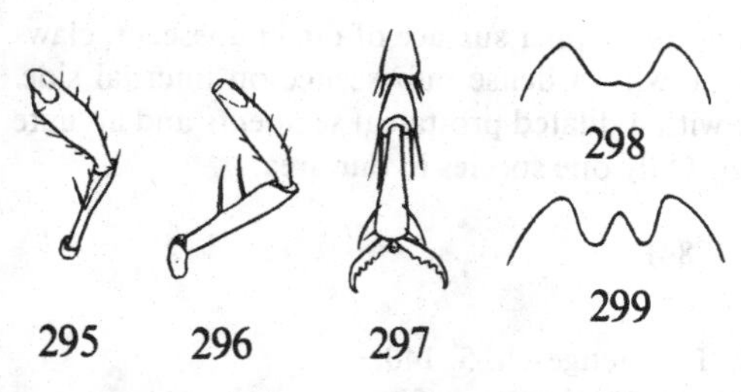

Figs 295, 296. Labial palp of 295: *Synuchus vivalis* (Ill.) and 296: *Calathus rotundicollis* (Dft.).
Fig. 297. Terminal tarsal segments of *Synuchus vivalis* (Ill.).
Figs 298, 299. Mentum of 298: *Olisthopus* and 299: *Agonum*.

what reminiscent of a *Pterostichus*. Pronotum (Fig. 292) strongly condiform with depressed lateral sides and side-margin crenulate basally. Elytral striae extremely fine, punctate anteriorly, without dorsal puncture. Legs very slender. Tarsi glabrous above; claws smooth. Wings full. Male with 3 dilated pro-tarsal segments and meta-trochanters prolonged into a sharp spine. Parameres as in *Calathus*.

206. ***Sphodrus leucophthalmus*** (Linnaeus, 1758)
Fig. 292; pl. 3: 11.

Carabus leucophthalmus Linnaeus, 1758, Syst. Nat. ed. 10: 413.
Carabus planus Fabricius, 1792, Ent. Syst. 1: 133.

20-26 mm. Piceous to almost black, dull; palpi, base of antennae and tarsi somewhat paler.

Distribution. Denmark: earlier recorded from a number of towns, now apparently missing. Latest record is from Nysted (LFM) in 1907. - Sweden: earlier, mainly in the 19th century, known from different localities in Sk., Hall., Sm., Vg., Upl. and Vstm. In the last 50 years only reported from Halmstad (Hall.) in a cellar. - No records from Norway. - Finland: old records from various parts of the country, latest record from Sb: Vehmersalmi in 1946. - Europe, N.Africa and W.Asia.

Biology. Exclusively synanthropous, occurring in cellars, stables, bakeries, mills, etc., sometimes in company with *Laemostenus terricola* and *Blaps*. Mainly in August-September.

Sphodrus has become increasingly rare during this century, undoubtedly due to improved hygienic conditions, and has now possibly disappeared from our area.

Genus *Laemostenus* Bonelli, 1810

Laemostenus Bonelli, 1810, Obs. Ent. 1 (Tab. Syn.).
Type-species: *Carabus janthinus* Duftschmid, 1812.
Pristonychus Dejean, 1828, Spec. Gén. Col. 3: 43.
Type-species: *Carabus terricola* Herbst, 1784.

Similar to *Pterostichus* but pronotum narrowly cordiform and with a metallic hue on upper surface. Pronotum (Fig. 293) with sides narrowly deplanate. Elytra with rather strong, punctate striae; no dorsal punctures. Upper surface of tarsi pubescent, claws with fine denticles at base. Meta-tibiae with a dense pubescence on internal side. Wings reduced into tiny scales. Male with 4 dilated pro-tarsal segments and arcuate meso-tibiae. Parameres as in *Calathus*. Only one species in our area.

207. ***Laemostenus terricola*** (Herbst, 1784)
Fig. 293; pl. 3: 9.

Carabus terricola Herbst, 1784, Arch. Insectengesch. 5: 140.
Carabus subcyaneus Illiger, 1801, Magasin Insektenk. 1: 57.

13-17.5 mm. Black, underside and appendages piceous brown; elytra with a bluish or violaceous lustre.

Distribution. Denmark: scattered records from all districts. Rare and apparently decreasing in number since 1950. - Sweden: scattered distributed from Sk. to Dlr.; has certainly decreased in number in this century except in eastern Sk., where apparently less rare. - Norway: a very scattered distribution in Ø, AK, VE, AAy, Ry, HOy and STi. - Finland: recorded from Ab: Korpo, N: Helsinki and Oa: Ylistaro (many specimens in potato cellars); also records from Vib and Kr in the USSR. - Europe north to 60°N, to Iceland, and south to Portugal, C.Italy and Yugoslavia, east to Volga. Introduced in North America.

Biology. A troglophilous species, in our area predominantly synanthropous, living in dark places in and around houses, e.g. in cellars and stables, often in company with *Blaps*. Occasionally remote from human habitations, for instance occurring in hollow trees with bird's nests; also in the burrows of rabbits and other mammals. I C.Europe it is often found among guano in caves. *Laemostenus* is night-active and most frequently encountered in August-September when breeding probably takes place.

Genus *Dolichus* Bonelli, 1810

Dolichus Bonelli, 1810, Obs. Ent. 1 (Tab. Syn.).
Type-species: *Carabus flavicornis* Fabricius, 1787 (= *Carabus halensis* Schaller, 1783).

All essential characters as in *Calathus*, but basal part of pronotum strongly constricted, hind-angles broadly rounded; hind-margin narrower than width of elytra over shoulders (Fig. 294).

A number of species in E.Asia, only one in our area.

208. ***Dolichus halensis*** (Schaller, 1783)
Fig. 294; pl. 3: 10.

Carabus halensis Schaller, 1783, Schrift. Naturf. Ges. Halle 1: 317.
Carabus flavicornis Fabricius, 1787, Mant. Ins. 1: 199.

15-18 mm. Flat and rather weakly sclerotized. Piceous, pronotum sharply yellow along side margins. Head usually with two small rufous spots. Elytra often with a large, mutual, triangular, rufo-testaceous spot in the anterior half (various other colour variants occur, especially in E.Asia). Appendages yellow. Pronotum as in fig. 294. Elytra dull from strong microsculpture. Wings full. Male with 3 rectangularly dilated protarsal segments.

Distribution. Denmark: very scattered records from the islands, in Jutland only recorded from SJ (Sønderborg before 1900). Very strongly decreasing in number; after 1950 only one record: NEZ (Hornbæk 1951). - Sweden: very rare during this century and decreasing in number. Only found in western Sk., except for one accidental record on the east coast (Stenshuvud 1966, 1 specimen on the seashore). Last record in western Sk.: Tygelsjö, 4 specimens 15.vii.1962. - Not recorded from Norway or East Fennoscandia. - C., E. and S.Europe, absent from the western and northern parts of the continent. East to the Amur and Ussuri regions.

Biology. On open, often cultivated fields, usually on clayey soil. It is a common species in agricultural regions of E.Europe and is a pronounced steppe element in our fauna. Probably only breeding in periods, population maintenance being dependent on immigration from the south or east. It is an autumn breeder; most numerous in August.

Genus *Synuchus* Gyllenhal, 1810

Synuchus Gyllenhal, 1810, Ins. Suec. Descr. 1 (2): 77.
Type-species: *Carabus nivalis* Panzer, 1797.

As in *Calathus* with serrate claws (Fig. 297) and liable to be confused with *C. piceus*. Easily recognized on the pear-shaped terminal segment of the labial palpi (Fig. 295). Pronotum convex with rounded sides and suggested hind-angles. Third elytral interval with 2 dorsal punctures. Elytra with faintly arcuate basal margin; microsculpture forming transversely arranged meshes. Wings dimorphic, full or reduced. Male with 3 dilated protarsal segments. Both parameres short and rounded, the right one being very small and fiddle-like.

209. ***Synuchus vivalis*** (Illiger, 1798)
Figs 295, 297; pl. 6: 3.

Carabus nivalis Panzer, 1797, Fauna Ins. Germ. 37: 19; *nec* Paykull, 1790.
Carabus vivalis Illiger, 1798, Käf. Preuss.: 197.

6-8.5 mm. Brownish black, without any trace of metallic hue, margin of pronotum and suture of elytra usually paler; appendages rufous.

Distribution. Denmark: rather distributed but uncommon, most scattered in western and northern parts of Jutland. - Sweden: generally distributed but uncommon in the south, north to Vrm., Dlr. and Gstr. Somewhat more scattered distributed in the northern half to Nb. and Lu. Lpm.; more records in the coastal areas than in the interior. - Norway: north to NTi. - Finland: comparatively common in the south, north to Ob; also in Vib and Kr of the USSR. - Most parts of Europe except the north, south to Portugal, Corse, C.Italy, Asia Minor; Siberia east to the Amur region.

Biology. Predominantly on rather dry ground in open country, living on sandy or gravelly, sometimes clay-mixed, soil with sparse vegetation of grasses, *Calluna,* etc. Often on cultivated land; also in forest edges and in light woods. Manley (1971) showed that the related *S. impunctatus* Say is phytophagous. The female of this species carries seeds of *Melampyrum* into underground storage chambers. Phytophagy thus seems possible also for the adult of *S. vivalis.* The larva is partly carnivorous. Breeding takes place in autumn; newly emerged imagines occur in summer.

Genus *Olisthopus* Dejean, 1828

Olisthopus Dejean, 1828, Spec. Gén. Col. 3: 176.
Type-species: *Carabus rotundatus* Paykull, 1790.
Odontonyx auct., *nec* Stephens, 1828.

Similar to *Synuchus,* but the hind-angles of pronotum are completely disappeared and pronotal base is strongly punctate. Upper surface with a metallic hue. Mentum (Fig. 298) without median tooth (cf. *Agonum*). Elytra with 3 dorsal punctures, striae finely punctate. The claws are smooth. Elytra with a subapical sinuation and rather abruptly truncate. Wings normally full, but quite reduced in some individuals. Male protarsi with 3 dilated segments. The two parameres are rounded, of rather similar shape, though the right one is smaller.

210. ***Olisthopus rotundatus*** (Paykull, 1790)
Fig. 298; pl. 6: 4.

Carabus rotundatus Paykull, 1790, Mon. Car. Suec.: 41.
Carabus rotundicollis Marsham, 1802, Ent. Brit.: 471.

6.4-7.8 mm. Brown to piceous, upper surface bronzed, base of antennae and legs pale.

Distribution. Denmark: rather distributed and common, known from all districts. - Sweden: generally distributed but usually uncommon, recorded from Sk. to southern Dlr. Isolated record in one locality in Vb. (Umeå). - Norway: restricted to AK and coastal areas from AAy to STy. - In East Fennoscandia comparatively rare in the south of Finland and in Vib of the USSR. - West Palaearctis; in Europe south to Portugal, C.Italy and Asia Minor.

Biology. A xerophilous species, living on open, dry ground, usually on sandy soil with sparse vegetation, for instance in grass areas (e.g. with *Corynephorus*) and *Calluna*-heaths. It is a common inhabitant of the alvar on Öl., occurring among grasses and scattered *Juniperus.* Less frequent in light pine forest. Reproduction takes place in autumn; young beetles emerge in summer.

Genus *Agonum* Bonelli, 1810

Agonum Bonelli, 1810, Obs. Ent. 1 (Tab. Syn.).
Type-species: *Carabus marginatus* Linnaeus, 1758.
Anchomenus Bonelli, 1810, Obs. Ent. 1 (Tab. Syn.).
Type-species: *Carabus prasinus* Thunberg, 1784 (= *Carabus dorsalis* Pontoppidan, 1763).

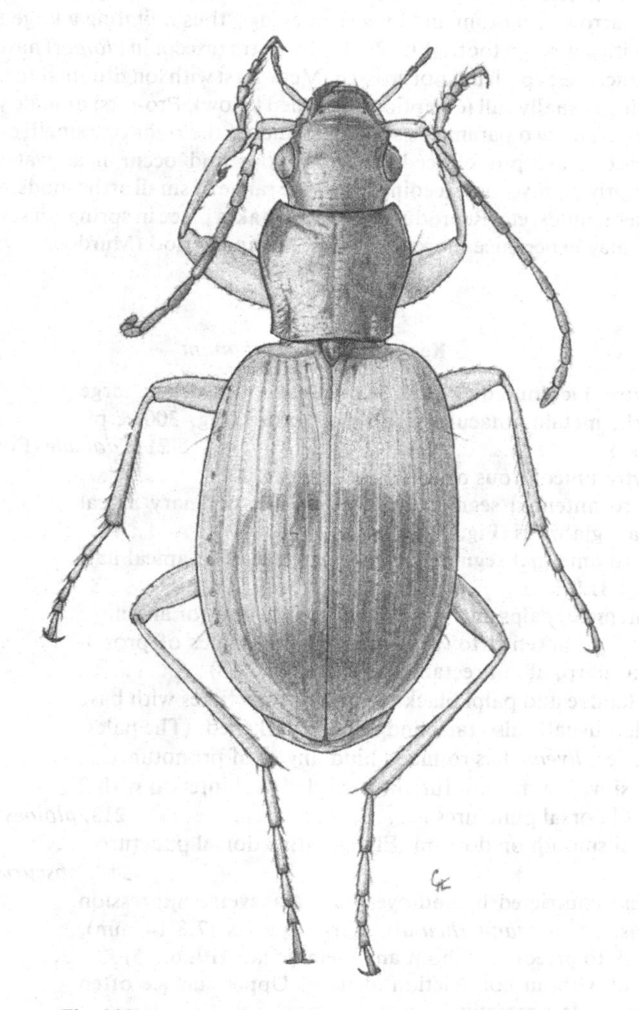

Fig. 300. *Agonum dorsale* (Pont.), length 6-8.2 mm.

Platynus Bonelli, 1810, Obs. Ent. 1 (Tab. Syn.).
Type-species: *Carabus angusticollis* Fabricius, 1801 (= *Carabus assimilis* Paykull, 1790).
Other names have been used for subgenera. *Platynus* is often considered a separate genus, including the subgenera *Anchomenus, Oxypselaphus* and *Paranchus.*

Agonum, as taken in its widest sense, is a very large genus and has a world-wide distribution. The species (cf. pl. 6: 11-15) are characteristic in their general habitus by the short and narrow pronotum and long slender legs, thus imitating a large *Bembidion.* Mentum with a median tooth (Fig. 299). The elytra (except in *albipes*) have more than 2 dorsal punctures; epipleura not crossed. Meta-tarsi with longitudinal furrows. Claws simple. Wings usually full (exceptions recorded below). Pro-tarsi of male with 3 dilated segments. The two parameres similar, rounded, the right one smaller.

Most species are pronouncedly hygrophilous and occur near water. They are predominantly carnivorous, feeding on a wide range of small arthropods, e.g. Collembola, Diptera, mites, etc. Reproduction usually takes place in spring. In several species the adults may experience more than one breeding period (Murdoch, 1966; Wasner, 1979).

Key to species of *Agonum*

1 Elytra bicoloured: bright rufo-testaceous with a large dark, metallic macula across the suture (Fig. 300 & pl. 6: 13) .... 211. *dorsale* (Pontoppidan)
– Elytra unicolorous or with pale margins 2
2(1) Third antennal segment, except for the ordinary apical setae, glabrous (Fig. 311) 3
– Third antennal segment pubescent, at least in apical half (Fig. 312) 23
3(2) Antennae, palps and legs unicolorous yellow (or antennae slightly darkened towards apex). Hind-angles of pronotum sharp, about rectangular (Figs 304, 308) 4
– Antennae and palpi black or brown, sometimes with base paler; usually also tarsi and femora darkened. (The palest species, *livens,* has rounded hind-angles of pronotum) 5
4(3) Tarsi with a median furrow. Third elytral interval with 2 small dorsal punctures .... 213. *albipes* (Fabricius)
– Tarsi smooth on dorsum. Elytra with 3 dorsal punctures .... 212. *obscurum* (Herbst)
5(3) Head contricted behind eyes by a transverse impression (least so in *mannerheimii*). Large species (7.8-14 mm), black to piceous, without any metallic hue (Pl. 6: 15) 6
– Head without constriction at neck. Upper surface often more or less metallic 10

6(5) Hind-angles of pronotum sharp, about right-angled or denticulate (Figs 302, 303) 7
– Hind-angles of pronotum rounded (Fig. 301) 9
7(6) Sides of pronotum sinuate in basal half to hind-angles (Fig. 302) .... 216. *assimile* (Paykull)
– Hind-angles of pronotum constitute a small prominent denticle; sides otherwise not sinuate (Fig. 303) 8

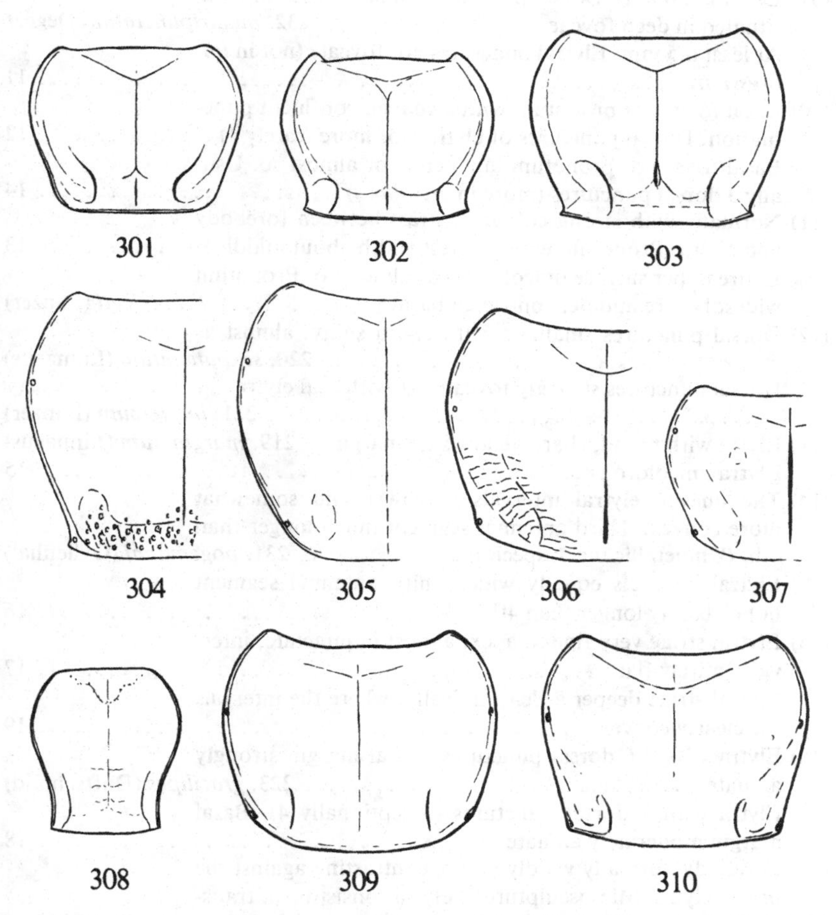

Figs 301-310. Pronotum of *Agonum*. – 301: *mannerheimii* (Dej.); 302: *assimile* (Payk.); 303: *krynickii* (Sperk); 304: *albipes* (F.); 305: *thoreyi* Dej.; 306: *bogemannii* (Gyll.); 307: *quadripunctatum* (Deg.); 308: *obscurum* (Hbst.); 309: *consimile* (Gyll.) and 310: *exaratum* (Mann.).

8(7) Base of pronotum (at least medially) without raised border. Elytra less than 3 times as long as pronotum.... 217. *krynickii* (Sperk)
– Base of pronotum with fine but evidently raised border. Elytra about 3.5 times as long as pronotum 218. *longiventre* (Mannerheim)
9(6) Base of pronotum with fine raised border (Fig. 301). First antennal segment piceous .... 215. *mannerheimii* (Dejean)
– Base of pronotum unmargined. At least 1st antennal segment rufous .... 214. *livens* (Gyllenhal)
10(5) Less than 6 mm. Dorsal punctures on 3rd elytral interval situated in deep foveae .... 232. *quadripunctatum* Degeer
– At least 6.5 mm. Elytral punctures not foveate (not in *impressum)* 11
11(10) Basal foveae of pronotum with a coarse, confluent punctuation. Dorsal punctures of elytra 5 or more (rarely 4) 12
– Basal foveae of pronotum inpunctate or almost so. Usually 3 dorsal punctures (more in *gracilipes*).......................... 14
12(11) Normally with strong colour contrast between forebody and elytra. Pronotum with greatest width about middle............... 13
– Entire upper surface unicolorous or almost so. Pronotum widest before middle, contricted basally............. 222. *ericeti* (Panzer)
13(12) Dorsal punctures small. ♀ with elytra shiny, almost as in the ♂ .... 220. *sexpunctatum* (Linnaeus)
– Dorsal punctures strongly foveate. ♀ with dull elytra .... 221. *impressum* (Panzer)
14(11) Elytra with broad, sharp yellow side-margin .. 219. *marginatum* (Linnaeus)
– Elytra unicolorous 15
15(14) The uneven elytral intervals narrower and somewhat more convex. Third antennal segment much longer than 4th. (Unmetallic black species) .... 231. *bogemanni* (Gyllenhal)
– Elytral intervals equally wide. Third antennal segment not or barely longer than 4th 16
16(15) Elytral striae very fine to apex, almost impunctate, intervals entirely flat 17
– Elytral striae deeper at least apically, where the intervals are clearly convex 19
17(16) Elytra with 4-6 dorsal punctures. Basal margin strongly arcuate .... 223. *gracilipes* (Duftschmid)
– Elytra with 3 dorsal punctures (exceptionally 4). Basal margin moderately arcuate 18
18(17) Forebody normally vividly green, contrasting against the brassy elytra. Microsculpture of elytra consisting of transverse meshes .... 224. *muelleri* (Herbst)
– Upper surface unicolorous, from green or coppery to almost black. Elytral microsculpture almost regularly

isodiametric .... 225. *sahlbergii* (Chaudoir)

19(16) All segment of meta- and meso-tarsi provided with a deep furrow on each side (Fig. 317). (Pure black, dull species) .... 230. *lugens* (Duftschmid)

– Last (and usually also penultimate) tarsal segments without furrows (Figs 315, 316). (Often with a metallic hue or more shiny black) 20

20(19) Hind-angles of pronotum protruding outside posterior seta as a small, obtuse denticle. Femora rufo-piceous, at least in inner half .... 226. *dolens* (Sahlberg)

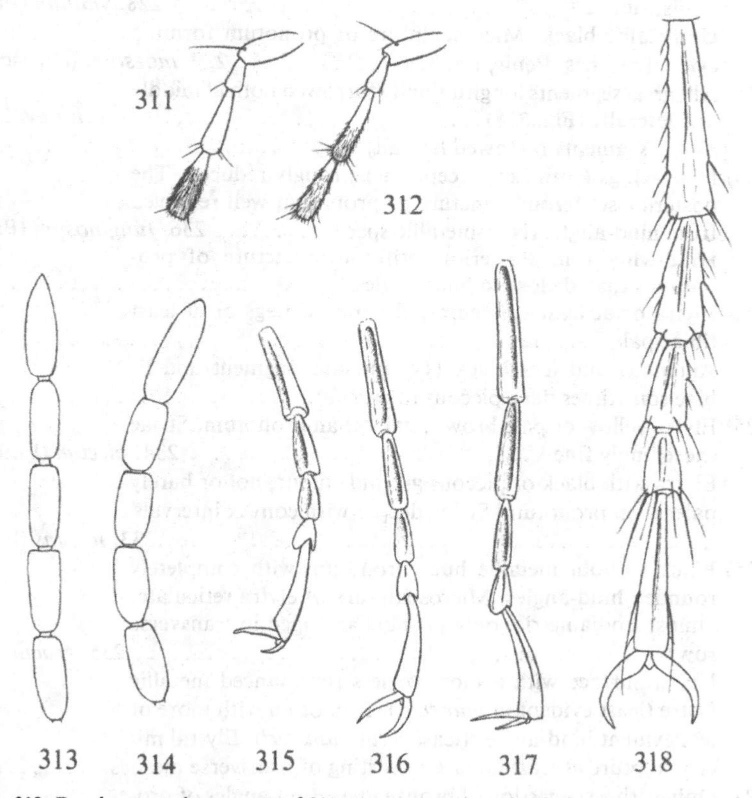

Figs 311, 312. Basal antennal segments of 311: *Agonum* s.str. and 312: Sg. *Europhilus* Chaud. Figs 313, 314. Antenna of *Agonum*. – 313: *consimile* (Gyll.) and 314: *exaratum* (Mann.). Figs 315-318. Meta-tarsus of *Agonum*. – 315: *versutum* Sturm; 316: *viduum* (Pz.); 317: *lugens* (Dft.) and 318: *thoreyi* Dej.

- No denticle at hind-angle of pronotum. Femora black or dark piceous, at most rufo-piceous in innermost part 21

21(20) First antennal segment dark rufous also above; tibiae and usually epipleura likewise somewhat pale. Microsculpture of elytra consists of transverse lines without regular meshes .... 227. *versutum* Sturm

- First antennal segment black, or pale underneath only; tibiae and epipleura black or piceous. Elytral microsculpture forming evident meshes, though often arranged in transverse rows 22

22(21) Upper surface more or less metallic (best seen along elytral sides). Microsculpture of pronotum obsolete on disc. Penis, fig. 320 .... 228. *viduum* (Panzer)

- Unmetallic black. Microsculpture of pronotum forming evident meshes. Penis, fig. 319. .... 229. *moestum* (Duftschmid)

23(2) All tarsal segments longitudinally furrowed both at middle and laterally (Fig. 318) .... 237. *thoreyi* Dejean

- Tarsal segments furrowed laterally only 24

24(23) Hind-wings (with rare exceptions) strongly reduced. The posterior setiferous puncture of pronotum well removed from hind-angle. (Non-metallic species) .... 236. *fuliginosum* (Panzer)

- Hind-wing full. Posterior setiferous puncture of pronotum situated close to hind-angle 25

25(24) Elytra or at least epipleura pale brown. Legs or at least tibiae pale 26

- Antennae and legs black (1st antennal segment and tibiae sometimes dark piceous in *gracile*). 27

26(25) Elytra yellow or pale brown, paler than pronotum. Striae exceedingly fine .... 234. *piceum* (Linnaeus)

- Elytra with black or piceous ground colour, not or barely paler than pronotum. Striae deeper with convex intervals .... 233. *micans* (Nicolai)

27(25) Black without metallic hue. Pronotum with completely rounded hind-angles. Microsculpture of elytra reticulate, almost isodiametric, only in spots arranged in transverse rows .... 235. *gracile* Sturm

- Upper surface with a more or less pronounced metallic lustre (least evident in *munsteri*). Pronotum with more or less evident hind-angles (least so in *munsteri*). Elytral microsculpture at least in part consisting of transverse meshes 28

28(27) Only with a suggestion of bronze hue. Hind-angles of pronotum very rounded (sometimes almost disappeared) .... 239. *munsteri* (Hellén)

- Upper surface with metallic, brassy, more rarely bluish

or greenish lustre (un-metallic specimens only quite exceptional). Pronotum with evident, though obtuse, hind-angles 29

29(28) Pronotum (Fig. 310) with well developed hind-angles. Terminal antennal segments short (Fig. 314) . 240. *exaratum* (Mannerheim)

– Hind-angles of pronotum (Fig. 309) only suggested. Antennal segments as in fig. 313 .... 238. *consimile* (Gyllenhal)

Subgenus *Anchomenus* Bonelli, 1810

Anchomenus Bonelli, 1810, Obs. Ent. 1 (Tab. Syn.).
Type-species: *Carabus prasinus* Thunberg, 1784 (= *Carabus dorsalis* Pontoppidan, 1763).
Clibanarius des Gozis, 1882, Mitt. Schweiz. Ent. Ges. 6: 295 (*nec* Dana, 1852).
Type-species: *Carabus dorsalis* Pontoppidan, 1763.
Idiochroma Bedel, 1902, Cat. rais. Col. NAfr. 1: 216.
Type-species: *Carabus dorsalis* Pontoppidan, 1763.

211. ***Agonum dorsale*** (Pontoppidan, 1763)
Fig. 300, pl. 6: 13.

Carabus dorsalis Pontoppidan, 1763, Danske Atlas 1: 678, pl. 29.
Carabus prasinus Thunberg, 1784, Nova Acta R. Soc. Scient. Upsal. 4: 74.

6.0-8.2 mm. A very characteristic species by the sharp bicoloured pattern. Forebody metallic green. Elytra bright rufo-testaceous with a large metallic green macula across the suture. Base of antennae and entire legs rufo-testaceous. Forebody narrow, pronotum only slightly wider than head. Wings full.

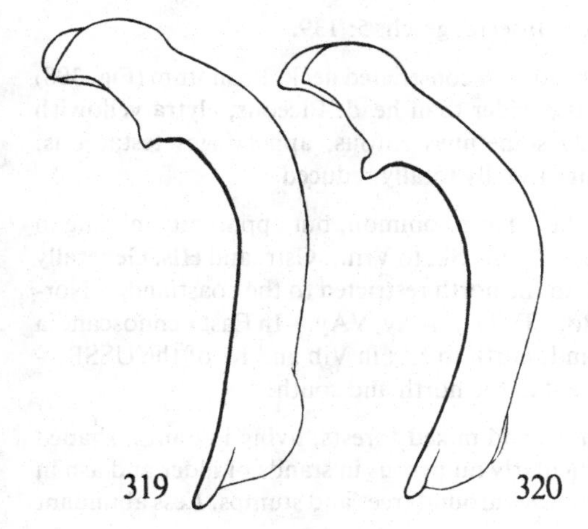

Figs 319, 320. Penis of *Agonum*. – 319: *moestum* (Dft.) and 320: *viduum* (Pz.).

Figs 321, 322. Apex of penis of *Agonum*. – 321: *munsteri* (Hell.) and 322: *consimile* (Gyll.).

Distribution. Denmark: very distributed and common. The records from the western and northern parts of Jutland are all from after 1950 and probably represent recent invasion. - Sweden: scattered distributed from Sk. to Vstm. Rather common in Sk., Öl., Gtl.; in other areas usually rare and very local. - Norway: Ø, VA, AAy and VAy. - Finland: Ab, N; locally common around Helsinki. - Entire Europe north to 60°N; also Morocco, W.Asia and W.Siberia.

Biology. The least hygrophilous of all *Agonum,* occurring in open meadows and grassland, usually on gravelly or clayey, often limestone soil. Also on arable land, particularly in winter crops on heavy soil. It feeds on aphids, insect larvae, etc. Predominantly night-active; during daytime it can be found gregariously under stones. The species is most numerous in May-June, which is the breeding season. Dicker (1951) observed large numbers of eggs deposited on the underside of strawberry leaves, each egg being enclosed in a sheath of soil. The young adults appear in August-September. They hibernate in hedges and open woodland, and are frequently found in huge aggregations in the winter quarters.

Subgenus ***Oxypselaphus*** Chaudoir, 1843

Oxypselaphus Chaudoir, 1843, Bull. Soc. Nat. Mosc. 16: 415.
Type-species: *Oxypselaphus pallidulus* Chaudoir, 1843 (= *Carabus obscurus* Herbst, 1784).
Anchus LeConte, 1854, Proc. Acad. Nat. Sci. Philad. 7: 38.
Type-species: *Anchus pusillus* LeConte, 1854 (= *Agonum puncticeps* Casey, 1920).

212. ***Agonum obscurum*** (Herbst, 1784)
Fig. 308.

Carabus obscurus Herbst, 1784, Arch. Insectengesch. 5: 139.

5.0-6.6 mm. Forebody very narrow, head with constricted neck. Pronotum (Fig. 308) with basal half very narrow, only little wider than head. Piceous, elytra yellowish brown, palest at shoulder, forebody sometimes rufous, appendages testaceous. Elytral striae strongly punctate. Wings usually totally reduced.

Distribution. Denmark: very distributed and common, but apparently missing in the larger part of WJ. - Sweden: reported from Sk. to Vrm., Gstr. and Hls. Generally distributed and common. From Upl. to the north restricted to the coastland. - Norway: only in the south-east: Ø, AK, Bø, VE, TEy, AAy, VAy. - In East Fennoscandia rather common in the south of Finland, north to Kb; in Vib and Kr of the USSR. - Holarctic; in Europe everywhere except in the north and south.

Biology. Predominantly in deciduous and mixed forests, living in damp, shaded sites among litter and moss. It is particularly numerous in stands of alder and ash in forest swamps, occurring among wet leaves around trees and stumps. Less abundant in peaty woods between pillows of *Sphagnum*. Also in densely vegetated marshes in open country. It is a nocturnal and predominantly carnivorous species which feeds on Collembola, mites, etc. Propagation normally takes place in spring (Larsson, 1939; Barndt, 1976), but findings of young imagines in April-May (Lindroth, 1945) indicate that autumn breeding may also occur.

Subgenus *Paranchus* Lindroth, 1974

Paranchus Lindroth, 1974, Handb. Ident. Brit. Ins. 4 (2): 81.
  Type-species: *Carabus albipes* Fabricius, 1796.
Anchomenus auctt. *nec* Bonelli, 1810.

213. ***Agonum albipes*** (Fabricius, 1796)
  Fig. 304.

Carabus ruficornis sensu Goeze, 1777, Ent. Beytr. 1: 663; *nec* Degeer, 1774.
Carabus albipes Fabricius, 1796, Ind. Alph. Ent. Syst.: 33.
Carabus pallipes Fabricius, 1801, Syst. Eleuth. 1: 187; *nec* 1787.

6.8-9.0 mm. Distinguished from all other species by the presence of only 2 dorsal punctures on elytra, and from all species except *thoreyi* by the medially furrowed tarsal segments. Piceous, sides and suture of elytra usually brown, all appendages pale testaceous; young individuals long retain a rufous colour. Base of pronotum strongly punctate (Fig. 304), hind-angles sharp. Elytral striae fine, impunctate, without subapical sinuation of sides. Hind-wings always with a reflexed apex but are somewhat varying in size; probably always too small to be functionary.

Distribution. Denmark: rather distributed and uncommon in eastern Jutland (SJ + EJ) and on the islands; also in western SJ (Rudbøl, Ribe å), NWJ (Vilsund); no records from WJ and NEJ. – Sweden: scattered distributed north to Dlr.; not rare but local. No records from large parts of Sm. Most finds along the coast and around the greater lakes. – Norway: in coastal areas from Ø to HO; moreover in STy at 64°N. – In East Fennoscandia only known from the south coast of Finland (locally common) and from Vib of the USSR. – Europe north to 60-61°N, south to S.Spain, S.Italy, Yugoslavia. Also in Asia Minor and N.Africa. In eastern N.America.

Biology. A shore-dwelling species, predominantly occurring on the banks of great, eutrophic at well as oligotrophic, lakes and on seashores; less often at small stagnant waters and on river banks. The adults usually live near the water-edge, in open sites on almost barren soil, notably clay-mixed sand or gravel. They are frequently found in abundance under stones or among plants washed ashore. It is a strictly nocturnal species. According to Larsson (1939) it has unstable breeding habits, the greater part of the population being autumn breeders, the rest reproducing in spring.

Subgenus *Platynus* Bonelli, 1810

Platynus Bonelli, 1810, Obs. Ent. 1 (Tab. Syn.).
Type-species: *Carabus angusticollis* Fabricius, 1801 (= *Carabus assimilis* Paykull, 1790).
Limodromus Motschulsky, 1864, Bull. Soc. Nat. Mosc. 37 (2): 316.
Type-species: *Carabus angusticollis* Fabricius, 1801.

214. *Agonum livens* (Gyllenhal, 1810)

Harpalus livens Gyllenhal, 1810, Ins. Svec. 1(2): 149.

7.5-10.5 mm. This and the following species are distinguished by the entirely rounded hind-angles of pronotum. Piceous, head almost constantly with two rufous spots on frons. Appendages dark rufous but apex of each antennal segment and often in part legs infuscated. Narrow, somewhat reminiscent of *Patrobus* in general habitus. Head with a transverse constriction behind eyes, visible also dorsally. Pronotum strongly narrowing towards base, hind-angles only suggested. Elytral striae fine, intervals faintly convex; 3 dorsal punctures. Wings full.

Distribution. Denmark: scattered occurrence; especially found on the islands; has become very rare after 1950. In Jutland only found in 2 localities after 1950: Gallehus and Draved skov (both SJ). - Sweden: rather distributed in the east from Sk. to Upl., Gstr.; also one locality in Hls. In the west to Hall., Vg. and southeastern Vrm. Usually rather rare. - Not in Norway. - In Finland widely distributed (north to Ob) but not common; also in Vib and southern Kr in the USSR. - Europe south to N.Spain, C.Italy and Serbia; W.Siberia.

Biology. In marshy deciduous forests, notably in stands of alder and birch, living among litter and moss. In Finland, Krogerus (1960) found the species to be especially typically of peaty wood, on slightly acid soil in *Sphagnum*-mosses, regularly occurring together with *Trechus rivularis.* Also on clayey soil without *Sphagnum,* occurring among leaves at the margin of pools. It is most numerous in May-June, when reproduction takes place, and in the autumn, when the young beetles emerge. Hibernating adults are sometimes found in great aggregations under bark of trees.

215. *Agonum mannerheimii* (Dejean, 1828)
Fig. 301.

Anchomenus Mannerheimii Dejean, 1828, Spec. Gén. Col. 3: 104.
Anchomenus morio Gebler, 1847, Bull. Soc. Nat. Mosc. 20 (1): 325.

10-11.5 mm. Much broader and more convex than *livens* but with similar pronotum (Fig. 301). Black or dark piceous, frons with two dark rufous spots, base of each antennal segment, tibiae and tarsi rufous brown. Basal transverse impression of head weaker than in *livens.* Elytra widening apically, striae deep, intervals strongly convex;

number of dorsal punctures varying from 3 to 6; the microsculpture as in *assimile*. Hind-wings seemingly full, with reflexed apex, but too small to be functionary.

Distribution. Not in Denmark. - Sweden: very rare and only found in the north in a few localities in Nb. and one locality in Ås. Lpm.; also isolated in Vrm.: Stömne (1951) - Norway: very rare, only Os. - In East Fennoscandia very rare but recorded from all over Finland (mostly old finds), and also from Vib and Kr of the USSR. - In NE.Europe and east to W.Siberia.

Biology. A strongly hygrophilous species, favouring *Sphagnum* swamps with *Comarum, Rubus chamaemorus,* etc. According to Krogerus (1960) it is especially typical of soligenous swamps in more or less dark spruce forest, occurring among moss and litter or under bark of trees and stumps. Reproduction takes place in spring.

216. ***Agonum assimile*** (Paykull, 1790)
Fig. 302; pl. 6: 15.

Carabus assimilis Paykull, 1790, Mon. Car. Suec.: 53.
Carabus angusticollis Fabricius, 1801, Syst. Eleuth. 1: 182.

8.7-12.3 mm. At once recognized by the sharp, protruding hind-angles of pronotum (Fig. 302). Shining black, mouth-parts, antennae and legs dark piceous. Head as in other members of the subgenus transversely impressed behind the eyes. Sides of pronotum broadly depressed, its base finely but continously bordered. Elytra very broad, widening apically; striae deep, intervals strongly convex; 3 dorsal punctures; microsculpture consisting of somewhat transverse, slightly irregular meshes, arranged in transverse rows. Wings with reflexed apex, small, and probably not functionary.

Distribution. Denmark: distributed and rather common in SJ, EJ, eastern NEJ, and all districts of the islands. Isolated records in WJ (Nørholm skov) and NWJ (Struer). - Sweden: rather common and distributed in the western part of southern Sweden, north to Dlr. In the eastern part very local or lacking, found north to southern Vb. - Norway: north to Ns, but mainly in the southern and eastern districts. - Finland: rather common in the south, scattered records north to Lk; also Vib and Kr of the USSR. - Most parts of Europe, north to 64-66°N, south to N.Spain, S.Italy and Bulgaria. Also W.Asia and Siberia, east to the Amur region.

Biology. A stenotopic species of deciduous forests on mull-rich soil, occurring in cool and wet, shaded habitats among litter, under bark of tree stumps, etc., often near water. It is particularly characteristic of stands of *Alnus* and *Fraxinus*. Also in damp and shaded sites in parks and gardens. Under experimental conditions, Thiele (1964, 1967) found the species to prefer dryness and to be very resistent to desiccation. Its choice of a cool, damp microclimate is determined by the preference for low temperatures. *A. assimile* is a nigh-active species having spring propagation. Hibernating adults usually aggregate in fallen trees and stumps.

217. ***Agonum krynickii*** (Sperk, 1835)
Fig. 303.

Anchomenus Krynickii Sperk, 1835, Bull. Soc. Nat. Mosc. 8: 151.
Anchomenus uliginosus Erichson, 1837, Käf. Mark Brandenburg 1: 107.

10.5-12.8 mm. Very similar to *assimile* but more dull black. Pronotum (Fig. 303) broader, sides hardly sinuate, except just in front of the denticulate hind-angles. The base, at least at middle, without marginal bead. Elytra somewhat more stretched, less widening posteriorly; their microsculpture is stronger, consisting of completely isodiametric meshes without transverse arrangement. Wings seemingly full, but probably not functioning.

Distribution. Denmark: very rare, only recorded from LFM (3 localities), SZ (2 localities), and NEZ (Malmmosen, Jægersborg dyrehave 1969). - Sweden: very rare and with a restricted distribution in the southeast. It has certainly been decreasing during this century. Earlier found in Sk., Sm. (one loc.), Öl. and Gtl. In recent years only on Öl. - Not in Norway or East Fennoscandia. - An East European species penetrating towards west to the Elbe and C.Italy, north to the Leningrad-area, east into Siberia.

Biology. Decidedly stenotopic, being confined to dark and marshy habitats in mull-rich deciduous forests. The species usually occurs among rich vegetation, e.g. of *Filipendula ulmaria*. It is found in litter and under moss and bark of tree stumps, etc. Often found together with *A. assimile*. It is a spring breeder.

218. ***Agonum longiventre*** (Mannerheim, 1825)

Platynus longiventris Mannerheim, 1825, *in* Hummel: Essais Ent. 4: 22.

12.4-14.0 mm. Largest species of the genus. Characteristic by the stretched, almost parallel-sided elytra (more than 3 times as long as pronotum). Coloured as *assimile,* and also with the same microsculpture and the same complete border of pronotum. But the hind-angles are small, denticulate as in *krynickii,* and form of pronotum is broader, less narrowing at base than in both the preceding species; its margins are conspicuously lifted basally. Wings full.

Distribution. In our area only found in Sweden: very rare with a very restricted distribution around the lower parts of river Dalälven (Upl., Dlr., Gstr.). Usually single or few specimens found. - An eastern species in Europe, found nearest to our area in Estonia and in the Leningrad-area. Absent from western Europe. Widespread in W.Siberia.

Biology. Like the two preceding species this carabid is confined to dark, moist deciduous woodland. In Sweden it has been encountered in old, temporarily inundated forest, notably in richly vegetated habitats, under the bark of dead trunks of *Populus tremula,* often together with *A. assimile.*

Subgenus *Agonum* s.str.

219. ***Agonum marginatum*** (Linnaeus, 1758)
Pl. 6: 12.

Carabus marginatus Linnaeus, 1758, Syst. Nat. ed. 10: 416.

8.8-10.4 mm. Metallic green or brassy with broadly yellow elytral sides. The only other similarly coloured Carabid is *Chlaenius vestitus* (Pl. 7: 18). Black, also underside with greenish reflection; 1st antennal segment and legs dark rufous, tibiae paler. Uneven intervals of elytra somewhat more convex. Wings full.

Distribution. Denmark: rather distributed and rather common; known from all districts. - Sweden: distributed from Sk. to Vrm.; uncommon, especially since 1960. Distribution more scattered in the east. - Norway: Ø, AK, Bø, VE, Ry. - Finland: only known from southernmost Finland, in a few places locally abundant; also Vib and southern Kr of the USSR. - In Europe north to 60-61°N, south to S.Spain, S.Italy and Greece; also N.Africa and W.Asia.

Biology. On clayey or clay-mixed sandy soil near water, in Scandinavia preferably in coastal salt meadows and on the seashore, usually occurring among grasses, *Carex, Juncus,* etc. Also inland, for instance at the margin of lakes and ponds and in clay-pits. In C.Europe the species occurs to a larger extent on river banks. It is a heat-preferent species, often running in warm sunshine. Breeding takes place in spring.

220. ***Agonum sexpunctatum*** (Linnaeus, 1758)
Pl. 6: 11.

Carabus 6-punctatus Linnaeus, 1758, Syst. Nat. ed. 10: 416.

6.5-8.0 mm. Normally with strong colour contrast between forebody and elytra. Black, lower surface greenish; head and pronotum vividly green, elytra coppery, usually with green margins; also femora and first antennal segments metallic. The colour of the forebody may be changed to bluish or almost black, that of the elytra into yellow-green or violet, with less contrast as a result. Quite black (probably old) specimens are very rare ("ab. *montanus*"). Sides of pronotum broadly flattened, elevated posteriorly; hind-angles completely rounded. Elytra broad with almost parallel sides; 3rd interval with (4-) 6-7 (-8) well-defined punctures. Elytral microsculpture regularly isodiametric. Wings full.

Distribution. Denmark: rather distributed and common in Jutland, northern Zealand, and Bornholm; more scattered in the rest of the area. - Sweden: rather common and widespread; known from all districts except Hrj. - Norway: mainly in the east and southeast; also in Ry, SFi, NTy and Fø. - In East Fennoscandia common all over the country of Finland, and in Vib and Kr of the USSR. - Palaearctic, in Europe only absent from the extreme north and parts of the south; east to the Ussuri region.

Biology. On open, sun-exposed, moist ground, usually near water, e.g. in meadows, grassland, damp heaths, and woodland glades. Predominantly on peaty, but also on clayey soil rich in mosses but otherwise sparsely covered with grasses, *Carex,* dwarf-shrubs, etc. It is strongly heliophilous, often running in the sunshine. The breeding period is in spring.

221. ***Agonum impressum*** (Panzer, 1797)

Carabus impressus Panzer, 1797, Fauna Ins. Germ. 37: 14.

8-10 mm. In general habitus very like *sexpunctatum* but with less colour contrast and elytra strongly foveate. Underside and femora greenish black, remaining appendages black without metallic reflection. Upper surface coppery or bronze; head, usually centre of pronotum and sidemargin of elytra, metallic green. Pronotum formed as in *sexpunctatum,* rather strongly punctate in the basal foveae. Elytral striae and intervals similar, except that the 5-7 dorsal punctures on 3rd interval are situated in large foveae. The ♂ has shiny, the ♀ quite dull elytra.

Distribution. In our area only in East Fennoscandia; USSR: southern Kr at lake Ladoga. - Palaearctic, in Europe mainly in the east, penetrating to the west into the Netherlands, NE.France and N. Italy.

Biology. At the margin of rivers and lakes, occurring on sandy or gravelly soil. On the shores of lake Ladoga the species has been found on fine, humus-mixed sand, in vegetation of *Scirpus, Agrostis* and *Juncus.*

222. ***Agonum ericeti*** (Panzer, 1809)

Carabus ericeti Panzer, 1809, Fauna Ins. Germ. 108: 6.
Harpalus bifoveolatus Sahlberg, 1827, Ins. Fenn. 1: 258.

6.5-8.5 mm. Entire upper surface unicolorous or almost so. Extremely variable: golden green, coppery, bluish, rarely unmetallic black. Appendages black. Pronotum narrower, more constricted basally, greatest width before middle; hind-angles usually suggested. Elytra more oviform, as a rule with more protruding shoulders. 4-7 dorsal punctures. Elytral microsculpture more irregular. Wings with reflexed apex, though more or less reduced (as least in our area), and not functionary.

Distribution. Denmark: very rare and only recorded from Jutland. Most records are from the central parts, but also SJ (Draved, Ø. Løgum), EJ (Svanemose), WJ (Vemb) and NEJ (Lundby bakker). - Sweden: rather rare and with scattered occurrence; especially very local in the north. - Norway: scattered in the southeast and southwest; also found in ST and Fø. - East Fennoscandia: rather common over most of Finland and in Vib and Kr of the USSR. - Scattered distribution in N., C. and E.Europe, south to N.Spain and N.Italy.

Biology. A stenotopic species, living almost exclusively in ombrotrophic *Sphagnum*-bogs (Mossakowski 1970a), both in wet *Sphagnum* and on fairly dry ground, for instance in hummocks with *Calluna, Vaccinium, Rubus chamaemorus,* etc.; often near peat-cuttings. According to Krogerus (1960) *A. ericeti* prefers strongly acid soil (pH 3.6-4.6) and a rather dry and warm microclimate. In bright sunshine the beetle often runs about on bare spots of dry, dark peat, where the temperature may become very high. In contrary to most other members of the genus *A. ericeti* has a poorly developed flying ability and seems to be rather stationary. It is most numerous in May-June, which is the breeding period.

223. ***Agonum gracilipes*** (Duftschmid, 1812)

Carabus gracilipes Duftschmid, 1812, Fauna Austriae 2: 144.

6.5-9 mm. A very slender species with long legs and antennae. Upper surface bronzed, 1st antennal segment and at least tibiae and tarsi reddish brown. Marginal depression of pronotum sharply delimited, hind-angles evident. Elytral striae extremely fine, intervals completely flat, basal margin strongly arcuate, dorsal punctures 4-6. Wings full.

Distribution. Denmark: very rare and decreasing in number since 1900. In Jutland only 2 records of single specimens; F: Langeland 1 specimen 1979. Most records are from LFM, SZ, NWZ, NEZ and B. - Sweden: very rare, only recorded from Sk., Bl., Hall., Öl., Gtl., G. Sand., Ög., Vg., Upl. and Hls. Usually only found singly; only on the south and east coast of Sk. sometimes numerous though always local. - Norway: rare, only On and TEi (both inland localities). - In East Fennoscandia only recorded from Al: Kökar and N: Tvärminne (in sea drift). - W., C. and E. Europe; W.Siberia.

Biology. *A. gracilipes* is a species of great vagility showing migratory tendencies, in Scandinavia rarely establishing populations of long duration. It is mostly found on sandy seashores, e.g. under seaweed; only sporadically occurring on inland shore localities. The species regularly flies at night and may come to light. It is a spring breeder.

Note. In periods of the 19th century the species regularly occurred, and in some periods reproduced, in our area. In the 20th century it has been found with strongly decreasing frequency; no permanent populations seem to exist.

224. ***Agonum muelleri*** (Herbst, 1784)

Carabus Mülleri Herbst, 1784, Arch. Insectengesch. 5: 139.
Carabus parumpunctatus Fabricius, 1801, Syst. Eleuth. 1: 199.

7.2-9.5 mm. Forebody normally vividly green, contrasting against the brassy elytra. Unicolorous green, brass, bronze or even bluish specimens are rare. First antennal segment rufous, tibiae and at least base of femora yellowish brown. Rather similar to *sexpunctatum,* but pronotum broader with narrower marginal deplanation and suggested hind-angles. Elytral striae fine to apex, almost impunctate, intervals flat; 3 (-4) dorsal punctures. Microsculpture of elytra consisting of transverse meshes. Wings full.

Distribution. Denmark: very distributed and common. - Sweden: common and distributed in the southern half of the country, in the northern parts lacking or very rare. - Norway: north to NTi. - Finland: rather common, north to Ks. - A widespread European species, north to about 64°N.

Biology. The most eurytopic member of the subgenus, living in habitats of widely differing humidity, usually on clayey soil on open, sun-exposed ground with sparse vegetation, e.g. on moderately dry meadows. It is a common inhabitant of agricultural land, mainly occurring in cereal fields on heavy soil. Also in gardens and in clear-felled areas in forests. The beetle is strongly heliophilous. The propagation period is in spring.

225. ***Agonum sahlbergii*** (Chaudoir, 1850)

Anchomenus Sahlbergii Chaudoir, 1850, Bull. Soc. Nat. Mosc. 23 (2): 117.
Anchomenus Archangelicus J. Sahlberg, 1875, Notis. Sällsk. Faun. Fl. Fenn. Förh. 14: 118.

7.5-8.3 mm. Upper surface unicolorous, from green or coppery to almost black. Appendages darker than in *muelleri*. Pronotum narrower and flatter, with lateral reflection more pronounced basally. Basal margin of elytra less arcuate, their microsculpture almost regularly isodiametric. Lateral furrows of basal meta-tarsal segments stronger. Wings full.

Distribution. In our area only at the White Sea in the Soviet part of East Fennoscandia. Also in Siberia. Formerly taken in Scotland at Glasgow (probably an introduction).

Biology. In our area confined to the sandy shores of the White Sea, for instance occurring under seaweed.

226. ***Agonum dolens*** (Sahlberg, 1827)

Harpalus dolens Sahlberg, 1827, Ins. Fenn. 1: 256.

6.7-8.3 mm. A rather small species, most like *versutum* but separated on colour, hind-angles of pronotum and elytral microsculpture. Black or piceous, upper surface with dark bronze or brassy lustre, rarely bluish or greenish, or quite black. First antennal segment slightly pale (but less so than in *versutum*), femora clear rufous brown (at least in inner half), usually also tibiae. Hind-angles of pronotum just outside the seta delimited as a small, blunt denticle. Marginal deplanation evident. Elytral microsculpture consisting of evident, almost isodiametric meshes, arranged in irregular transverse rows. The meta-tarsal segments keeled as in *viduum* (cf. fig. 316).

Distribution. Denmark: very rare and certainly not breeding. Only 5 scattered records, mostly from coastal localities. - Sweden: distributed all over the country but local. Most abundant in the north, rare in the southern areas. - Norway: mainly in the southeast; moreover in STi and NTi. - East Fennoscandia: rather rare but distributed all over Finland as well as in the Soviet parts. - NE.Europe, south to the Netherlands and C.Germany, east to the Amur region and Kamchatka.

Biology. Strictly riparian, occurring on muddy lake shores and river banks, usually in moderately dense *Carex* vegetation. The beetles often live gregariously near the water edge; sometimes in company with *Pelophila borealis*. It is most numerous in spring, when breeding takes place.

227. ***Agonum versutum*** Sturm, 1824
Fig. 315.

Agonum versutum Sturm, 1824, Deutschl. Fauna Ins. 5 (5): 191.

7.0-8.6 mm. Similar to a small *viduum* but with a different elytral microsculpture. Black, upper surface dark bronzed, often touching upon green or blue, seldom unmetallic, epipleura usually pale. First antennal segment, tibiae and sometimes innermost part of femora, dark rufous. Pronotum shorter than in *viduum*, usually clearly wider in posterior half. Elytral intervals flat anteriorly, or almost so. Their microsculpture consisting of transverse lines without evident meshes. Lateral furrows of meta-tarsi shallow (Fig. 315). Wings full.

Distribution. Denmark: scattered distributed in Jutland, NEZ and B. Very scattered or missing in the rest of the country. - Sweden: rather common and widespread in the south, north to Dlr. and Hls. Rare in northern Sweden and almost missing in Lapland, Jmt. and Hrj. - Norway: in the southeast, from VA north to Os. - Finland: rather common and recorded north to Ob; also Vib and Kr in the USSR. - In Europe south to C.France, N.Italy and Bosnia; through Siberia east to the Amur region.

Biology. A very hygrophilous species, living on soft muddy surface at the border of standing or slowly running fresh waters. It occurs on more or less acid soil poor in nutrients as well as on rich clayey soil; usually in exuberant vegetation of *Carex*, *Glyceria*, etc., notably on mossy ground. The species prefers open country, but is also found in moderately shaded sites, e.g. in open forest. Often in company with *A. viduum*. It is most numerous in spring and early summer, when breeding takes place. The adults have been found hibernating in decaying trunks.

228. ***Agonum viduum*** (Panzer, 1797)
Figs 316, 320.

Carabus viduus Panzer, 1797, Fauna Ins. Germ. 37: 18.
Carabus obscurus Paykull, 1790, Mon. Car. Suec.: 75; *nec* Herbst, 1784.

7-9 mm. Upper surface metallic, best seen laterally behind shoulder, and thus best separated from the otherwise very similar *moestum*. Deep black, upper surface with greenish or bronze, rarely bluish lustre; specimens with reduced metallic reflection are no doubt old and worne. First antennal segment and tibiae black or dark piceous. Pronotum large with sides more rounded than in *versutum*, widest before middle.

Elytral intervals more or less convex. Microsculpture obsolete on disc of pronotum, on the elytra forming transverse meshes arranged in transverse rows. Dorsal keel of basal meta-tarsal segments (Fig. 316) strong only basally. Wings full and probably functional in all individuals. Penis (Fig. 320) evenly arcuate.

Distribution. Denmark: rather distributed and rather common; known from all districts. - Sweden: distributed in about the same area as *versutum,* but more common than this species, especially in the south up to river Dalälven. - Norway: north to NTi, but apparently lacking in HO and SF. - Finland: rather common, distributed north to Ks; also in Vib and Kr of the USSR. - In Europe south to N.Spain and C.Italy; also Siberia east to river Lena.

Biology. A eurytopic shore-dweller, living at the margin of all kinds of fresh, even brackish waters, on different types of muddy soil, usually among rich vegetation of e.g. *Carex.* In the shore zone of eutrophic lakes it is often found together with *A. moestum,* at oligo- and dystrophic waters frequently with *Bembidion doris* and *Pterostichus diligens.* It prefers open, sun-exposed ground, but also occurs in somewhat shaded sites, for instance in open woodland. *A. viduum* in most numerous in May-June which is the breeding period.

229. ***Agonum moestum*** (Duftschmid, 1812)
Fig. 319.

Carabus moestus Duftschmid, 1812, Fauna Austriae 2: 138. Primary homonym of *Carabus moestus* Gmelin *in* Linnaeus, 1790.
Harpalus emarginatus Gyllenhal, 1827, Ins. Suec. 1 (4): 450.

8-9.5 mm. Constantly unmetallic black. First antennal segment and tibiae often dark piceous, as in *viduum.* It was earlier regarded as a dark variety of this. Also, the depressed lateral part of pronotum is somewhat less pronounced, the elytral intervals a little less convex. The microsculpture of the pronotal disc is forming evident meshes, and on the elytra it usually shows a less clearly transverse arrangement. The dorsal keel on the two basal meta-tarsal segments is sharper. Penis (Fig. 319) less arcuate ventrally.

Distribution. Denmark: rather distributed in eastern Jutland and on the islands including B. Obviously absent from large areas of western and northern Jutland. - Sweden: found from Sk. to southern Hls. Rather common in Sk., Öl, Gtl. and around Mälaren; in the rest of the area sparse or lacking. - Norway: only Ø and AK. - In Finland rather rare, recorded north to Om; also records from Vib and Kr in the USSR. - Europe north to 61°N, south to C.Spain, C.Italy and Greece, east to the Amur region.

Biology. More stenotopic than the preceding species, living in more or less shaded sites on clayey or muddy shores of eutrophic lakes and ponds, e.g. among *Phragmites.* Also in marshy deciduous forest, notably in alder swamps with a rich vegetation of *Carex, Lysimachia vulgaris, Solanum dulcamara,* etc. It is a spring breeder; most numerous in May-June.

Note. *A. emarginatum* (Gyll.) was described as a distinct species and is slenderer with more parallel-sided elytra. It is apparently a macropterous form of *moestum*, which otherwise has somewhat reduced and probably non-functional wings.

230. ***Agonum lugens*** (Duftschmid, 1812)
Fig. 317.

Carabus lugens Duftschmid, 1812, Fauna Austriae 2: 139.

8.6-9.8 mm. Deep black, pronouncedly dull due to the strong microsculpture. Appendages not paler. Elytra more stretched with sides more parallel than in the four preceding species. Pronotum with hind-angles suggested as a very small denticle (as in *dolens*). The microsculptural meshes of elytra are exactly isodiametric without any transverse arrangement. All meta-tarsal segments, also the last, with a median keel (Fig. 317). Wings always full and functionary.

Distribution. Denmark: very rare, and with strongly decreasing abundance. Since 1950 only one record: SZ, Tystrup sø 1958. A number of older records, especially from LFM and SZ, probably represent immigrating specimens. - Sweden: distributed but rare around Mälaren (Sdm., Upl., Vstm.); in other areas (Sk., Öl., Gtl.) usually very rare and decreasing in number. In Sk. only one record of a breeding population in this century: Ivö 1968. - Not in Norway. - In East Fennoscandia only recorded from Al (Finström and Föglö). - From N.Africa and the Iberian Peninsula over C. and E.Europe to the Urals.

Biology. On soft, muddy soil, notably at the margin of eutrophic lakes, occurring among rich and tall vegetation of e.g. *Phragmites, Scirpus* and *Glyceria,* often in somewhat shaded sites under bushes and in sparse tree growth. Occasionally found on seashores. The species normally propagates in spring and is most numerous in May-June. The adults hibernate in tree stumps, fallen trunks, etc.

Subgenus *Sericoda* Kirby, 1837

Sericoda Kirby, 1837, *in* Richardson, Fauna Bor. Amer. 4: 14.
Type-species: *Sericoda bembidioides* Kirby, 1837.
Agonodromius Reitter, 1908, Fauna Germ. 1: 239.
Type-species: *Carabus quadripunctatus* Degeer, 1774.

This is no doubt the most primitive subgenus of *Agonum,* above all manifested by the precense of 2-4 stiff terminal setae on both parameres of the ♂.

231. ***Agonum bogemannii*** (Gyllenhal, 1813)
Fig. 306.

Harpalus Bogemannii Gyllenhal, 1813, Ins. Suec. 1(3): 697.

6.4-8 mm. Flat, with parallel-sided elytra, in general habitus like a *Dromius.* Black, underside more piceous, but appendages not paler. Antennae stout, shorter than half body length. Head as broad as pronotum over front-angles. Pronotum (Fig. 306) with hind-angles strongly obtuse, the disc with fine, irregular transverse lines. Elytra with very fine, slightly irregular striae, with the even intervals somewhat broader; apex almost truncate. Microsculpture of elytra isodiametric, almost granulate. Three faint dorsal punctures.

Distribution. Not in Denmark. - Sweden: extremely rare, only single specimens known from the 19th century in Sm., Vg., Boh., Hls. and "Lapponia". - Finland: extremely rare; one old record from Ab; further records from Lk: Kittilä and Lk: Pelkosenniemi (in the 1970's), also from "Tavastia" and "Lapponia". - Holarctic with disrupted distribution; from western N.America to Scandinavia; also found in the mountains of C.Europe. Apparently a species on its way towards extinction.

Biology. Predominantly an inhabitant of the high boreal coniferous region, occurring in burnt forest, usually under the bark of trees damaged by fire. It has been observed flying towards a forest fire and settle on the hot ashes (Lindroth, 1972). During the Second World War it appeared in burnt forests in eastern Karelia.

232. ***Agonum quadripunctatum*** (Degeer, 1774)
Fig. 307.

Carabus quadripunctatus Degeer, 1774, Mem. Hist. Ins. 4: 102.

4.5-5.8 mm. A small, *Bembidion*-like species, distinguished by (3-) 4 (-5) deep dorsal foveae on the 3rd interval of elytra. Dull black with bronze hue, tibiae more or less piceous. Pronotum short (Fig. 307), hind-angles obtuse. Microsculpture very strong; on the elytra consisting (on each interval) of alternating patches of parallel, differently oblique strigulae; hence the impression of a pattern of silvery and black spots, changing according to light angle. To the human eye this attire is extremely similar to a piece of charcoal.

Distribution. Denmark: very rare and usually only found in single specimens. Taken in numbers 1979-81 at Gadbjerg near Vejle (EJ) in a burned section of a wood. - Sweden: widespread throughout the country but rare and lacking over large areas. - Norway: mainly in the eastern districts: AK, HEs, Os, TE, AAy, ST, NTi. - In East Fennoscandia widely distributed, but rarely observed, over the entire area. - In E.Europe and eastwards to Siberia, Kamchatka; N.America.

Biology. Like the preceding species and *Pterostichus quadrifoveolatus* this species is attracted to sites of fire. It occurs in forests, notably in coniferous and mixed stands, and near human habitations. Often found under charred branches on the ground and under the bark of tree stumps damaged by fire, sometimes in company with *quadrifoveolatus.* The species has a high power of dispersal and has repeatedly been found in drift material on the seashore. It is a spring breeder; newly emerged beetles have been encountered in late summer and autumn.

Subgenus *Europhilus* Chaudoir, 1859

Europhilus Chaudoir, 1859, Stett. Ent. Zeit. 20: 124.
Type-species: *Agonum micans* Nicolai, 1822.

Dorsal punctures at least 4. Wings full, except normally in *fuliginosum*.

233. *Agonum micans* Nicolai, 1822

Agonum micans Nicolai, 1822, Diss. Col. Agr. Hal.: 19.
Carabus pelidnus sensu Duftschmid, 1812, Fauna Austriae 2: 144; *nec* Paykull, 1792.

6.2-7.4 mm. Elytra black or dark piceous, often with somewhat paler ground-colour, their epipleura yellowish brown; upper surface almost constantly bronzed. First antennal segment, palpi and legs dark rufous with paler tibiae. Pronotum with reflexed margin strongly widened inside hind-angles, which are clearly suggested. Elytra widening behind middle, striae moderately deep. Their microsculpture as in *piceum*.

Distribution. Denmark: very rare in Jutland: SJ (Højer 1886), EJ (near Horsens 1978), and on Funen (Tranekær 1980, Sønder Broby 1937). Breeding populations seem only to exist in a few localities in LFM, SZ and NWZ. No records from NEZ and B. - Sweden: scattered distributed all over the country and absent from large areas. Usually rare, only common around Stockholm and around the lower parts of the river Dalälven. Probably increasing in number. - Norway: AK, HEs, O, Bø, VE, TE, STi, NTi. - Finland: distributed north to Lk but nowhere common; also Vib in the USSR. - Europe north to the Polar Circle, south to C.France, N.Italy and Greece, east to W.Siberia.

Biology. On eutrophic lake shores and river banks, occurring on muddy clay-soil in sparse but often tall vegetation of *Scirpus, Carex,* etc.; often in rather shaded sites under *Salix*-bushes or in sparse growth of *Alnus* and *Fraxinus.* In Norway it is especially typical of the large rivers (Andersen, 1982). The species is regularly encountered together with *Bembidion dentellum.* It is a spring breeder; adults hibernate under bark.

234. *Agonum piceum* (Linnaeus, 1758)
Pl. 6: 14.

Carabus piceus Linnaeus, 1758, Syst. Nat. ed. 10: 416.

5.5-7.3 mm. The palest species of the subgenus. Piceous, upper surface almost constantly with a faint bronzed hue; forebody darker than the pale brown or ferrugineus elytra (including epipleura). First antennal segment paler than the following segments, legs testaceous with darker tarsi. Body flatter and narrower than in *micans.* Pronotum with barely suggested hind-angles; elytral striae extremely fine, intervals as a rule completely flat; subapical sinuation evident. Elytral microsculpture forming narrow meshes, arranged in irregular transverse rows. (See also the pale form of *thoreyi).*

Distribution. Denmark: very distributed and rather common. - Sweden: recorded from all districts except P. Lpm. Rather distributed and common, only at higher altitudes rare or absent. - Norway: mainly in coastal areas from Ø to HOi, and in ST and NTi. - In East Fennoscandia rather common, north to Le in Finland and to the Kola peninsula in the USSR. - Total distribution as preceding species.

Biology. At the margin of lakes, ponds and slow-running rivers, occurring on clayey or muddy soil with a rich vegetation of *Carex, Equisetum, Glyceria,* etc. Preferably on open ground, but also in moderately shaded situations under bushes and in scattered tree growth; often in company with *A. thoreyi.* It has a good dispersal power and is frequently found in sea drift. Breeding takes place in spring. The adults hibernate under bark of fallen trees, etc.

235. *Agonum gracile* Sturm, 1824

Agonum gracilis Sturm, 1824, Deutschl. Fauna Ins. 5 (5): 197.

6.0-7.3 mm. Habitus as *piceum,* but less stretched and much darker. Entire body black without any trace of metallic hue. First antennal segment and tibiae sometimes piceous. Hind-angles of pronotum completely rounded. Elytra with slightly rounded sides and obsolete subapical sinuation, striae a little more evident. Microsculpture of elytra consisting of almost isodiametric meshes, slightly irregular, only in spots forming suggested transverse rows.

Distribution. Denmark: scattered throughout the country; uncommon, seems most frequent in WJ. - Sweden: distributed and common all over the country; found in all districts. - Norway: scattered throughout most parts of the country. - Finland: common all over the country and also in the adjacent parts of the USSR. - Europe, south to C.France and N.Italy, east to the Amur region.

Biology. Very hygrophilous, predominantly occurring in quagmires with *Sphagnum,* but also on the shores of oligo- and dystrophic lakes and ponds, living among *Carex, Menyanthes,* etc.; often in company with *Bembidion doris* and *Pterostichus diligens.* Less frequent in the shore zone of eutrophic lakes. It is most abundant at low and medium altitudes, rare in the alpine zone. The species has a good dispersal power and is regularly found in shore-drift. It is a spring breeder.

236. *Agonum fuliginosum* (Panzer, 1809)

Carabus fuliginosus Panzer, 1809, Fauna Ins. Germ. 108: 5.

5.5-7.8 mm. General outline characteristic: pronotum, as compared with the elytra, broad. Microsculpture diagnostic. Black, entirely unmetallic, elytra sometimes brown or reddish; base of antennae and at least tibiae rufo-piceous. Pronotum much wider than one elytron; its posterior setiferous puncture well removed from hindangle. Elytra short, oviform, convex, widest behind middle, preapical sinuation evident. The microsculpture is more regularly "brick-like" than in other *Europhilus*-species. Hind-wings, except in single full-winged specimens, strongly reduced.

Distribution. Denmark: very distributed and common. - Sweden: a common and well distributed species. - Norway: fairly common throughout the country. - Also commonly distributed over the territory of East Fennoscandia. - In Europe from the far north south to C.France, N.Italy and Bosnia; east to W.Siberia.

Biology. A hygrophilous species, living on different kinds of moist, often shaded ground, notably in marshy woodland, e.g. in alder swamps. Also in fens and the border zone of eutrophic lakes and rivers, together with *A. thoreyi;* less numerous in oligotrophic bogs. Regularly encountered in sea-drift. Reproduction takes place in spring.

237. ***Agonum thoreyi*** Dejean, 1828
Figs 305, 318.

Carabus pelidnus Paykull, 1792, Mon. Curc. Suec., p. 149: *nec* Herbst, 1784.
Agonum thoreyi Dejean, 1828, Spec. Gén. Col. 3: 165.
Agonum puellum Dejean, 1828, Spec. Gén. Col. 3: 166.
Agonum micans auctt.; *nec* Nicolai, 1822.

6-8 mm. At come recognized on the elytral microsculpture and the structure of the meta-tarsi. A stretched species, especially the pronotum (Fig. 305), with slender legs. Piceous black to dark brown; first antennal segment and legs usually reddish brown. Elytral microsculpture consisting of isodiametic meshes without transverse arrangement. All tarsal segments with a median furrow on dorsum (Fig. 318). On the continent and the British Isles lives a pale form, in which the elytra are yellow or light brown, usually clouded along the suture; also the appendages are paler. This is the true *thoreyi* Dej., whereas our form belongs to var. (or subsp.) *puellum* Dej.

Distribution. Denmark: very distributed and rather common. - Sweden: distributed from Sk. to Hls.; also found in Vb. and Nb. Common in Sk. and in the lowlands around the great lakes. - Norway: scattered in Ø, AK, VE, Ry, Nsi and Fø. - East Fennoscandia: common in the south of Finland, north to Lk; also in the adjacent parts of the USSR in Vib and Kr. - Europe, south to S.France and C.Italy, east to the Amur and Ussuri regions.

Biology. In swamps on clayey soil with rich and tall vegetation, preferably at the margin of eutrophic lakes and large rivers, occurring near the water edge among *Phragmites, Typha,* etc., frequently together with *Odacantha melanura* and the staphylinid *Paederus riparius* (L.). It is often found in great number in floating heaps of reed. Rarely occurring in oligotrophic bogs. Sometimes encountered in drift material on the seashore. It is a spring breeder. According to Wasner (1979) the egg production is much greater, and the breeding period considerably longer, than in *fuliginosum, piceum* and *gracile.* This fact enables the species to exist under the extreme living conditions of a reed swamp. The adult beetles hibernate in leaf-sheaths and hollow stems of plants.

238. ***Agonum consimile*** (Gyllenhal, 1810)
Figs 309, 313, 322.

Harpalus consimilis Gyllenhal, 1810, Ins. Suec. 1(2): 161.

5.3-6.5 mm. This and the two following species form a group of small, northern species, almost constantly with a pronounced metallic lustre. Black, appendages not perceptibly paler, upper surface with brassy, sometimes bluish or greenish lustre (exceptionally almost unmetallic). *A. consimile* has the pronotum (Fig. 309) more dilated anteriorly than *gracile*, and clearly suggested, though very obtuse hind-angles. Elytra flatter, intervals more or less convex. Antennae (Fig. 313) slender, segments 4-10 with straight sides, 6-7 at least twice as long as wide (cf. *exaratum*). First meta-tarsal segment without or with only suggested furrow. Microsculpture of elytra forming somewhat irregular but clearly transverse meshes, at least arranged in transverse rows. Penis (Fig. 322) with subapical ventral sinuation.

Distribution. Not in Denmark. - Sweden: distributed and abundant in the high mountains of northern Sweden, from Hrj. north to T. Lpm. - Norway: mainly in N, TR and F and in the central mountain areas in the southern part of the country. - Finland: locally common in the north, single finds from the southern parts; also in Lr in the USSR. - A northern palaearctic species.

Biology. Only in the fjelds, especially in the birch region, occurring in mires rich in mosses (usually not *Sphagnum*) and with a field layer dominated by *Carex*- and *Eriophorum*-species. Krogerus (1960) experimentally showed that *A. consimile* prefers higher temperatures and a less acid substrate (pH 6.0-7.6) than *munsteri*. The species is sometimes found together with *Elaphrus lapponicus*. Reproduction probably takes place in spring.

239. ***Agonum munsteri*** (Hellén, 1935)
Fig. 321.

Europhilus Munsteri Hellén, 1935, Notul. Ent. 15: 88.

5.3-6.1 mm. Easily confused with *consimile* and only differing in the following points: the metallic lustre is fainter bronzy. The entire body is more convex (see elytral profile) and the later-basal foveae of pronotum therefore deeper. The pronotum is larger, proportionally both broader and longer, the hind-angles are virtually disappeared. Microsculpture similar. Penis (Fig. 321) straight ventrally.

Distribution. Denmark: SJ, Draved Kongsmose, in numbers July 1985 (P. Jørum & V. Mahler). - Sweden: distributed from Vrm. to Nb. and Lu. Lpm.; rare and hitherto only found in about 15 localities. Isolated in the south: Ög., Simonstorp. - Norway: HEs and NTi, two records close to the Swedish border. - In East Fennoscandia rather rare but widely distributed over the entire area. - A species of the northern conifer zone, south to the Netherlands and N.Germany.

Biology. In ombrotrophic bogs, usually at the border of small dystrophic lakes and ponds, occurring in wet *Sphagnum*. The field layer of the habitat is dominated by e.g. *Carex, Oxycoccus, Andromeda* and *Rhynchospora alba*. According to Krogerus (1960) it is a rather cold-preferent and decidedly acidophilous species (pH 4.0-5.0). It probably breeds in spring.

240. ***Agonum exaratum*** (Mannerheim, 1853)
Figs 310, 314.

Anchomenus exaratus Mannerheim, 1853, Bull. Soc. Nat. Mosc. 26: 143.
Platynus aldanicus Poppius, 1906, Öfvers. Finska Vetensk.Soc. Förh. 48 (3): 36.

5.2-6.5 mm. Best recognized on the form of pronotum. Coloured as *consimile*, though more constantly with brassy lustre above; tibiae always quite black. Antennae (Fig. 314) with shorter and more rounded outer segments. Pronotum (Fig. 310) with more pronounced hind-angles and sides before them straight or slightly sinuate; basal foveae with a more or less pronounced tubercle. Elytral striae usually deeper and intervals rather convex. First meta-tarsal segment with evident internal furrow. Microsculpture similar. Penis without ventral sinuation (as in *munsteri*).

Distribution. Not found in Denmark, Sweden, Norway or Finland. On the tundra from Hudson Bay through Siberia to Kola Peninsula.

Biology. Barely below the forest limit. On soft marshy ground at the margin of ponds and pools, with mosses, *Carex, Eriophorum,* etc.

Tribe Amarini

Genus *Amara* Bonelli, 1810

Amara Bonelli, 1810, Obs. Ent. 1 (Tab. Syn.).
Type-species: *Carabus vulgaris* Linnaeus *sensu* Panzer, 1797 (= *Amara lunicollis* Schiødte, 1837).

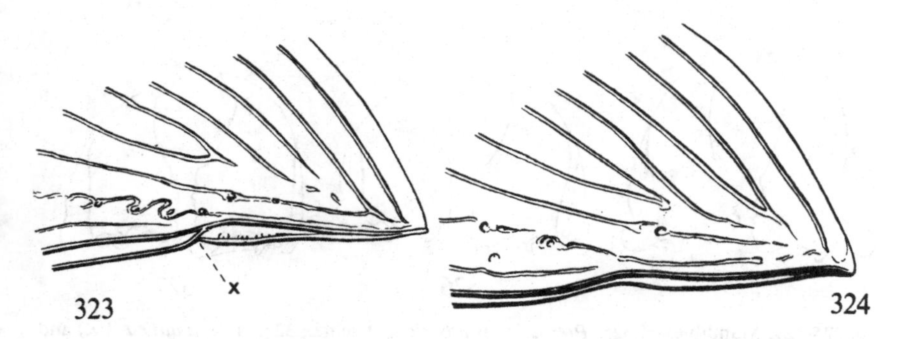

Figs 323, 324. Apex of left elytron of 323: *Amara* (x = "crossed" epipleura) and 324: *Harpalus*.

This is a large genus of medium-sized Carabids (cf. pl. 6: 16-18). The body is stout and the legs short, resulting in a marked similarity with *Harpalus.* However, the elytra of *Amara* have "crossed" epipleura (Fig. 323), the head has two supra-orbital punctures (except in *hyperborea*), and a posterior puncture is present at hind-angle of pronotum. Finally, the elytra lack dorsal punctures, which is only exceptionally the case in *Harpalus.* Seventh elytral stria has, with a few exceptions, one to five punctures near apex (Figs 338-340); the posterior puncture may be very small and removed to the level of the second stria.

In consequence of the mainly vegetarian mode of life, the mandibles (Fig. 326) are short and blunt. Pronotum is broad, on each side provided with two latero-basal foveae or striae which, however, may be more or less obliterated, especially the outer one. The wings are full, but dimorphic in a few species. The pronotum of the male is often broader than in female and is also more shiny (with less microsculpture). The meso- and meta-tibiae of the male are arcuate; the latter may be pubescent on innerside in some species. The pro-tarsi have constantly 3 dilated segments. Right paramere long, styloid, usually hooked at apex. Penis simple, the internal sac without sclerites. For this reason and because of the generally high intraspecific variation, many species-groups provide great difficulties in identification.

All species are more or less xerophilous and usually restricted to open country in places where the vegetation is short. Species with metallic sheen are usually diurnal, sun-loving animals; those with more dull coloration are predominantly active at night and spend the bright hours in the soil at plant roots, etc. The food of the adults consists to a great extent of seeds of especially Compositae and Cruciferae, but also of other vegetable matter. The larvae seem to be mainly carnivorous.

Key to species of *Amara*

1	Elytra with a seta-bearing pore-puncture at base of the abbreviated scutellar stria	2
–	Elytra without such basal pore-puncture	11
2(1)	Terminal spur of pro-tibia trifid (Fig. 328)	3

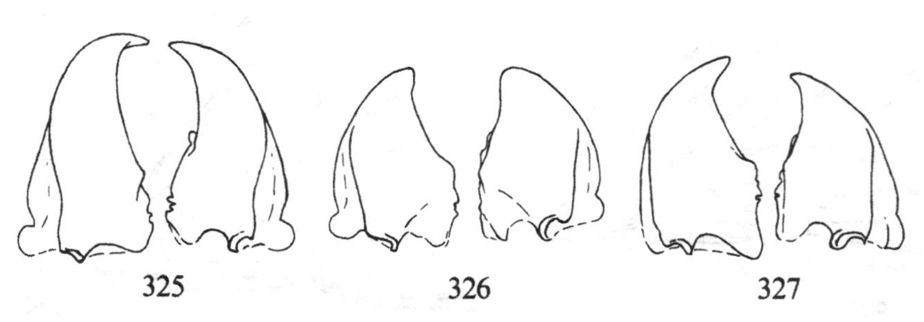

Figs 325-327. Mandibles of 325: *Pterostichus adstrictus* Eschtz.; 326: *Amara aulica* (Pz.) and 327: *Harpalus aeneus* (F.).

– Terminal spur of mid-tibia simple (Fig. 329) 5

3(2) Less than 8 mm. Pronotum with sides obliquely depressed posteriorly and anterior angles protruding .... 243. *plebeja* (Gyllenhal)

– Usually more than 8 mm. Pronotum not depressed laterally, convex to hind-angles; anterior angles less protruding 4

4(3) Three basal antennal segments rufo-testaceous. Two basal abdominal segments more or less punctate laterally. Pronotum with hind-angles rounded at tip . 241. *strenua* Zimmermann

– Usually four basal antennal segments pale, but the fourth segment may be infuscated at apex. Two basal abdominal segments rugulose and without or with only suggested punctuation laterally. Hind-angles sharp. Elytral striae

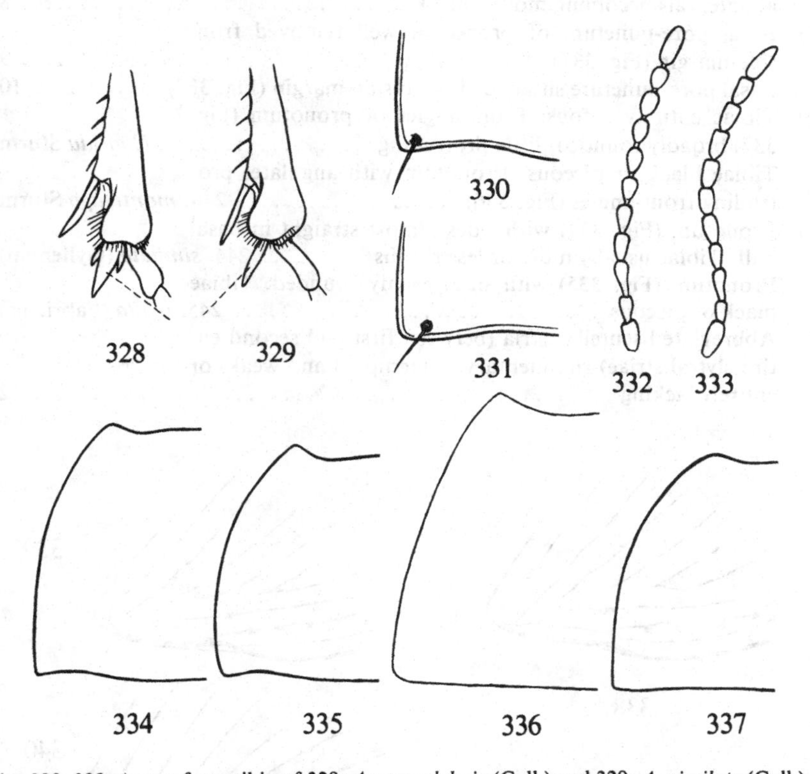

Figs 328, 329. Apex of pro-tibia of 328: *Amara plebeja* (Gyll.) and 329: *A. similata* (Gyll.).
Figs 330, 331. Hind-angle of pronotum of 330: *Amara ovata* (F.) and 331: *A. nitida* Sturm.
Figs 332, 333. Antenna of 332: *Amara infima* (Dft.) and 333: *A. tibialis* (Payk.).
Figs 334-337. Pronotum of *Amara*. – 334: *similata* (Gyll.); 335: *ovata* (F.); 336: *montivaga* Sturm and 337: *nitida* Sturm.

deeper .... 242. *tricuspidata* Dejean

5(2) Antennae entirely pale. Upper surface not metallic or with brassy hue. Pronotum coarsely punctate at base .... 271. *praetermissa* (Sahlberg)

– Antennae infuscated, only three or four basal segments pale. Upper surface almost constantly with metallic lustre. Base of pronotum smooth or with fine punctures 6

6(5) Less than 7 mm. Legs entirely pale .... 259. *anthobia* Villa & Villa

– More than 7 mm. Femora more or less infuscated 7

7(6) Almost constantly more than 10 mm. Elytral striae fine to apex .... 257. *eurynota* (Panzer)

– 10 mm or less. Elytral striae deepened towards apex with all intervals becoming more convex 8

8(7) Basal pore-puncture of pronotum well removed from side-margin (Fig. 331) 9

– Basal pore-puncture situated close to side-margin (Fig. 330) 10

9(8) Tibiae entirely rufous. Front-angles of pronotum (Fig. 337) broadly rounded, little protruding .... 247. *nitida* Sturm

– Tibiae black or piceous. Pronotum with angulate, protruding front-angles (Fig. 336) .... 246. *montivaga* Sturm

10(8) Pronotum (Fig. 334) with sides almost straight in basal half. Tibiae usually more or less rufous .... 244. *similata* (Gyllenhal)

– Pronotum (Fig. 335) with sides evenly rounded. Tibiae black or piceous .... 245. *ovata* (Fabricius)

11(1) Abbreviated scutellar stria (between first and second entire elytral striae) rudimentary, interrupted and weak, or entirely lacking 12

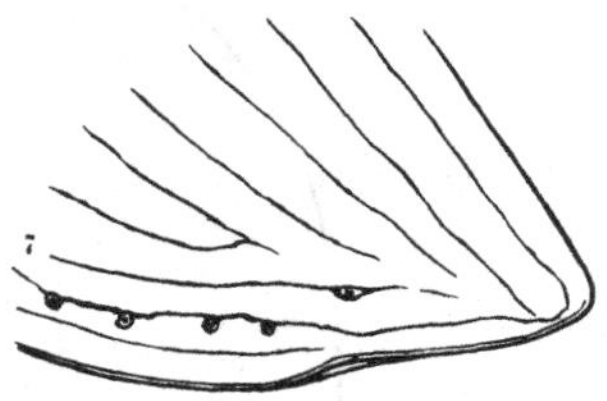

338

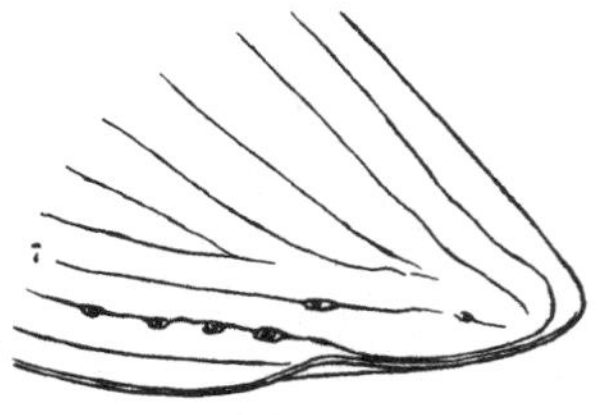

339

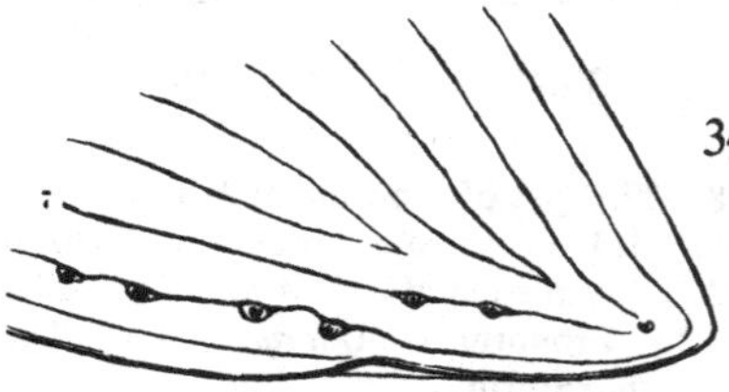

340

Figs 338-340. Elytral apex with preapical punctures of 338: *Amara apricaria* (Payk.); 339: *A. spreta* Dej. and 340: *A. lunicollis* Schiødte.

– Abbreviated scutellar striae well developed, not interrupted and not shallower than adjacent striae 14

12(11) Distal antennal segments oviform (Fig. 332). Elytral microsculpture regularly isodiametric .... 270. *infima* (Duftschmid)

– Distal antennal segments virtually conical, as usual (Fig. 333). Elytral microsculpture consisting of transverse meshes 13

13(12) Legs entirely pale, rufous. Outer basal fovea of pronotum more or less obliterated .... 260. *lucida* (Duftschmid)

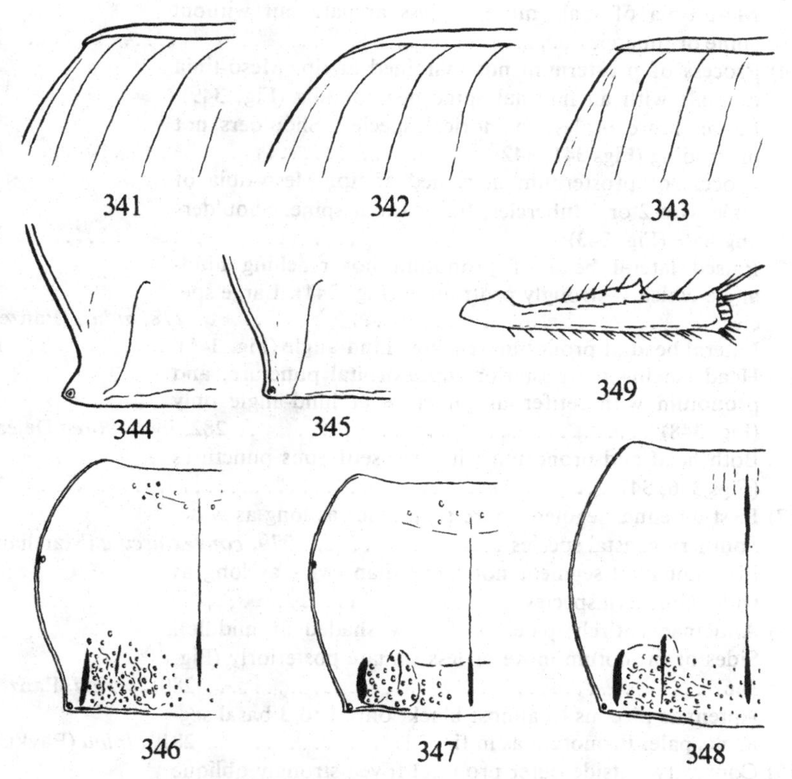

Figs 341-343. Shoulder of 341: *Amara aulica* (Pz.); 342: *A. convexiuscula* (Marsh.) and 343: *A. fulva* (Müll.).

Figs 344, 345. Hind-angle of pronotum of 344: *Amara aulica* (Pz.) and 345: *A. convexiuscula* (Marsh).

Figs 346-348. Pronotum of 346: *Amara torrida* (Pz.); 347: *A. alpina* (Payk.) and 348: *A. hyperborea* Dej.

Fig. 349. Meso-tibia of *Amara alpina* (Payk.).

- Femora more or less infuscated. Outer pronotal fovea small but deep and sharply defined .... 261. *tibialis* (Paykull)

14(11) Pronotum with a convexity outside outer fovea. Seventh elytral stria with a single preapical puncture (Fig. 338) Meso-tibia of male either with a spine or with 2-3 small tubercles internally 15

- Pronotum without strong lateral convexity (except in *equestris*). Seventh elytral stria with two or more preapical punctures (Figs 339,340), the posterior of which may be very small and removed to the level of second stria. Meso-tibia of male more or less arcuate but without spine or tubercles 24

15(14) Process of prosternum not margined at tip. Meso-tibia of male with an internal spine near middle (Fig. 349). Large, more or less cylindrical species. Shoulders not protruding (Figs 341, 342) 16

- Process of prosternum margined at tip. Meso-tibia of male with 2 or 3 tubercles, but without spine. Shoulders angulate (Fig. 343) 20

16(15) Raised lateral bead of pronotum not reaching hind-angle which is strongly protruding (Fig. 344). Large species .... 278. *aulica* (Panzer)

- Lateral bead of pronotum reaching hind-angle (Fig. 345) 17

17(16) Head lacking the posterior supra-orbital puncture, and pronotum with setiferous puncture at hind-angle only (Fig. 348) .... 282. *hyperborea* Dejean

- Both head and pronotum with two setiferous punctures (Figs 346, 347) 18

18(17) First antennal segment more than twice as long as wide. Southern coastal species .... 279. *convexiuscula* (Marsham)

- First antennal segment not more than twice as long as wide. Northern species 19

19(18) Antennae entirely pale (or faintly shaded at middle). Sides of pronotum more or less sinuate posteriorly (Fig. 346) .... 280. *torrida* (Panzer)

- Antennae piceous to almost black, only 1 to 3 basal segments pale. Pronotum as in fig. 347 .... 281. *alpina* (Paykull)

20(15) Convexity outside outer pronotal fovea strongly oblique (Fig. 350), interrupted before base by the basal pore-puncture. Very stout species with pronotum nearly twice as wide as long 21

- Basal pronotal convexity not or slightly oblique. Pronotum at most 1.5 times as wide as long 22

21(20) Testaceous or pale brown (usually with greenish hue).

Eyes flat. Sides of pronotum sinuate in posterior half (Fig. 350) .... 275. *fulva* (Müller)

– Piceous to almost black. Eyes semicircular. Hind-angles of pronotum denticulate but sides not sinuate (Fig. 370) .... 277. *consularis* (Duftschmid)

22(20) Mentum simple. Frons devoid of microsculpture anteriorly. Elytral striae punctate to apex .... 273. *crenata* Dejean

– Mentum almost constantly double. Entire frons with wawy microsculpture. Punctuation of striae obliterating apically 23

23(22) Meta-tibia of male not pubescent. Convexity outside pronotal fovea not quite reaching base .... 276. *majuscula* (Chaudoir)

– Meta-tibia of male densely pubescent internally. Pronotum with sides slightly sinuate posteriorly, the convexity usually fusing with basal bead (Fig. 351) .... 274. *apricaria* (Paykull)

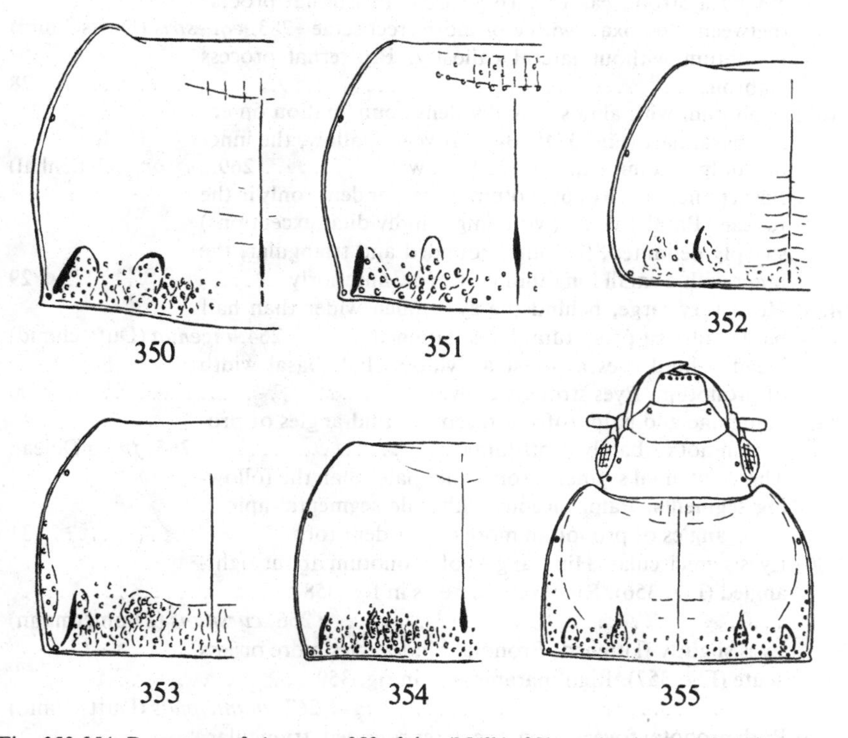

Figs 350-354. Pronotum of *Amara*. – 350: *fulva* (Müll.); 351: *apricaria* (Payk.); 352: *brunnea* (Gyll.); 353: *equestris* (Dft.) and 354: *bifrons* (Gyll.).
Fig. 355. Forebody of *Amara quenseli* (Schönh.).

24(14) Antennae entirely pale or very slightly infuscated towards apex 25
– Antennae piceous to black with 1 to 3 (rarely 4) basal segments pale, or at least underside of first segment pale 32
25(24) Eyes flat. Pronotum with protruding front-angles and obliquely depressed sides (Fig. 355) .... 268. *quenseli* (Schönherr)
– Eyes convex. Pronotum with rounded front-angles (except in *brunnea*) and not or barely depressed sides 26
26(25) Pronotum (Fig. 352) with strongly, evenly rounded sides and hind-angles, and obliterated outer fovea. Smaller, 5.2-6.8 mm. .... 272. *brunnea* (Gyllenhal)
– Pronotum with only slightly rounded sides, but hind-angles and outer fovea (except sometimes in *bifrons*) evident. Larger 27
27(26) Outer basal fovea of pronotum separated from side-margin by a strong callosity (Fig. 353). Prosternal process (between pro-coxae) with 6 or more erect setae 283. *equestris* (Duftschmid)
– Pronotum without lateral callosity. Prosternal process glabrous 28
28(27) Pronotum with almost equally dense punctuation on entire basal part (Fig. 354). Basal foveae shallow, the inner one only seldom with a linear furrow .... 269. *bifrons* (Gyllenhal)
– Basal punctuation of pronotum sparse, or dense only in the foveae. Basal foveae (with single individual exceptions) sharply delimited, the outer rounded and triangular, the inner with a small longitudinal furrow anteriorly 29
29(28) Head very large, behind the eyes much wider than half basal width of pronotum. Eyes flattened ...... 264. *ingenua* (Duftschmid)
– Head behind eyes at most as wide as half basal width of pronotum. Eyes strongly convex 30
30(29) Antennae and palpi rufo-testaceous. Hind-angles of pronotum not or barely protruding...................... 265. *fusca* Dejean
– Three antennal segments somewhat paler than the following segments. Palpi piceous with pale segmental apices. Hind-angles of pronotum more or less dentiform 31
31(30) Eyes semicircular. Hind-angles of pronotum about right-angled (Fig. 356). Right paramere as in fig. 358 .... 266. *cursitans* (Zimmermann)
– Eyes slightly flattened. Pronotal hind-angles more or less acute (Fig. 357). Right paramere as in fig. 359 .... 267. *municipalis* (Duftschmid)
32(24) Both pronotal foveae deep, the outer rounded, triangular, sharply delimited externally. The posterior pore-puncture almost touching the lateral bead 33

– Outer basal fovea obliterated, shallow, or very small; in dubious cases the posterior pore-puncture is well removed from side-margin 35

33(32) Eyes flat (Fig. 352). Pronotum obliquely flattened laterally .... 268. *quenseli* (Schönherr)

– Eyes convex. Pronotum not obliquely flattened 34

34(33) Eyes semicircular. Hind-angles of pronotum about right (Fig. 356). Right paramere: fig. 358 .... 266. *cursitans* (Zimmermann)

– Eyes slightly flattened. Pronotal hind-angles more or less acute (Fig. 357). Right paramere: fig. 359 .. 267. *municipalis* (Duftschmid)

35(32) Antennae dark with 3 or 4 sharply contrasting pale basal segments; segment 3 may be shaded apically 36

– Antennae with only one or 2 paler basal segments; segment 3 as dark as segment 4, or a little paler just at base 43

36(35) Legs entirely pale 37

– At least femora more or less infuscated 38

37(36) Front-angles of pronotum protruding, angulate (Fig. 360) .... 258. *familiaris* (Duftschmid)

– Front-angles of pronotum rounded, little protruding (Fig. 361) .... 260. *lucida* (Duftschmid)

38(36) Elytral striae fine throughout, intervals quite flat. Inner basal fovea of pronotum forming a sharp streak (Fig. 363) .... 254. *aenea* (Degeer)

– Elytral striae deepened towards apex and intervals becoming more convex. Inner pronotal fovea less sharp 39

39(38) Setiferous puncture at pronotal hind-angle situated close to lateral bead. Seventh elytral stria with 2 subapical punctures 40

– Posterior setiferous pronotal puncture more removed

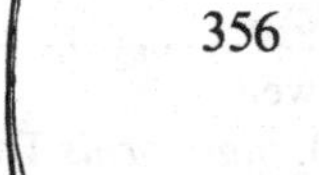

356

357

358

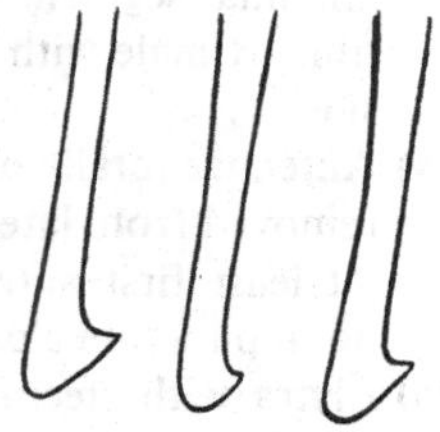

359

Figs 356, 357. Hind-angle of pronotum of 356: *Amara cursitans* (Zimm.) and 357: *A. municipalis* (Dft.).
Figs 358, 359. Apex of right paramere of 358: *Amara cursitans* (Zimm.) and 359: *A. municipalis* (Dft.).

from side-margin. Elytra with 3 preapical punctures, one at apex of second or first stria 41

40(39) Antennae with 3 basal segments and basal part of segment 4 pale rufous. Pronotum punctate around inner fovea and sparsely so also outside it .... 253. *littorea* Thomson

– Antennal segment 3 very seldom entirely pale, segment 4 always dark. Base of pronotum impunctate or with a few punctures near inner fovea .... 252. *curta* Dejean var.

41(39) Front-angles of pronotum broadly rounded (Fig. 337). Antennae with entire third and base of fourth segment pale.. 247. *nitida* Sturm var.

– Front-angles of pronotum angulate, protruding (Fig. 362) Entire fourth antennal segment dark, and also apex of third segment usually darkened 42

42(41) Lateral row of punctures (in eighth elytral stria) somewhat more open at middle but not interrupted. Penis: fig. 369 .... 249. *convexior* Stephens

– Lateral row of punctures with more or less wide interruption at middle. Penis: figs 367, 368 .... 248. *communis* (Panzer)

43(35) Pronotum laterally broadly deplanate (Fig. 364), from hind- to front-angles. Uneven intervals of elytra more or less convex .... 263. *interstitialis* Dejean

– Lateral deplanation of pronotum non-existing or faint, never reaching anterior third; only in *erratica* sometimes with faintly uneven convexity of intervals.......................... 44

44(43) Front-angles of pronotum rounded, little protruding (Fig. 365). Second antennal segment always black. Meta-tibia of male with a row of long setae on inside. Northern species .... 262. *erratica* (Duftschmid)

– Front-angles of pronotum acute and protruding. Second antennal segment often pale, at least underneath. Metatibia of male with both pubescence and long setae on inside.. 45

45(44) Antennae totally black. Posterior setiferous puncture well removed from lateral pronotal bead .... 250. *nigricornis* Thomson

– At least first antennal pale underneath. Posterior pronotal puncture closer to lateral bead 46

46(45) Elytra with steep stripe at apex (as in *communis),* where the striae are deep and intervals convex 47

– Elytra flat, striae fine, intervals flat................................ 48

47(46) Pronotum very convex, foveae deep (Fig. 366). 7.3-9 mm .... 251. *lunicollis* Schiødte

– Pronotum rather flat, with little pronounced foveae. 5.8-7.4 mm .... 252. *curta* Dejean

48(46) First and second antennal segments bright rufous. Base of pronotum more or less punctate, at least near inner fovea .... 255. *spreta* Dejean

– At least second antennal segment black or infuscated. Pronotal base impunctate or with a few punctures near inner fovea .... 256. *famelica* Zimmermann

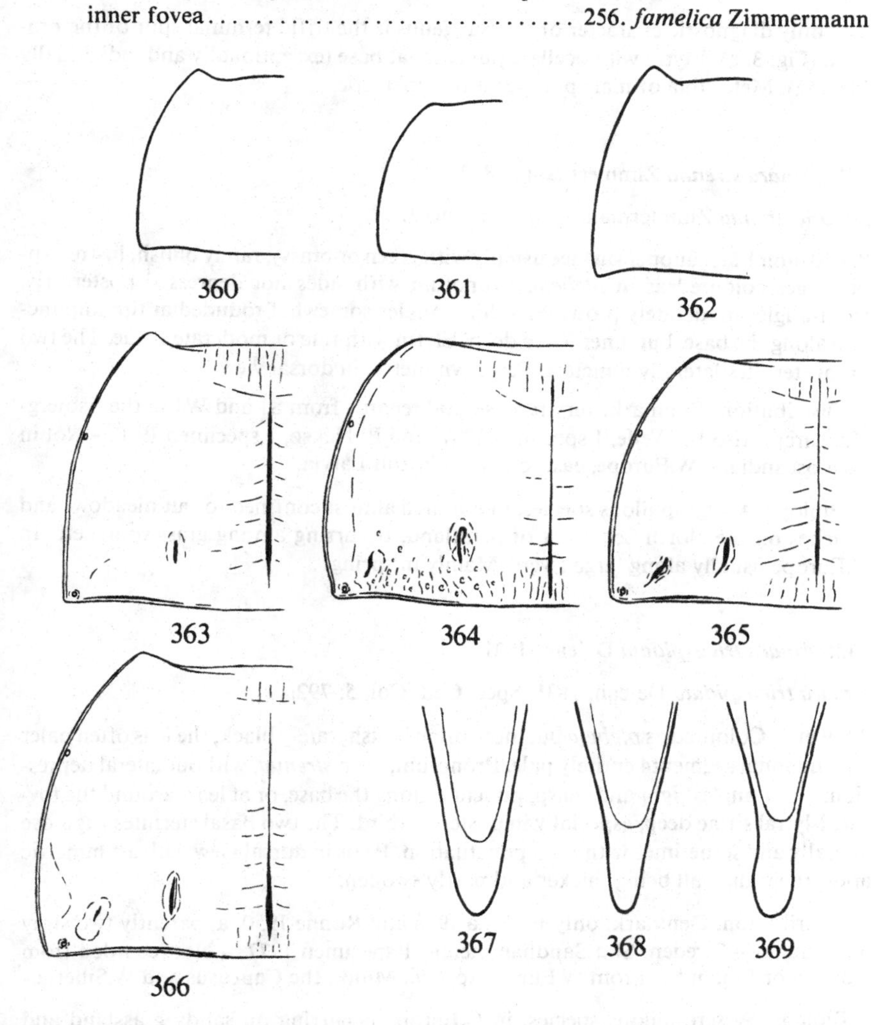

Figs 360-366. Pronotum of *Amara*. – 360: *familiaris* (Dft.); 361: *lucida* (Dft.); 362: *communis* (Pz.); 363: *aenea* (Deg.); 364: *interstitialis* Dej.; 365: *erratica* (Dft.) and 366: *lunicollis* Schiødte. Figs 367-369. Apex of penis of 367, 368: *Amara communis* (Pz.) and 369: *A. convexior* Stph.

Subgenus *Zezea* Csiki, 1929

Zezea Csiki, 1929, Coleopt. Catal. 104: 402.
Type-species: *Amara angustata* Say, 1833.
Triaena LeConte, 1848, *nec* Hübner, 1818.

The only diagnostic character of this subgenus is the trifid terminal spur of the protibia (Fig. 328). Elytra with ocellate puncture at base (exceptionally and individually lacking). Meta-tibia of male pubescent on inner side.

241. ***Amara strenua*** Zimmermann, 1832

Amara strenua Zimmermann, 1832, Faunus 1: 32.

8.5-10 mm. Black, upper surface usually with green or brassy, rarely bluish, lustre. Appendages coloured as in *plebeja*. Pronotum with sides not depressed posteriorly, front-angles moderately protruding; hind-angles somewhat rounded at tip; impunctate along the base, but inner fovea deep. Elytra with fine or moderate striae. The two basal sternites laterally punctate. Penis symmetric in dorsal view.

Distribution. Denmark: very rare; several records from SJ and WJ in the Esbjerg-Ribe area. Also EJ: Vejle, 1 specimen 1887, and B: Neksø, 1 specimen 1918. - Not in Fennoscandia. - W.Europe, east to Rumania and Latvia.

Biology. A hygrophilous species, in our area almost confined to salt meadows and beaches on the North Sea coast of S.Jutland, occurring among grass roots, etc. In C.Europe usually along large rivers. Mainly in spring.

242. ***Amara tricuspidata*** Dejean, 1831

Amara tricuspidata Dejean, 1831, Spec. Gén. Col. 5: 792.

7.5-9 mm. Coloured as *plebeja* but more often bluish, rarely black, the legs often paler and antennal segment4 entirely pale. Pronotum, as in *strenua*, without lateral depression. Hind-angles right and sharp, punctate along the base, or at least around the foveae. Elytral striae deep, especially in posterior third. The two basal sternites rugulose laterally and sometimes with faint punctuation. Penis in dorsal view with asymmetric apex, the right wall being thicker and partly swollen.

Distribution. Denmark: only B: Bagå 1948 and Rønne 1950, apparently two stray individuals. - Sweden: Sk.: Sandhammeren, 1 specimen 1977. - Not recorded from Norway or Finland. - From W.Europe to Asia Minor, the Caucasus and W.Siberia.

Biology. A xerophilous species, in C.Europe occurring on sandy grassland and dunes; also in cultivated fields. In our area only accidentally in drift material on seashores.

243. ***Amara plebeja*** (Gyllenhal, 1810)
Fig. 328.

Harpalus plebejus Gyllenhal, 1810, Ins. Suec. 1 (2): 141.

6.3-7.8 mm. Easily recognized on the form of the pronotum. Black, upper surface with brassy, sometimes greenish or bluish, lustre. Antennae with 3 basal segments and base of fourth rufo-testaceous. Tibiae yellowish brown. Eyes flat. Pronotum with oblique depression posteriorly along side-margin, base sinuate laterally and more or less punctate, foveae well impressed; hind-angles acute, while front-angles are strongly protruding. Elytral striae deepened towards apex. The two anterior abdominal sternites not or faintly punctate.

Distribution. Denmark: very distributed and rather common. - Sweden: generally distributed and rather common in the south, more local in the north up to the polar circle. - In Finland common in the southern and central parts, found north to Obs. Also in the districts of Vib and Kr of the USSR. - Europe and western Siberia.

Biology. Rather eurytopic, occurring on different types of moderately humid soil. Notably on firm, clayey soil with meadow grasses and sedges, often near water. Also on arable land. The adults feed mainly on grass seeds and often climb the forage plants (e.g. *Alopecurus, Deschampsia, Festuca* and *Poa*). The species has a high dispersal power and is regularly found in drift material on seashores. Breeding takes place in spring.

Subgenus *Amara* s. str.

This subgenus and *Celia* are the two largest subgenera in *Amara*. They have been separated differently, but usually on the absence or presence in the male of a dense pubescence on meta-tibia. This is an unstable character, as also demonstrated in the obviously monophyletic subgenus *Bradytus,* where both character states occur. The species of *Amara* s.str. possess this meta-tibial pubescence. The only exceptions are *erratica* and *interstitialis,* for which reason they have often been arranged in *Celia.* However, especially the structure of the folds of the internal sac of penis shows them to belong in *Amara* s. str. despite the lacking meta-tibial pubescence.

The ground colour is black, but the upper surface has a metallic, often very bright lustre. The antennal colour is surprisingly constant and often a good separating character. The wings are invariably full.

244. ***Amara similata*** (Gyllenhal, 1810)
Figs 329, 334.

Harpalus similatus Gyllenhal, 1810, Ins. Suec. 1(2): 138.
Amara curvicrus Thomson, 1872, Opusc. Ent. 4: 362.

7.8-10 mm. Black, upper surface with, sometimes faint, brassy or greenish, rarely bluish, reflection. Three basal antennal segments and base of fourth segment rufo-testaceous. Tibiae yellowish brown, at least basally, as are tarsi. Pronotum (Fig. 334) more narrowed forwards than backwards, sides, especially posteriorly, little rounded. Hind-angles, in connection with the laterally sinuate hind-margin, more or less acute. Base never quite impunctate, at least inner fovea evident, though sometimes small and streak-like; posterior pore-puncture close to side-margin (as in fig. 330). Elytra constantly with a basal pore-puncture present. Striae deepened towards apex, seventh stria with 3 (or 4) preapical punctures.

Distribution. Denmark: rather distributed and common; more scattered in western Jutland. - Sweden: rather distributed and common in the southern half of the country, rare and local in the north; not found at high elevations. Probably only recently established north of 63°N and today found at least as far north as in Lu. Lpm.: Vuollerim. - Norway: somewhat scattered distributed over the entire country except for the extreme north-west; usually rather rare. - Finland: widely distributed but not common, found north to Obs. Also in Vib and Kr of the USSR. - From Europe to eastern Siberia.

Biology. On different kinds of moderately dry soil, predominantly in open, sun-exposed country, e.g. meadows and grassland; often on arable land, notably in fallow fields, also in open forest. The vegetation of the habitat is usually dominated by grasses and weeds, e.g. crucifers, the seeds of which form part of the diet of the imago. The beetles are often found on the forage plants. They are predominantly diurnal. Breeding occurs in spring.

245. ***Amara ovata*** (Fabricius, 1792)
Figs 330, 335.

Carabus ovatus Fabricius, 1792, Ent. Syst. 1: 154.
Carabus obsoletus Duftschmid sensu Dejean, 1828, Spec. Gén. Col. 3: 460, *nec Carabus obsoletus* Rossi, 1790.
Amara sarsi Munster, 1927, Nyt Mag. Naturv. 65: 287.

8-9.5 mm. Closely related to *similata* and best separated on the shape of the pronotum. Body broader. Coloration the same, but the metallic reflection often less pronounced and more bluish. Pronotum (Fig. 335) broader, especially in the male, with sides more arcuate; hind margin usually less sinuate inside the almost rectangular hind-angles. Base of pronotum impunctate or almost so, foveae more or less obliterated. Elytra with sides somewhat more rounded, and with 3 or 4 preapical punctures.

Distribution. Denmark: known from all districts, but with a rather scattered occurrence and usually uncommon. - Sweden: somewhat scattered distributed over the entire country, lacking only in the extreme north-west; usually rather rare. - Norway: recorded from three separate areas: Ø, AK and HEs; VAy, Ry, and HO; and Ns. - Finland: widely distributed and mostly rather rare, found north to Li. Also in Vib, Kr and Lr of the USSR. - Transpalaearctic, east to Japan, south to Iran.

Biology. In drier habitats than *A. similata,* usually occurring in open country on rather dry, gravelly soil with sparse but often tall vegetation of grasses, crucifers, etc., for instance in gravel pits and on railway embankments, also in arable fields. The imago is partly phytophagous, feeding on plant seeds, etc. It is a typical spring breeder.

246. ***Amara montivaga*** Sturm, 1825
Fig. 336.

Amara montivaga Sturm, 1825, Deutschl. Fauna Ins. 5 (6): 45.

7.8-9.2 mm. Very broad and convex species. Black, upper surface with a usually strong green reflection. Antennae coloured as in *similata,* but legs entirely black, or at least tibiae very dark piceous. Pronotum (Fig. 336) very broad, shorter than in related species and with greatest width near hind-angles; sides strongly rounded and converging forwards; front-angles protruding; base impunctate with almost obsolete foveae; basal pore-puncture well removed from side-margin. Elytra with greatest width near shoulder, more convex than in *ovata.*

Distribution. Not in Denmark. - Sweden: a rare and local species with limited occurrence in Vg., Boh., Dlsl., Nrk., Vrm. and Dlr. Lately also recorded from Nb. - In Norway rare and local, restricted to the south-eastern part; also one isolated record in Ry. - Fairly common in southern and central Finland, north to Kb. Also in Vib and southern Kr. - Europe and western Siberia.

Biology. In dry, open country on gravel and sand with rich weed vegetation. It is pronouncedly synanthropic, usually occurring near human habitations, on ruderal places, in gravel pits, etc. In C.Europe predominantly in mountains. Mainly in spring.

247. ***Amara nitida*** Sturm, 1825
Figs 331, 337.

Amara nitida Sturm, 1825, Deutschl. Fauna Ins. 5 (6): 35.

7.2-8.5 mm. Best recognized on the form of pronotum (Fig. 337). Black, upper surface with brassy, sometimes greenish, rarely bluish, reflection. Antennae coloured as in *similata,* but tibiae paler, reddish brown. Pronotum (Fig. 337) with anterior margin almost truncate, whereby best separated from *communis;* base almost constantly more or less punctate; foveae usually evident, pore-puncture well removed from lateral margin (as in *montivaga).* Specimens without basal pore-puncture on elytra have been called *"imbella"* Reitt.

Distribution. Denmark: very rare; EJ (Vejle, Horsens), NEJ (Frederikshavn), LFM (3 localities), NWZ (Asnæs) and NEZ (3 localities). - Sweden: found in most of the provinces from Sk. to Jmt. but rare and local. Also a single record from Nb. - Norway: not a common species but found in most provinces north to ST. - Finland: rare but found over most of the country, north to ObN. Also in Vib and Kr of the USSR. - Europe and Siberia.

Biology. In open country on moderately dry, usually clay-mixed gravelly soil with sparse vegetation, e.g. in gravel pits. Also on light cultivated soil (e.g. in the Stockholm area). Mainly in spring.

248. ***Amara communis*** (Panzer, 1797)
Figs 362, 367, 368.

Carabus communis Panzer, 1797, Fauna Ins. Germ. 40: 2.
Amara pulpani Kult, 1949, Ent. Listy 12: 77.
Amara pseudocommunis Burakowski, 1957: 343.

6-6.8 mm. A rather small and variable species, best recognized on the front-angles of pronotum. Black, upper surface usually brassy, sometimes greenish or bluish. Antennae with 3 basal segments rufous, the third always darkened apically; the fourth segment not pale at base. Tibiae more or less reddish. Head large with convex eyes, thereby separating the species from immature *familiaris*. Pronotum (Fig. 362) somewhat broader in the male, rather varying in form, punctuation and basal foveae, but always with triangular produced front-angles. Elytral striae clearly deepened apically. Eighth stria more or less interrupted at middle, and seventh stria with 3 preapical punctures. Penis: figs 367, 368. Female with 2 setiferous punctures on last sternite.

Distribution. Denmark: rather distributed and rather common. - Sweden: very distributed and common over most of the country, unrecorded only from G. Sand., P. Lpm. and T. Lpm. - Norway: fairly common in all provinces except F. - Finland: very common and found all the way to Li. - USSR: also in Vib, Kr and Lr. - Europe and Siberia, south to the Caucasus.

Biology. A eurytopic species, occurring in open country on almost every kind of moderately dry soil, e.g. in meadows and fields, also in light forests. Often under moss and dry leaves. The adult beetle feeds on plant seeds. It is a diurnal spring breeder.

249. ***Amara convexior*** Stephens, 1828
Fig. 369.

Amara convexior Stephens, 1828, Ill. Brit. Ent. Mand. 1: 131.
Amara continua Thomson, 1873, Opusc. Ent. 5: 529.

7-8.2 mm. Closely related to *communis*. More parallel-sided, as the pronotum is narrower compared to the elytra, the difference being most pronounced in the male. Basal foveae of pronotum usually better delimited and more coarsely punctate. The row of punctures in eighth stria somewhat more open at middle but not actually interrupted (more as in *lunicollis*). Penis (Fig. 369) constricted at apex.

Distribution. Denmark: scattered occurrence on the eastern coast of Jutland, north to Randers and including Samsø. Also SJ (Rømø, Stensbæk), WJ (Esbjerg, Fanø), NWJ (Øster Assels, Fur), and NEJ (Rødhus). A few records from F, LFM, SZ, NWZ; more frequent in NEZ; no records from B. - Sweden: rare and local, mainly in the south; north to Upl., Dlr. and Vstm. - Norway: not common, but found in most provinces north to Ns. - Finland: only known in a few specimens from the Helsinki area. - In the USSR a few finds from Vib and at the river Svir in Kr. - From W.Europe to the Urals and Asia Minor.

Biology. Apparently more xerophilous than *A. communis,* preferring rather dry, sandy or gravelly soil. For instance in dry grassland with growth of *Anthoxanthum odoratum, Ameria vulgaris* and *Dianthus deltoides,* occurring in company with *A. equestris, lunicollis* and *aenea* (Barndt, 1976). Also in gravel pits. Pronouncedly diurnal. Breeding takes place in spring.

250. ***Amara nigricornis*** Thomson, 1857

Amara nigricornis Thomson, 1857, Skand. Col. 1: 39.
Amara melanocera Tschitschérine, 1899, Abeille 29: 274.
Amara Natvigi Csiki, 1929, Col. Catal. 104: 422.

7-7.8 mm. Best recognized on the entirely black antennae; only first segment with an indication of piceous underneath. Most similar to *lunicollis.* Black, upper surface more or less brassy or coppery. Pronotum somewhat narrower than in *lunicollis,* with posterior pore-puncture more removed from side-margin. Basal foveae small, but the inner one deep. Elytra with row of punctures in eighth stria clearly interrupted at middle. Seventh stria with 2 preapical punctures.

Specimens with elytral striae hardly deepened at apex may be separated from *famelica* on the still darker antennae, the position of the posterior pronotal puncture, and the more convex elytra, notably at apex.

Distribution. Not in Denmark. - Sweden: rare to very rare in the interior of the northern provinces, south to Dlr., unrecorded from Gstr., Med. and Ång. - In Norway rare and local, the distribution being very discontinuous: HEn, On, VAi, STi, Nsi, TRi, Fn and Fø. - Finland: very rare, scattered records from the northern provinces; one specimen from Ab: Nousiainen 29.iv.1978 (leg. T. Lammes, det. F. Hieke). Also in Kr and Lr of the USSR. - Boreoalpine, in C.Europe in the Cevennes, the Alps and the Carpathians.

Biology. An inhabitant of the northern coniferous region, usually occurring in open forest and at forest edges, on moderately dry, gravelly soil with sparse vegetation of grasses, etc. Often on burned clear-felled areas of forests, where a moss vegetation has been established (Lundberg, 1984). Singly in the alpine region. Mainly occurring in spring.

251. ***Amara lunicollis*** Schiødte, 1837
Figs 340, 366.

Amara lunicollis Schiødte, 1837, Naturh. Tidsskr. 1: 164.
Amara vulgaris auctt., *nec* (Linnaeus, 1758).

7.3-9 mm. In this and all the following species of the subgenus the male meta-tibia possesses a dense mat of pubescence on the inner surface. A strongly convex but rather stretched species, with markedly broad pronotum (Fig. 366). Black, upper surface with brassy or green, rarely bluish, lustre, strongest in the male. Appendages dark except for one or two basal segments of antennae which are piceous or rufous. Also base of tibiae may be lighter. Pronotum (Fig. 366), notably in the male, with sides strongly rounded and somewhat obliquely depressed inside hind-angles. Basal foveae usually more impressed than in *communis* and *convexior*, the outer usually consisting of an oblique streak. Elytral striae clearly, and usually strongly, deepened towards apex; the row of punctures in eighth stria as in *convexior*. Seventh stria (Fig. 340) with 3 preapical punctures.

Distribution. Denmark: rather distributed and rather common. - Sweden: rather common to common, widespread throughout the country, absent only from the high altitude areas of the north-west. - Norway: fairly common and widespread throughout the country, except for the inner parts of the south. - Finland: found north to Li, but not very common. Also Vib, Kr and Lr in the USSR. - Holarctic, south to Iran.

Biology. A rather eurytopic species which occurs on more or less dry ground, notably on peaty or sandy soil in open country. It is a typical inhabitant of *Calluna* heaths, regularly occurring together with *Carabus arvensis*, *Bradycellus ruficollis* and *B. caucasicus*. Also on not too dry grassland, in meadows and arable land, under moss carpets on rocks, in open woodland, etc. It is predominantly diurnal. Breeding takes place in spring.

252. ***Amara curta*** Dejean, 1828

Amara curta Dejean, 1828, Spec. Gén. Col. 3: 468.
Amara cyanocnemis Thomson, 1857, Skand. Col. 1: 37.

5.8-7.4 mm. A small species, similar to *familiaris* in general habitus but with darker appendages. Black, upper surface dull bronze, rarely bluish. Antennae with 1 or 2 basal segments rufous (rarely also with part of third segment pale). Legs brown with black femora. Head and eyes smaller than in *communis*; eyes more convex than in *familiaris*. Antennae, notably outer segments, strikingly short. Basal foveae of pronotum small, linear, or almost evanescent; the laterobasal seta closer to side-margin than in *communis*. Elytra with 2 preapical punctures. Last sternite in female with 4 punctures.

Distribution. Denmark: very rare; LFM: Møn, Langebjerg, in numbers 1864; Møn, Jydelejet, in numbers 1942-57, also 1 specimen 1975; Møn, Ulfshale, 1 specimen 1951; NEZ: København, 1 specimen before 1900. - Sweden: rare and local from Sk. to southern Hls. - Norway: local in the south-east; also records from SFi, MRi and NTi. - East Fennoscandia: rare in the south of Finland, and in Vib and southern Kr of the USSR. - Europe and Siberia.

Biology. In dry, open country on stony sand or gravel with sparse and short vegetation, for instance in gravel pits and on sand hills, often on southern slopes near the coast. The beetles occur under stones and among plant roots. Mainly in spring.

253. ***Amara littorea*** Thomson, 1857

Amara littorea Thomson, 1857, Skand. Col. 1: 21.
Amara Kodymi Jedlička, 1936, Čas. Čzl. Spol. Ent. 33: 4.

7.2- 8.6 mm. Similar to *curta,* notably in the pronotum, but larger and more convex, especially the head. Black, upper surface dull with faint bronzy, rarely bluish, lustre. Three first antennal segments and base of fourth pale rufous. Tibiae reddish brown, at least basally. Front-angles of pronotum less prominent than in *communis.* Basal forveae small and flat, streak-like, with faint or lacking punctuation (as in *curta*). Posterior puncture closer to side-margin than in *communis.* Elytra with striae clearly deepened apically. Seventh elytral stria with 2 preapical punctures. Female with 4 coarse punctures on last sternite.

Distribution. Denmark: B, Rosmandsbæk, 1 specimen 1951. - Sweden: very rare and very scattered, most records being old. Recorded from Sk. to T. Lpm. but with large gabs in C. and N.Sweden. One of the latest records is from the Stockholm area, where it was found rather frequently. In Sk. the latest record is from 1881 except for two stray individuals on seashores. - Norway: rare and local in the south-east, north to 62°N. - In Finland rare, with occasional finds north to LkW; also in Vib and Kr of the USSR. - In Central Europe from the Rhine east to Ukraine; also in E.Siberia.

Biology. In moderately dry, open country on clayey or clay-mixed gravelly soil. Preferably on cultivated land, e.g. occurring among weeds on waste-places, sometimes together with *A. aulica, eurynota* and *Harpalus distinguendus.* Single specimens are found in drift on seashores. During daytime the beetles are concealed in the soil. Newly emerged imagines occur in the autumn.

254. ***Amara aenea*** (Degeer, 1774)
Fig. 363.

Carabus aeneus Degeer, 1774, Mém. Hist. Ins. 4: 98.
Harpalus trivialis Gyllenhal, 1810, Ins. Suec. 1 (2): 140.

6.2-8.8 mm. Agreeing with the two following species through the elytral striae being fine throughout (as they are in *eurynota*), but in all 3 species the basal pore-puncture of the elytra is lacking. Black, metallic lustre of upper surface usually strong, brassy or green, rarely somewhat bluish. Antennae with segments 1-3 and base of segment 4 clear rufo-testaceous. Tibiae always pale. Head smaller than in *communis,* with flat eyes. Front-angles of pronotum (Fig. 363) less pointed, the outer basal fovea obliterated or wanting, the inner forming a short, straight, sharp streak. Elytra with 3 preapical punctures. Last abdominal segment of female with 4 setae.

Distribution. Denmark: very distributed and very common. - Sweden: generally distributed in the south, north to southern Dlr. and Upl.; also two localities in Hls. Often a very common species where it occurs. - Norway: rather common in the south and south-east, mainly along the coast. - In Finland fairly common, found north to

Ok; also Vib and Kr in the USSR. - Europe, Siberia and North Africa, including the Azores, the Canary Islands and Madeira; south to the Himalayas. Introduced in North America.

Biology. A xerophilous species living in open country on dry, sandy or clayey soil with rather dense but short vegetation. Notably on heaths and grassland; also on cultivated land, often on lawns in parks and gardens. The beetle is pronouncedly diurnal, often seen running on the surface in bright sunshine. It feeds on plant seeds and has been reported to be harmful to winter cereals. Breeding takes place in spring.

255. ***Amara spreta*** Dejean, 1831
Fig. 339.

Amara spreta Dejean, 1831, Spec. Gén. Col. 5: 791.

7.8-9.5 mm. Similar to *aenea* but broader and with a different antennal colour. Body with same coloration but only two basal antennal segments bright rufo-testaceous (segment 2 exceptionally slightly clouded), segment 3 clearly black. Head broader. Pronotum with sides depressed posteriorly, punctate along base which is sinuate laterally; hind-angles therefore more or less acute; basal foveae deeper. Elytra (Fig. 339) almost constantly with only 2 preapical punctures. Last sternite of female with 2 punctures.

Distribution. Denmark: distributed and common along the western and northern coast of Jutland; also in C.Jutland; otherwise very scattered in Jutland. On the islands and Bornholm fairly distributed along the coasts. - Sweden: distributed and locally common along the coasts in Sk., Bl. (Mjällby), Hall. and Vg. (Göteborg), in Sk. also in several inland localities. - Norway: rare and very local, only recorded from Ø, VAy and Ry. - Finland: known from a few localities in the south. Apparently somewhat more frequent in Vib and in Kr at Lake Onega. - From W.Europe to the Caucasus, Siberia and Mongolia.

Biology. A stenotopic species, living in open country on dry, sandy soil with sparse vegetation. In our area almost restricted to shifting sand and duneland on the coast, occurring among *Ammophila* etc., often together with *Calathus ochropterus*. Rarer in the inland. Mainly in spring.

256. ***Amara famelica*** Zimmermann, 1832

Amara famelica Zimmermann, 1832, Faunus 1: 36.

6.6-9 mm. As preceding species recognized on antennal colour. Body somewhat narrower than in *spreta* and also more convex, eyes flatter, and structure of pronotum different. Metallic colour of body less bright and specimens without metallic lustre occur. Only first antennal segment quite pale, second segment darker, usually entirely black. Tibiae black. Pronotum with sides less flattened basally, base more or less punctate, and little sinuate laterally; hind-angles about right; foveae usually deeper.

Elytra almost constantly with 2 subapical punctures. Last sternite of female with 4 punctures.

Distribution. Denmark: very rare. In Jutland 13 scattered records; no records from EJ. Also two old records from NEZ (Amager, Tisvilde). - Sweden: rare and local in the southern half of the country, but apparently lacking in the south-east. Many of the records are old. - Norway: rare and local, restricted to the south-eastern part of the country. - Finland: fairly common and found all over the country except for the northernmost parts; also Vib, Kr and Lr in the USSR. - From W.Europe to North Africa, Asia Minor and Central Siberia.

Biology. In Sweden repeatedly encountered in moderately humid habitats, usually on clay-mixed sand or gravel, e.g. near lake shores among grass and carices, often together with *A. plebeja;* singly in sand pits and on heather ground. In Denmark and in C.Europe it is mainly associated with *Calluna* heaths (Mossakowski, 1970b), sometimes occurring together with *Bembidion nigricorne* and *Cymindis vaporariorum.* Predominantly in spring.

257. ***Amara eurynota*** (Panzer, 1797)

Carabus eurynotus Panzer, 1797, Fauna Ins. Germ. 37: 23.
Carabus acuminatus Paykull, 1798, Fauna Suec. Ins. 1: 166.

9.5-12.6 mm. The largest and also the flattest of all species of *Amara* s. str. Black, upper surface brassy, coppery or greenish. Antennae with 3 basal segments and base of fourth segment rufo-testaceous. Inner basal fovea of pronotum small but deep, the outer obliterated. Elytra with a basal bore-puncture; elytral striae very fine, not at all deepening towards apex; uneven intervals of elytra more or less convex.

Distribution. Denmark: rather distributed but not common, being very sparse in western parts of Jutland. - Sweden: rather common and distributed in the southern half of the country, more scattered in the northern half and unrecorded from several districts (see catalogue). - Norway: scattered and rather local, but found in most provinces north to Ns. - Finland: fairly common and found all over the country except in the northernmost parts; also in Vib, Kr and Lr of the USSR. - From W.Europe to North Africa, Asia Minor and Central Siberia.

Biology. In open country, usually on clay-mixed gravelly soil with sparse but often tall vegetation. It is clearly favoured by human activity, often occurring in fallow fields and on waste land among weeds (e.g. *Rumex, Polygonum, Capsella*). The food of the adult consists partly of plant matter, e.g. seeds of *Capsella.* Mainly in spring and autumn.

258. ***Amara familiaris*** (Duftschmid, 1812)
Fig. 360.

Carabus familiaris Duftschmid, 1812, Fauna Austriae 2: 119.

5.6-7.2 mm. This and also the two following species are the only members of the subgenus with entirely pale, bright rufous legs. *A. familiaris* is separated on the triangularly protruding front-angles of pronotum (Fig. 360). Piceous, upper surface with brassy, greenish or bluish lustre. Antennae with 3 or 4 basal segments pale. Eyes much flatter than in *communis*. Basal foveae of pronotum obsolete, impunctate or almost so. Elytral striae deepening towards apex. Male with right paramere without hook.

Distribution. Denmark: very distributed and common. - Sweden: recorded from all districts except Hrj.; very common in the south. - Norway: rather common in all provinces north to TR. - Finland: very common, lacking only in the northernmost provinces. Also in Vib, Kr and Lr of the USSR. - From W.Europe to North Africa, the Caucasus, Central Siberia and Mongolia. Introduced in North America.

Biology. Very eurytopic, occurring in open, sun-exposed country on almost every kind of soil, both on moderately humid and on dry ground, even in dunes. It is especially numerous in meadows, on waste land among weeds and on cultivated soil, often in fallow fields. The adults feed on plant seeds etc. The species has a high power of dispersal and is frequently found in drift material on seashores. It is a pronounced spring species.

259. ***Amara anthobia*** Villa & Villa, 1833

Amara anthobia Villa & Villa, 1833, Col. Eur. Dupl.: 33.

6-7 mm. Extremely similar to *familiaris*, but with a setigerous pore-puncture at the base of the abbreviated scutellar stria (as in *similata, ovata, montivaga, nitida* and *eurynota*). Eyes more convex than in *familiaris*, and the pronotal front-angles somewhat less produced. Male with right paramere having a small hook at tip. Penis smaller, slenderer and more arcuate.

Distribution. In Denmark a few scattered records; repeatedly captured only in Jægersborg Dyrehave (NEZ) since 1942. - Sweden: very rare and so far only 3 records: Sk., Vitemölle (1947), Öl., Ekerum (1972) and Sk., Hallands Väderö (1980). Now probably established in the country. - Norway and East Fennoscandia: no records. - C. and S.Europe, the Caucasus and Asia Minor.

Biology. Both in open country, usually on sandy, sparsely vegetated, often cultivated soil, and in woodland. Most Danish specimens have been found in compost heaps near deer shelters. Accidentally in drift material on seashores.

260. ***Amara lucida*** (Duftschmid, 1812)
Fig. 361.

Carabus lucidus Duftschmid, 1812, Fauna Austriae 2: 121.

4.6-6.4 mm. Extremely similar to *familiaris* and separated on the only slightly protruding front-angles of pronotum (Fig. 361). Pronotum therefore has an almost truncate front-margin. Smaller and a little narrower than *familiaris*. Coloured as this, but eyes more convex. Most species have a rudimentary scutellar stria on elytra and may be confused with immature specimens of *tibialis*, but are separated, among other things, on the obsolete pronotal foveae.

Distribution. In Denmark rather distributed and rather common. - Sweden: restricted to coastal areas in the south, from Boh. to Öl., Gtl. and G. Sand. Rather common on the south and east coast of Sk., and on Öl. and Gtl. - Norway: very rare, only recorded from VAy. - In Finland only recorded from Ab: Lojo (Krogerus); also one record from Vib: Zaporožskoe. - C. and S.Europe, south to Asia and Iran, east to Kazan.

Biology. In dry open country, notably on sandy soil with sparse, short grass vegetation. It is pronouncedly coastal in our area, mainly occurring in dunes and on commons near the sea; rarer in the inland. Mainly in spring and late summer.

261. ***Amara tibialis*** (Paykull, 1798)
Fig. 333.

Carabus tibialis Paykull, 1798, Fauna Suec. Ins. 1: 168.

4.4-5.7 mm. A small, parallel-sided species. Because of the small size only to be compared with *infima*, though at once separated on the antennae (Fig. 333). Black, upper surface with brassy or greenish, rarely bluish, lustre. Antennae with three basal segments, and usually also base of fourth segment, pale; segment 3 often clouded apically. Femora always dark. Eyes strongly convex. Pronotum with well marked front-angles, basal foveae small but deep, at least the outer one. Elytra parallel-sided, striae fine, more or less distinctly punctate. Abbreviated scutellar stria more or less incomplete, sometimes missing.

Distribution. In Denmark rather distributed and rather common. - Sweden: generally distributed and common in the south up to river Dalälven. In the north (Hls. - Lu. Lpm.) only a few scattered records, and no records at all from the high altitude areas of the north-west. - Norway: scattered along the southern coast and in the inner parts of the country north to NT. - In Finland rare but found north to ObS. Also in Vib and Kr of the USSR. - From W.Europe to W.Siberia, Mongolia and Central Asia.

Biology. On open, sun-exposed dry ground, usually on sandy or gravelly soil with short but often rather dense vegetation of grasses and dwarf-shrubs, often under *Calluna;* preferably in coastal regions. The species is frequently found in abundance in sea drift. It is most numerous in May-June.

262. ***Amara erratica*** (Duftschmid, 1812)
Fig. 365.

Carabus erraticus Duftschmid, 1812, Fauna Austriae 2: 120.

6.5-8.4 mm. Short and convex species. Upper surface exceptionally without metallic hue (probably old specimens), but usually with more or less lively metallic lustre, especially in the male. The lustre may be brassy, coppery, golden, green or bluish. Antennae with first segment usually somewhat paler, piceous or dark rufous, on the underside. Legs black. Eyes protruding. First antennal segment swollen towards base. Pronotum (Fig. 365) not, or very faintly, deplanate inside hind-angle; front-angles quite rounded; basal foveae small but deep. Elytral striae fine to apex, uneven intervals individually with a tendency to be more convex, sometimes with indicated transverse corrugation. Seventh stria with 2 subapical punctures. Male without pubescence internally on meta-tibia.

Distribution. Not in Denmark. - In Sweden only found in the very north, from Vb. to T. Lpm. (unrecorded from Ås. Lpm.), and here rather distributed and usually not rare. - Norway: scattered in the north: N, TR and F. - Rare in northern Finland, south to Ok; also in Lr and northern Kr in the USSR. - Circumpolar, in Europe boreomontane, found south to N.Italy and the Caucasus.

Biology. Confined to the mountains of northern Fennoscandia, occurring in open, not too dry moraine country with coherent vegetation, especially of grass. Both on meadows in the alpine region, and in open places within the birch and conifer regions. Imagines occur during the whole summer; newly emerged beetles have been found in late July and August and are supposed to hibernate (Lindroth, 1945). In the Alps of Tyrol the species has spring propagation and hibernates as larva (De Zordo, 1979).

263. ***Amara interstitialis*** Dejean, 1828
Fig. 364.

Amara interstitialis Dejean, 1828, Spec. Gén. Col. 3: 472.

7.4-10 mm. Broader and flatter than *erratica,* less shiny due to stronger microsculpture. Upper surface (except is overwintered specimens) vividly coloured in blue, green or brass. Antennae with one or two segments rufous, at least underneath. Legs black. Pronotum (Fig. 364) characterized by the flattened part along side-margin (similar to that of *quenseli*), continuing, although narrowing, towards front-angles. Basal foveae usually deeper than in *erratica.* Front-angles identical in the two species. Elytra with uneven intervals more or less convex, striae fine to apex. Seventh stria almost constantly with 3 subapical punctures. Meta-tibia of male as in *erratica.*

Distribution. Not in Denmark. - Sweden: distributed and rather common in the northern parts, south to Hls. and Dlr. - Norway: rather common from ST and northwards; also scattered in the interior southern parts. - Finland: rare but found over most of the country, excluding Al. Also in Vib, Kr and Lr in the USSR. - From N.Europe over Siberia to Alaska and NW.Canada.

Biology. An inhabitant of the high boreal coniferous region, living in moderately dry, open country, notably clay-mixed gravelly soil with grass and weeds, often on cultivated ground. Not occurring above the timber line. Mainly in spring and summer.

Subgenus *Celia* Zimmermann, 1832

Celia Zimmermann, 1832, Faunus 1: 20.
Type-species: *Harpalus bifrons* Gyllenhal, 1810.

The members of this subgenus in the narrower sense as discussed above under the subgenus *Amara* s. str. have certain colour features in common: the body is never quite black, at least the margin of pronotum are diaphanously paler and metallic lustre is either absent or obscured. Appendages more or less pale, the antennae at most with outer segments piceous. Seventh elytral stria with 2 subapical punctures. Wings full, except dimorphic in *quenseli* and *infima.*

264. ***Amara ingenua*** (Duftschmid, 1812)

Carabus ingenuus Duftschmid, 1812, Fauna Austriae 2: 110.

9.2-11.2 mm. Largest member of *Celia,* stout with notably broad head and flat eyes. Piceous, upper surface more or less bronze, extreme margin of pronotum transparent. Appendages dark rufous, but terminal palpal segments, antennae apically and femora sometimes infuscated. Pronotal foveae small but deep. Hind-angles not protruding but delimited forwards by a minute incision. Elytra convex with markedly rounded sides, broader posteriorly in the female. Meta-tibia of male with sparse, erect hairs in addition to the usual row of setae. Wings full.

Distribution. Denmark: known from all districts, but very scattered and usually collected singly; most generally distributed in NEZ. - Sweden: recorded from most of the provinces but uncommon, e.g. at highter altitudes. The frequency seems to be on the decrease in the last decades. - Norway: fairly common along the southern coast, otherwise scattered north to Ns. - Finland: fairly common, found north to Ks. Also Vib and Kr in the USSR. - From C.Europe to Asia Minor, the Caucasus, Mongolia and E.Siberia; isolated occurrence in S.France.

Biology. A pronouncedly synanthropic species, living on cultivated fields, on waste ground among weeds, at garbage deposits among ruderal plants, etc. Both on sandy and clayey soil, in C.Europe also on saline ground. The adult consumes plant seeds, preferably of *Polygonum aviculare* (Lindroth, 1945). It is most numerous in spring, when breeding takes place and in autumn when the young beetles appear.

265. ***Amara fusca*** Dejean, 1828

Amara fusca Dejean, 1828, Spec. Gén. Col. 3: 497.
Amara complanata Dejean, 1828, Spec. Gén. Col. 3: 496.

8-8.8 mm. Similar to a small *ingenua,* but easily separated on the strongly convex eyes, as well as on the paler colour. Piceous, upper surface more or less bronze, all appendages rufo-testaceous; femora may be slightly infuscated. Pronotum with hind-angles at most slightly prominent. Punctuation and development of basal foveae somewhat varying, as also in the two following species. Elytra parallel-sided, convex. Male as in *ingenua.*

Distribution. Denmark: very scattered occurrence, but with increasing abundance; first Danish record: B, Arnager, 1912. Present status: EJ: Samsø, Helgenæs, Mols, Femmøller; NWJ: Mønsted; NEJ: Års, Sæby, Læsø; F: Ristinge; several records in each of the districts LFM, SZ, NWZ, NEZ and B. No records from SJ and WJ. - In Sweden restricted to Sk. from where it is now recorded from at least 15 localities; sometimes abundant in sandy areas; earliest record is Sk.: Ven, 1934. - Not recorded from Norway or East Fennoscandia. - From W.Europe to the Caucasus and W.Siberia; primarily in coastal areas.

Biology. In dry open country on sandy or gravelly soil, e.g. on sandy commons and on steep slopes and hills, both inland and in coastal habitats; often under *Artemisia campestris.* Mainly in spring and autumn, seemingly with autumn reproduction. A late immigrant in our area.

266. ***Amara cursitans*** (Zimmermann, 1832)
Figs 356, 358.

Celia cursitans Zimmermann, 1832, Faunus 1: 22.
Celia fuscicornis Zimmermann, 1832, Faunus 1: 20.
Amara municipalis s. Schiødte, 1841; *nec* (Duftschmid, 1812).

7-8.8 mm. Very closely allied to *municipalis* but apparently constantly with a stronger parameral hook. Other differences are relative: larger on an average. Ground colour more brownish, with less pronounced metallic hue, antennae somewhat paler. Eyes a little larger and more convex. Pronotum slightly broader and shorter (hardly more than 1/3 of elytra), sides more rounded, particularly towards base. Hind-angles (Fig. 356) less prominent and acute, but rather variable. Elytra less widening towards apex. Male with meta-tibia as in *ingenua.* Hook of right paramere: fig. 358.

Distribution. Denmark: very scattered and rare. - Sweden: usually rare, with scattered distribution north to southern Dlr. (unrecorded from Ög.); also absent at high elevations. - Norway: rare, only recorded from Ø, AK, TEy and VAy. - Finland: very rare, found only in a few localities but there occasionally in numbers: Ab: Turku, N: Helsinki area; Ta: Nokia and Pälkäne; Sa: Mikkeli. Also in Kr. - Central Europe, mainly in mountainous areas, south to C.Italy and Bulgaria.

Biology. On dry, sandy or gravelly soil, usually on open ground with sparse vegetation, often near human habitations, e.g. on arable land, in gravel-pits, among ruderal plants, etc. In Denmark and C.Europe also in clear-felled areas in forests, occurring together with e.g. *Carabus arvensis, C. problematicus, Bradycellus verbasci,* and on burnt soil *Pterostichus quadrifoveolatus* (cf. Lauterbach, 1964). Breeding takes place in autumn, young adults emerge in spring.

267. ***Amara municipalis*** (Duftschmid, 1812)
Figs 357, 359.

Carabus municipalis Duftschmid, 1812, Fauna Austriae 2: 113.
Amara melancholica Schiødte, 1837, Naturh. Tidsskr. 1: 145.

5.5-7.8 mm. Considerably smaller than preceding species and with sharp hind-angles of pronotum. Black or piceous, upper surface more or less bronze, sometimes greenish, extreme margins of pronotum slightly paler. Antennae piceous or almost black, with first segment and base of the four following segments dark rufous. Palps more or less infuscated. Legs brown, often with darker femora. Eyes smaller than in *fusca* and less convex. Hind-angles of pronotum (Fig. 357) acute and prominent, since side-margins and basal margin of pronotum both are more or less sinuate. Basal foveae well developed, the outer triangular and rounded. Punctuation at base of pronotum variable but always fainter at middle than in *bifrons*. Elytra short, widest behind the middle, especially in the female. Male with meta-tibiae as in *ingenua*. Right paramere (Fig. 359) with a hook at apex.

Distribution. Denmark: fairly distributed in eastern Jutland and northern Zealand; in other districts a very scattered occurrence. No records from large areas in northern and western Jutland, from district F, and from Lolland in district LFM. Only one old record (1872) from B. - Sweden: scattered distributed over the country, sometimes common, but still unrecorded from large areas. - Norway: scattered and local in the south and south-east; northernmost records in STi and NTi. - Finland: rare but found north to ObS; also Vib and Kr in the USSR. - From France to W.Siberia and the Caucasus.

Biology. In dry, open country on sandy or gravelly soil with sparse vegetation. It is pronouncedly synanthropic, usually occurring among weeds on waste ground, on ruderal places and in gravel-pits, often together with *A. ingenua;* also on sandy fields and commons. During daytime under plants, e.g. *Artemisia campestris*. It is an autumn breeder, most numerous in September; young adults occur in spring.

268. ***Amara quenseli*** (Schönherr, 1806)
Fig. 355.

Carabus Quenseli Schönherr, 1806, Syn. Ent. 1: 201.
Celia silvicola Zimmermann, 1832, Faunus 1: 26.
Amara maritima Schiødte, 1841, Gen. Spec. Danm. Eleuth. 1: 178.

6.4-8.8 mm. Easily recognized on the structure of the pronotum (Fig. 355). Piceous to brown, with bronze, sometimes greenish or bluish, hue, notably in the more shiny male. Appendages rufous or pale brown, antennae rarely slightly infuscated apically. Eyes flat. Prosternum between pro-coxae with 2 or 4 setae (absent in all related species). Pronotum (Fig. 355) with sides obliquely depressed, front-angles protruding; immediately in front of hind-angle is a minute incision (as in *bifrons*). Elytral striae very fine. Wings dimorphic, either full or reduced into a rudiment. Male with arcuate meso-tibiae.

Distribution. Denmark: scattered distributed along the coasts; also some records from C.Jutland; no records from NWJ. - Sweden: rather distributed and often common in the northern half of the country. In the south very local and usually not common, most abundant in the coastal regions. - Norway: common in the north and the inner parts of the south; also scattered in the coastal areas of the south-west. - Finland: fairly common over most of the area (not Al). Also Vib, Kr and Lr of the USSR. - Circumpolar, in Europe boreoalpine; also in Iceland and Scotland.

Biology. A xerophilous species, occurring in open country on sandy or gravelly soil with sparse vegetation. It is a characteristic inhabitant of the grass and dwarf-shrub heaths of the alpine and subalpine regions, often occurring together with *Miscodera arctica.* Also in dunes and shifting sand areas on the seashore (f. *silvicola*), usually in *Ammophila* and *Elymus* vegetation. *A. quenseli* is a nocturnal species which hides under stones or among plant roots during daytime. It feeds on both animal and plant matter. Breeding takes place in the autumn, young adults emerging in June-July. In northern Fennoscandia development lasts at least two years.

Note. *A. silvicola* Zimmermann has been regarded as a distinct species. It is broader, flatter, and usually without metallic lustre. It has a coastal occurrence, confined to dune-sand. Also *A. fulva* occurs in a similar form. Neither of these forms are sharply delimited from the *forma typica* and should probably not even be regarded as subspecies. *A. quenseli* occurs in both forms also in N.America.

269. ***Amara bifrons*** (Gyllenhal, 1810)
Fig. 354.

Harpalus bifrons Gyllenhal, 1810, Ins. Suec. 1 (2): 144.
Amara livida s. Schiødte, 1841; *nec* (Fabricius, 1801).

5.3-7.4 mm. Living or fresh specimens are paler in colour than any other *Amara.* Ground colour yellowish or reddish brown, upper surface with faint bronze hue; all appendages testaceous, almost transparent. Pronotum (Fig. 354) with base more densely and equally punctate, foveae shallower than in related species, and sides with a minute incision close to hind-angle (as in *quenseli*). Elytra parallel-sided, their striae rather shallow, never strongly punctate. Wings full. Male without special marks, except for pro-tarsi.

Distribution. In Denmark rather distributed and common. - Sweden: generally distributed and common in the south, scattered in the north, and only 2 records north of 66°N; unrecorded in Hrj. and P. Lpm. - Norway: fairly common in most districts north to TRy. - Finland: common, north to ObN. Also Vib and Kr of the USSR. - Europe, the Caucasus and W.Turkestan.

Biology. A xerophilous species, living on sun-exposed, sandy soil with scattered vegetation, e.g. in dry grassland and on cultivated fields, frequently occurring together with *A. fulva.* It is predominantly nocturnal, during daytime hidden under leaf rosettes, e.g. of Boraginaceae and *Knautia arvensis.* It is a good flier, often coming to light and regularly occurring in sea drift. Reproduction takes place in the autumn.

270. ***Amara infima*** (Duftschmid, 1812)
Fig. 332.

Carabus infimus Duftschmid, 1812, Fauna Austriae 2: 114.

4.9-5.7 mm. Together with *tibialis* the smallest member of the genus, but broader and more convex than that species. At once recognized on the moniliform antennae (Fig. 332). Piceous, with faint metallic hue. Appendages reddish brown. Abbreviated scutellar stria of elytra absent or barely suggested. Pronotum with faint outer fovea. Elytral striae evidently punctate, not deepened apically. Elytral microsculpture consisting of absolutely isodiametric meshes. Wings dimorphic, full or quite reduced. Male with prosternum punctate at middle.

Distribution. Denmark: very scattered distributed in Jutland, along the northern coast of Zealand, and on Bornholm. Also two old records from F (Fåborg, Alsø Bakker before 1900), and from NEZ (Solrød Strand 1915). - Sweden: very scattered and rare in the south, north to 61°N; more generally distributed only along the coasts of Halland and eastern Skåne. - Norway: very local, a few records only in some southern districts. - Finland: found in a few localities - N: Tvärminne and Hyvinkää, Ta: Ruovesi (in numbers), Oa: Ylistaro; Ok: Vaala. Also in Vib and Kr in the USSR. - From W.Europe to C.Siberia.

Biology. A very xerophilous species occurring in open country on sandy or gravelly soil with scattered vegetation, predominantly on dry grassland with *Corynephorus, Nardus, Agrostis,* etc., also on heaths with *Calluna* and *Cladonia.* The beetles have been observed climbing *Spergula arvensis* to feed on the seeds (Schjøtz-Christensen, 1957). According to Schjøtz-Christensen (1965) *A. infima* is a pronounced winter-active species, having its period of reproduction in very late autumn and early spring; eggs laid in the autumn are supposed to hibernate. Newly emerged adults appear in late summer. During the summer the beetles aestivate in the soil, usually among grass roots, which is presumably an adaptation to the extreme temperature conditions of the habitat. A great proportion of beetles experience two or more breeding periods, thus surviving up to three years. *A. infima* has a poorly developed flying ability and seems to be a very stationary species.

271. ***Amara praetermissa*** (Sahlberg, 1827)

Harpalus praetermissus Sahlberg, 1827, Ins. Fenn. 1: 246.
Harpalus rufo-cinctus Sahlberg, 1827, Ins. Fenn. 1: 249.

6.2-8.2 mm. The member of *Celia* in our fauna with a pore-puncture at the base of the abbreviated scutellar stria. Otherwise similar to *bifrons*, but shorter, with broader pronotum, notably in the male. Piceous, upper surface with faint greenish hue. All appendages pale. Base of pronotum coarsely punctate except at middle; hind-angles almost rectangular. Microsculpture of elytra consisting of very densely arranged transverse lines, which are strong and tend to fuse into transverse meshes in the female.

Distribution. Denmark: scattered distributed in Jutland (records from all districts) and on Bornholm, but usually occurring singly or in small numbers. Also LFM: Gedser, 1 specimen 1903; Bøtø, 3 specimens before 1900; Høvblege on Møn, collected several times c. 1840-1981; NEZ: Tisvilde, 3 specimens: before 1900, 1903 and 1978. - Sweden: distributed over the entire country and rather common; only being more scattered in central Sweden; unrecorded from Gstr. - Norway: widespread and rather common, except in the western coastal districts. - In Finland rare, found in the southernmost and the northern parts. Also Vib, Kr and Lr of the USSR. - From W.Europe to C.Siberia, Mongolia and the Caucasus.

Biology. Mainly in dry, open country on gravelly or sandy, sometimes chalky or clayey soil with scattered vegetation, e.g. on dry grassy hills and in gravel pits; also among litter in thin deciduous forest. The species is widespread in the mountains up to the lower alpine region, e.g. occurring on dwarf-shrub heaths and grass meadows. It is an autumn breeder, being most numerous in June-August. In the arctic and subarctic climate of the Fjelds development probably lasts two years.

272. ***Amara brunnea*** (Gyllenhal, 1810)
Fig. 352.

Harpalus brunneus Gyllenhal, 1810, Ins. Suec. 1 (2): 143.

5.2-6.8 mm. A small species, easily recognized in that sides and hind-angles of pronotum both are completely rounded (Fig. 352). Piceous brown, appendages pale. Eyes smaller and more protruding than in *praetermissa*. Pronotum (Fig. 352) with more prominent front-angles. The elytral microsculpture sharp and almost isodiametric in the female; in the male more or less obliterated.

Distribution. Denmark: rather distributed in NWZ, NEZ and B; very scattered in the rest of the country. - Sweden: generally distributed and common except more scattered in the northern coastland; known from all districts. - Norway: widespread and common in the north and in the inner southern parts; scattered in coastal areas. - East Fennoscandia: common all over the area. - Europe south to N. Italy; Siberia; Alaska and Yukon.

Biology. Less xerophilous than most other species of *Amara*, living in more or less shady habitats, usually on gravelly soil. In the lowlands predominantly in open deciduous forests, notably of birch, occurring among moss and leaves; also in open country, e.g. in the drier parts of peat bogs. In the mountains mainly in the birch forests, but also above the timber line, e.g. in dwarf-shrub heaths, often in company with *A. praetermissa*. Also on the tundra of the Kola Peninsula. Sometimes found in drift material on seashores. It is a night-active species, breeding in autumn.

Note. Aberrant specimens with a basal elytral pore-puncture (as in *praetermissa*) have long been known from C.Europe, but were first recently discovered in Scandinavia: Sweden, Gotland 1976 and Småland 1978.

The species has been referred to a separate subgenus *Acrodon* Zimmermann, characterized by the undivided mentum tooth. However, this character is not as constant in *Celia* as has been presumed.

Subgenus *Bradytus* Stephens, 1828

Bradytus Stephens, 1828, Ill. Brit. Ent. Mand. 1: 136.

Type-species: *Carabus ferrugineus* Rossi, 1790 (= *Carabus fulvus* Müller, 1776).

Stout species of medium size. Prosternum margined at tip (as also in the following subgenera). Shoulders angulate. Seventh stria (as in *Curtonotus*) with a single preapical puncture. Wings full. Prosternum of male with a large punctate or impressed area at middle, the meso-tibia with 2 or more small tubercles internally near apex, and the meta-tibia (except in *majuscula* and *crenata*) has a dense internal pubescence.

273. *Amara crenata* Dejean, 1828

Amara crenata Dejean, 1828, Spec. Gén. Col. 3: 507.

7.0-8.5 mm. The species is very similar to *apricaria* (see next species) but, as in *majuscula*, without meta-tibial pubescence in the male. It is narrower than both these species, with more stretched and parallel-sided elytra. The pronotum has more rounded sides. The elytral striae are deeper with stronger punctures, notably towards apex. Mentum with a simple tooth (in *apricaria* and *majuscula* it is bifid). Upper surface more shiny due to weaker microsculpture, especially disappearing on frons. In the male the meso-tibia has more than 2 small tubercles.

Distribution. Denmark: extremely rare. LFM: Bøtø, 1 specimen 1981 under seaweed; B: Sose, 3 specimens 1958, 2 newly emerged specimens 1959, and 9 specimens 1960; since not re-captured at Sose. - Sweden: extremely rare. Sk.: Vitemölle, 1 specimen 1969; Gtl.: Sundre, 1 specimen 1973. - Not recorded from Norway. - In Finland only found in Al: Kökar, one stray individual. - From France to Yugoslavia, the Caucasus and Iran.

Biology. The habitat preference of this species is imperfectly known. The specimens from Bornholm were found in dry open country, on clayish gravelly-mixed soil with sparse vegetation; the beetles occurred under stones. In C.Europe often on saline ground (Hieke, 1976).

274. *Amara apricaria* (Paykull, 1790)
Figs 338, 351.

Carabus apricarius Paykull, 1790, Mon. Car. Suec.: 125.

6.5-9 mm. Cylindrical and very convex, distinguished on the cordiform pronotum (Fig. 351) and the anteriorly strongly punctate striae. Piceous to brown, upper surface darker, usually bronzed; outer margins of pronotum paler, all appendages reddish. Pronotum narrow, lateral convexity not or little oblique, pore-puncture situated lateral of this. The basal fovea deep, coarsely punctate, hind-angles sharp. Frons with wavy microsculptural lines.

Distribution. Very distributed and common over the entire area. - Holarctic, south to Asia Minor and Kashmir.

Biology. Very eurytopic, living in open country on almost every kind of moderately dry soil. It is favoured by human activity, usually occurring in weedy vegetation in fields, gardens, ruderal sites, etc. Also in the mountains, reaching the lower alpine region. The species is a good flier, often coming to light; frequently found in sea drift. It is an autumn breeder.

275. ***Amara fulva*** (Müller, 1776)
Figs 343, 350.

Carabus fulvus Müller, 1776, Zool. Dan. Prodr., p. 77.

8-10.4 mm. Broad and flat. Entirely yellow or pale brown, upper surface usually with greenish hue. Eyes flat. Sides of pronotum sinuate behind middle, hind-angles sharp, outer fovea oblique (Fig. 350). Shoulder with sharp, protruding tooth (Fig. 343). Male more shiny than female.

Distribution. Denmark: very distributed and common. - Sweden: distributed over most of the country and at least common in the south. Few or no records from the high elevated areas in central and northern Sweden; unrecorded from P. Lpm. - Norway: common in the southern and eastern parts, north to STi. - Finland: common north to Li; also in Vib, Kr and Lr of the USSR. - Europe and Siberia, south to Asia Minor. Introduced in North America.

Biology. A xerophilous species living in open country on dry sand, sometimes mixed with gravel or clay. It is confined to habitats with sparse vegetation, mainly of grasses, often occurring in shifting sand regions on the coast, e.g. in tufts of *Elymus,* on sandy river banks, in sand-pits, and also on cultivated soil. The beetles are night-active, during daytime buried in the soil. They reproduce in autumn.

Note. A special form (called *"arenaria"* by R. Krogerus) is found on locations with loose dune-sand. It is very pale, without metallic hue and has wider pro-tibiae, but cannot be regarded as a subspecies.

276. ***Amara majuscula*** (Chaudoir, 1850)

Bradytus majusculus Chaudoir, 1850, Bull. Soc. Nat. Mosc. 23 (2): 148.

8.3-9.2 mm. The male is at once separated from *apricaria* by the lack of the dense pubescence on the inner side of meta-tibiae. Otherwise deviating from that species only through the following more or less relative characters: larger on an average; broader, especially the pronotum which has the sides more rounded; the basal foveae usually shallower and more finely and densely punctate; the lateral convexity somewhat flatter, less sharply delimited and not quite reaching basal bead.

Distribution. Denmark: very scattered and rare. Before 1950 only 3 records of single individuals: EJ, Anholt 1949; NEZ, Saunte 1944, and B, Randkløve 1949. After 1950 recorded with increasing frequency. - Sweden: recorded for the first time in 1917. Now widespread but local and uncommon in the south, north to river Dalälven. In the north very rare and only a few captures in Hls., Jmt. and Lu. Lpm., all after 1944. Often observed in sea-drift in SE.Skåne, sometimes in great numbers. - Norway: very rare, only one record from HEs. - A recent addition also to the fauna of Finland, first

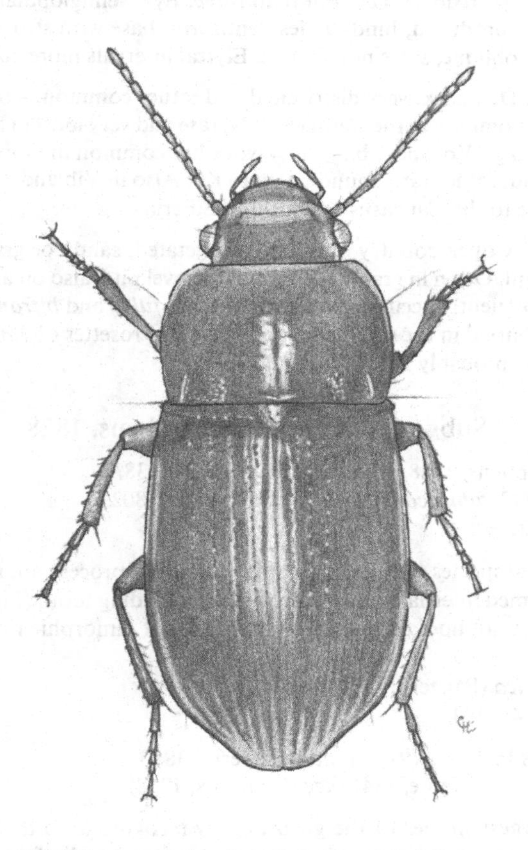

Fig. 370. *Amara consularis* (Dft.), length 8-9.4 mm.

record being from 1939; now rare but found north to Sb. Also in Vib and southern Kr in the USSR. - From C.Europe to E.Siberia and N.China.

Biology. Almost confined to arable and ruderal land, living on both sandy and clayey soil, often together with *A. apricaria.* It is an excellent flier, often coming to light and regularly occurring in drift material on seashores in late July and August. The breeding period is in autumn.

277. ***Amara consularis*** (Duftschmid, 1812)
Fig. 370.

Carabus consularis Duftschmid, 1812, Fauna Austriae 2: 112.

8-9.4 mm. Piceous to black (except from immaturity), not metallic. Underside dark brown, appendages reddish. Less flat than *fulva.* Eyes semiglobular. Pronotum with front-angles less produced, hind-angles dentiform, base with stronger punctuation, outer fovea less oblique, sides not sinuate. Elytral intervals more convex at apex.

Distribution. Denmark: very distributed and rather common. - Sweden: generally distributed and common in the south to 61°N, rare and very local in the north, recorded from Jmt., Ång., Vb. and Nb. - Norway: rather common in most provinces north to STi. - Finland: rather rare, found north to Kb. Also in Vib and Kr of the USSR. - From W.Europe to the Caucasus and Central Siberia.

Biology. In dry open country on sparsely vegetated, sandy or gravelly soil, sometimes rich in mull. Often in grassy fields and in gravel pits; also on arable land, e.g. in fallow fields, frequently occurring together with *A. fulva* and *bifrons.* In the daytime the beetles are buried in the soil, often under the leaf rosettes of *Knautia* etc. Mainly in April-August, probably breeding in autumn.

Subgenus *Curtonotus* Stephens, 1828

Curtonotus Stephens, 1828, Ill. Brit. Ent. Mand. I: 138.
Type-species: *Carabus convexiusculus* Marsham, 1802.
Cyrtonotus auct.

Large, cylindrical species (not under 7 mm). Prosternal process not margined. Mesotibia of male armed internally at middle with a protruding tooth (Fig. 349). The long right paramere is not hooked at apex. Wings full, but dimorphic in *alpina.*

278. ***Amara aulica*** (Panzer, 1797)
Figs 326, 341, 344.

Carabus aulicus Panzer, 1797, Fauna Ins. Germ. 38: 3.
Amara spinipes s. Schiødte, 1841; *nec* (Linnaeus, 1758).

11-14.3 mm. Largest species of the genus, easily recognized on the structure of the pronotum. Piceous, under surface paler, upper surface usually faintly bronzed. Antennae, palpi and often also leg rufous. Pronotum more or less heart-shaped; raised lateral bead not reaching hind-angles, which are strongly protruding (Fig. 344). Shoulder-tooth blunt but evident (Fig. 341). Elytra broadest behind middle. Meso-tibia of male internally with one tooth and 3 tubercles.

Distribution. Denmark: fairly distributed and rather common; very scattered or missing in large parts of western and northern Jutland. - Sweden: rather distributed and common, missing only in the extreme north-west of the country. - Norway: widespread and common, north to TRy. - In Finland common and found north to Li. Also in Vib, Kr and Lr of the USSR. - Europe, the Caucasus and W.Siberia. Introduced in North America.

Biology. On different kind of moderately humid, often clayey soil rich in mull, notably on meadows with tall vegetation; also on agricultural land, e.g. in corn and potato fields. The beetles frequently climb the vegetation at night in search for seeds, mainly of Compositae (*Carduus, Cirsium,* etc.), also of Umbelliferae, corn, etc. The main breeding period is in autumn, young adults emerge chiefly in early summer.

279. ***Amara convexiuscula*** (Marsham, 1802)
Figs 342, 345.

Carabus convexiusculus Marsham, 1802, Ent. Brit. 1: 462.

10.8-12.8 mm. Almost as large as *aulica* but slenderer and more cylindrical. Piceous, upper surface with evident bronzy hue. Legs paler brownish or yellowish rufous; otherwise coloured as *aulica.* Lateral bead of pronotum thinner, reaching apex of hind-angle, which is about right (Fig. 345). Elytra almost parallel-sided; shoulder quite rounded without tooth (Fig. 342). Meso-tibia of male internally with 2 spines.

Distribution. Denmark: fairly distributed along the coasts of eastern Jutland and the islands, and around the Limfjord; along the western coast of Jutland only a few records. - Sweden: distributed along the coasts from Boh. to Öl., Gtl. and G. Sand. Two inland localities are known from Sk.: Lund and Kristianstad (close to Hammersjön), in both these places in numbers in cultivated fields. In 1980-83 found rather abundantly in Närke (Hallsberg) in a garbage deposit. - Norway: rare, a few records from the south-eastern part: Ø, Bø, VE and TEy. - In Finland found in a few places along the south coast but locally common; also Vib in the USSR. - Europe and Siberia, mainly along the coast and in saline localities.

Biology. A halobiontic species, in our area almost confined to sandy, sometimes marshy seashores, occurring among sparse vegetation, under seaweed and shells, etc. In C.Europe frequently on inland saline localities as well as on ruderal sites, e.g. in refuse dumps. It is an autumn breeder, mainly occurring in June-August.

280. ***Amara torrida*** (Panzer, 1797)
Fig. 346.

Carabus torridus Panzer, 1797, Fauna Ins. Germ. 38: 173.

9.0-11.2 mm. Smaller than the two preceding species, like a small *convexiuscula* with abbreviated elytra. Closely related to *alpina.* Black to piceous, upper surface more or less bronzed. Antennae and palpi normally rufo-testaceous, very rarely with middle segments faintly shaded. Legs somewhat variable in colour. Rufinistic specimens with more or less rufous elytra are less frequent than in *alpina.* The sides of pronotum (Fig. 346) always clearly sinuate before base, but to a varying degree; hind-angles about right. Shoulder-tooth much reduced. Male as in *convexiuscula.*

Distribution. Not in Denmark. - Sweden: widespread and rather common in the north but not found south of 65°N. - In Norway a northern species, common in N, TR and F. - Finland: rather common in the north, found south to ObS. Also in Lr and northern Kr in the USSR. - Circumpolar, in our area a postglacial immigrant from the north-east. It was abundant in the British Isles during the glacial and early postglacial time.

Biology. Almost restricted to the conifer region of northernmost Fennoscandia, living in open sites on fairly humid, gravelly or clayey soil with grass or weeds, often on cultivated soil. Very rare in the birch region and above the timber line. Singly on the tundra of the Kola Peninsula. The beetles occur mainly in June-August and reproduce in July; development lasts two years (Forsskåhl 1972).

281. ***Amara alpina*** (Paykull, 1790)
Figs 347, 349.

Carabus alpinus Paykull, 1790, Mon. Car. Suec.: 119.

8-11 mm. Easily confused with *torrida* but separated on the colour of the appendages and the pronotal form. Extremely variable in colour: ground-colour usually black, often with upper surface vividly green or bronzy; in Scandinavia usually rufinistic, with elytra rufous, except for side-margin and suture which are black. Palpi and antennae black or piceous, but 1 to 3 basal antennal segments pale rufous. Legs varying in colour from reddish to black. Pronotum (Fig. 347) much varying in form with sides not or barely sinuate before the denticulate hind-angles; front-angles somewhat less rounded than in *torrida.* Hind wings sometimes reduced to varying extent. Male as in *convexiuscula.*

Distribution. Not in Denmark. - Sweden: rather generally distributed and not rare in the high mountains of the north, from Dlr. to T. Lpm. Postglacial finds have been made in Sk. - Norway: common in the north and in the mountains of the south. - Finland: very common in the arctic/alpine zone, found south to Ks; also Lr. - Circumpolar; in Europe south to Scotland.

Biology. A characteristic inhabitant of the alpine zone of the Fjelds up to 1700 m, notably occurring on dwarf-shrub heaths with *Empetrum,* but also in grassy meadows and on poor high mountain ground without continuous plant cover; often in very late snow-free sites. Less common in the birch region. Imagines occur during the entire short arctic summer, mainly in July when breeding takes place. According to Forsskåhl (1972) development probably lasts two years.

282. ***Amara hyperborea*** Dejean, 1831
Fig. 348.

Amara hyperborea Dejean, 1831, Spec. Gén. Col. 5: 800.
Amara peregrina Morawitz, 1862, Mélang. Biol. Bull. Acad. Sci. St. Petersbg. 4: 219.
Amara tumida auct.; *nec* Morawitz, 1862.

9-13 mm. Easily separated from all other species of *Amara* by the lack of the posterior supra-orbital puncture and the anterior pronotal seta.

Piceous to dark brown, elytra sometimes slightly rufinistic. Appendages, except palpi, a little paler; femora often infuscated. Head narrow. First antennal segment thick, widening apically. Pronotum (Fig. 348) with broad base, sides straight or extremely weakly sinuate in basal third; hind-angles obtuse. Basal fovea with two impressions separated by a convexity, latero-basal carina poorly developed. Elytra stretched, widening behind middle, shoulders angulate but with little pronounced callosity. Meso-tibiae of male only slightly modified. Wings large.

Distribution. Not in Denmark. - In Sweden extremely rare, only P. Lpm., between Adolfström and Peljekajse, 1 specimen 1965. - Unrecorded in Norway. - Finland: many records from Li since the 1920ies; in Lr recorded already in the 19th century. - Circumpolar.

Biology. A species of the northern coniferous region. In our area a recent immigrant. It is on the whole a vagrant species with a good ability of dispersal.

Subgenus *Percosia* Zimmermann, 1832

Percosia Zimmermann, 1832, Faunus 1: 17.
Type-species: *Carabus equestris* Duftschmid, 1812.

The diagnostic character is the presence of 6 or more erect setae on the prosternal process (between the pro-coxae). The outer basal pronotal fovea is separated from side-margin by a strong callosity (Fig. 353). One species.

283. ***Amara equestris*** (Duftschmid, 1812)
Fig. 353.

Carabus equestris Duftschmid, 1812, Fauna Austriae 2: 109.
Carabus patricius Duftschmid, 1812, Fauna Austriae 2: 110.

8.2-10.5 mm. Brown to almost black, under surface paler, sides of pronotum rufescent, appendages rufous. Eyes flat. Pronotum (Fig. 353) with lateral bead coarse, sides almost straight in basal half; the foveae deep, punctate. Elytra with 2 preapical punctures. Wings full and, though comparatively small, probably functionary. Male with shiny elytra and arcuate meso-tibiae.

Distribution. Denmark: scattered and uncommon; known from all districts but apparently absent from many areas. - Sweden: rather distributed but usually rare in the south of the country, north to about 62°N in Dlr. and Hls. One record from Vb. - Norway: rare, restricted to the lowland of the south-east. - In Finland rare, found north to Kb. Also in Vib and Kr of the USSR. - From W.Europe to the Caucasus and W.Siberia.

Biology. Preferably on open or faintly shaded ground, on dry sandy soil with scattered, usually grassy vegetation; for instance in gravel pits, along forest edges, in grassland and fields; often on recently abandoned cultivated land. During daytime occurring at roots of grass, under dry leaves, etc. It is a pronounced autumn breeder, having its main period of egg-laying in late July-September. The larvae hibernate, and the young beetles emerge in June. Only a few old adults survive the winter (Schjøtz-Christensen, 1965).

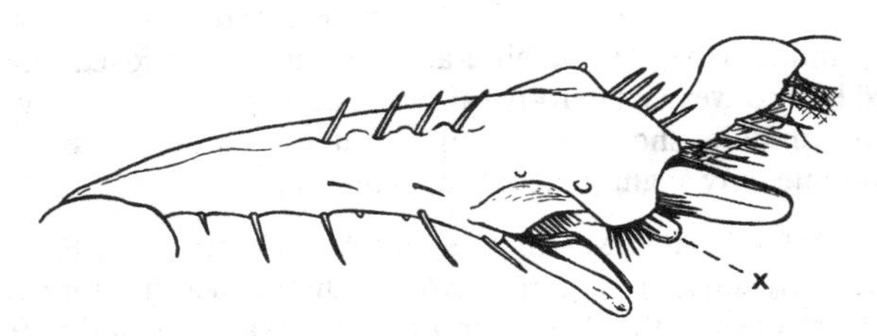

Fig. 371. *Zabrus tenebrionides* (Gz.), left pro-tibia from above, x = extra apical spur.

Genus *Zabrus* Clairville, 1806

Zabrus Clairville, 1806, Ent. Helv. 2: 80.
Type-species: *Carabus gibbus* Fabricius, 1794.

The single species of this genus in our area reminds of a big, stout, very convex *Harpalus* or of *Amara equestris.* It agrees with *Amara* in the "crossed" elytral epipleura and the lack of any dorsal puncture in third interval; with *Harpalus* in the presence of a single supra-orbital puncture and the lack of a seta at hind-angle of pronotum. A good diagnostic character is the small extra-tibial spur inside the ordinary apical spur (Fig. 371). Wings full. Male with pro-tarsi having 3 dilated segments.

284. ***Zabrus tenebrioides*** (Goeze, 1777)
Fig. 371, pl. 3: 13.

Carabus tenebrioides Goeze, 1777, Ent. Beytr. 1: 665.
Carabus gibbus Fabricius, 1794, Ent. Syst. 4 App.: 442.

14-16 mm. Piceous black, appendages rufo-piceous with darker femora. Pronotum strongly narrowing forwards, lateral bead thick basally, base densely punctate.

Distribution. Denmark: in Jutland a few more than 60 years old records from EJ;

also WJ: Holstebro, in numbers 1927. Earlier fairly distributed in the island districts, especially before 1900, but after 1950 only 2 records: LFM: near Stubbekøbing 1957, and SZ: Næstved in numbers 1977. Also some records from B, latest in 1942. - Sweden: very rare, confined to western Sk. and possibly Öl. Earlier a few captures from Trälleborg to Hälsingborg and sometimes frequent. In this century only found in a few localities: Kungstorp, Uppakra, Dalby, Lund. Latest records: Vellinge 1 specimen 1951, Lund 1 specimen 6.vii.1952 dead on a street. Also 2 specimens in Museum Lund labelled "Öland M"; labels may be incorrect. - Not in Norway or East Fennoscandia. - S. and C.Europe, the Caucasus, Asia Minor, Syria and Turkestan.

Biology. Predominantly in cereal fields and near-by grassland, both on sandy and clayish soil. The species is highly phytophagous, both in the larval and adult stage, regularly causing damage to cereals (e.g. wheat, rye and barley), especially in SE.Europe, occasionally also in Denmark. In summer the beetles climb the plants, usually after sunset, in search for grains. They have been observed to feed also on seeds of *Bromus;* animal food is also accepted. In autumn *Zabrus* eats young leaves of winter cereals. The larvae devour the tender shoots. *Zabrus* is an autumn breeder, its peak of activity is in July-August. Larvae hibernate, and young adults emerge in early summer.

Tribe Harpalini

An exceedingly large and multiformous group of world-wide distribution, reaching its maximum in the subtropical areas. Size varying from 2 to 17 mm. Stature stout, with rather short appendages.

Genus *Harpalus* Latreille, 1802

Harpalus Latreille, 1802, Hist. Nat. Crust. Ins. 3: 92.
Type-species: *Carabus ruficornis* Fabricius, 1775.

The subgenera *Ophonus* and *Pseudoophonus* have often been treated as separate genera.

A very large genus with species of moderate to rather large size (cf. pl. 3: 12 & 7: 1,2). A general similarity with *Amara* is evident. The best diagnostic characters are the presence of only one, or several irregular, supra-orbital puncture, the absence of a seta at hind-angle of pronotum, and the "non-crossed" epipleura (Fig. 324).

The mandibles (Fig. 327) are heavy, but rather plump, somewhat intermediate between *Pterostichus* and *Amara.* Third elytral interval has at least one "dorsal puncture" in apical half except in species with more or less pubescent elytra, in the two smallest species, and in *rufus.* Microsculpture stronger in the female than in the male; the elytra thus more dull. Antennae pubescent from segment 3 onwards. First metatarsal segment short. Both pro- and meta-tarsi have four dilated segments in the male and carry two rows of scale-like adhesive hairs underneath (Fig. 416). Parameres short and rounded. Penis more or less asymmetric in dorsal view, inner armature usually complicated and spiny.

Certain species-groups, notably in the subgenus *Ophonus*, can cause problems with the identification. A reliable identification requires dissection of the male genitalia.

Most *Harpalus* are pronouncedly xerophilous and confined to open, often sandy, country. Many are nocturnal, buried in the soil during daytime. Their food, to a large extent, consists of vegetable matter (seeds, pollen, fruits, etc.). The phenology of a number of species is described by Schjøtz-Christensen (1965, 1966a).

Key to species of *Harpalus*

1 All elytral intervals punctate and pubescent, the inner ones sometimes less densely 2

– At least inner elytral intervals entirely smooth and glabrous 15

2(1) Head with frons and temples pubescent, at least distinct when seen in profile. Pronotum with coarse puncture on disc 3

– Frons glabrous. Pronotum smooth on disc, or nearly so 13

3(2) Both pronotum and elytra with a distinct, bluish or greenish, metallic lustre 4

– Pronotum always without metallic lustre, elytra more or less so 5

4(3) Pronotum with raised, well delimited basal bead; hind-angles obtusely rounded (Fig. 379) .... 292. *azureus* (Fabricius)

– Pronotum without raised basal bead; hind-angles sharp, about right-angled .... 286. *nitidulus* (Stephens)

5(3) Pronotum with a fine, at middle indistinct, raised bead along basal margin (best seen from in front) 6

– Pronotum without such bead 9

6(5) Hind-angles of pronotum obtuse and rounded (Fig. 379); sides in front of hind-angles not or barely sinuate .... 292. *azureus* (Fabricius) var.

– Hind-angles right or slightly obtuse without rounded tip; sides more or less sinuate posteriorly 7

7(6) Pronotum broad, cordate (Fig. 373), with sides strongly rounded anteriorly and with posterior sinuation long and deep. Punctuation on disc sparse (as in *rufibarbis* but much coarser). Penis: fig. 385 .... 287. *puncticollis* (Paykull)

– Sides of pronotum less sinuate posteriorly, tip of hind-angle less sharp (Figs 374, 375, 378). Punctuation on disc denser 8

8(7) Pronotum (Fig. 378) narrow, broader near base which is faintly arcuate just inside hind-angle. Penis: fig. 389 .... 291. *puncticeps* (Stephens)

– Pronotum (Figs 374, 375) as broad as elytra over shoulders, its base virtually straight. Penis: fig. 386 .... 288. *melletii* Heer

9(5) Sides of pronotum (Fig. 372) with short but evident sinuation. Shoulders rounded. Punctuation of pronotum

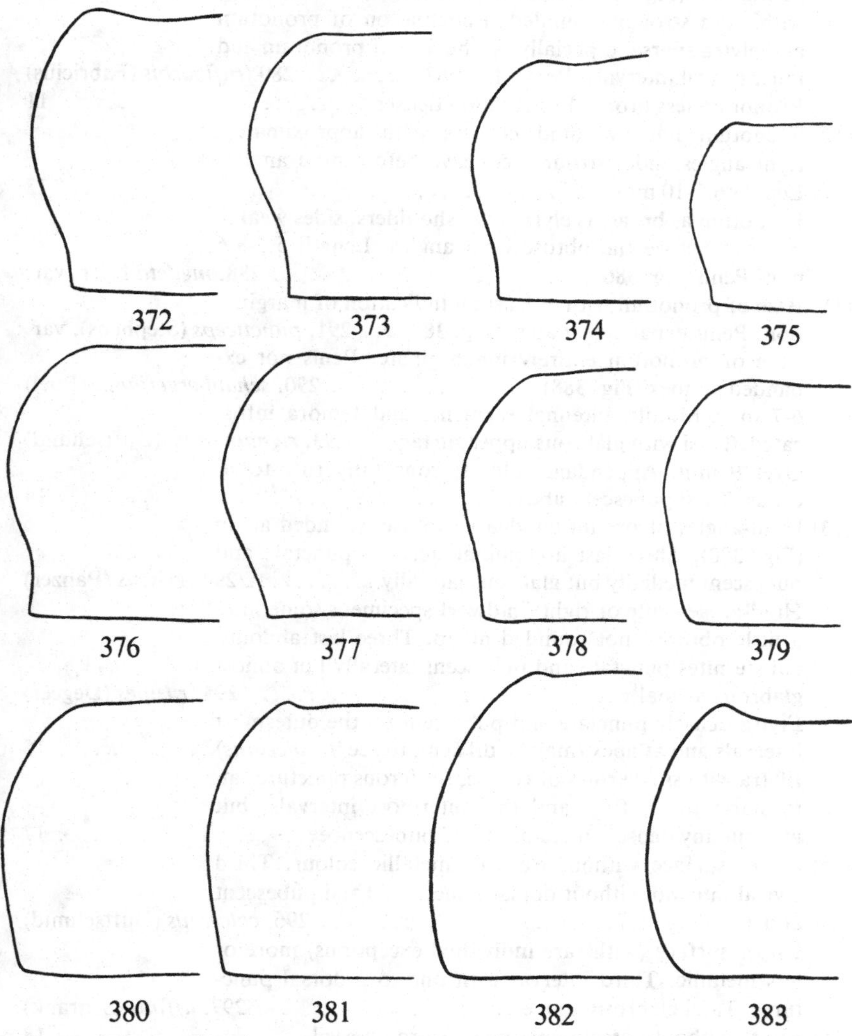

Figs 372-383. Pronotum of *Harpalus*. - 372. *rupicola* Sturm; 373: *puncticollis* (Payk.); 374: *melletii* Heer; 375: same; 376: *rufibarbis* (F.); 377: *schaubergerianus* (Puel); 378: *puncticeps* (Stph.); 379: *azureus* (F.); 380: *griseus* (Pz.); 381: *calceatus* (Dft.); 382: *serripes* (Quens.) and 383: *solitaris* Dej.

and elytra sparse but very strong .... 285. *rupicola* Sturm
- Shoulders angulate, often protruding as a denticle. Pronotum with lateral sinuation longer or shallower 10

10(9) Pronotum (Fig. 376) considerably broader than long, with sides strongly rounded. Punctuation of pronotum and elytra sparse, especially on the disc of pronotum and inner elytral intervals. Penis: fig. 387 .... 289. *rufibarbis* (Fabricius)
- Pronotum less broad. Punctuation denser 11

11(10) Pronotum narrow, hind corners with approximately right angles, sides strongly concave before hind angles. Length 6.5-10 mm 12
- Pronotum as broad as elytra over shoulders, sides weakly concave before the obtuse hind angles. Length 5.5-8.6 mm. Penis: fig. 386 .... 288. *melletii* Heer, var.

12(11) Base of pronotum with at least an indication of margination. Penis expanded at apex (Fig. 389) ... 291. *puncticeps* (Stephens), var.
- Base of pronotum entirely immarginate. Penis not expanded at apex (Fig. 388) .... 290. *schaubergerianus* (Puel)

13(2) 6-7 mm. Middle antennal segments and femora infuscated. Tarsi with glabrous upper surface .. 293. *signaticornis* (Duftschmid)
- Over 9 mm. Appendages almost constantly rufo-testaceous. Tarsi pubescent above 14

14(13) Hind-angles of pronotum clearly obtuse, rounded at tip (Fig. 380). Three last abdominal sternites punctate and pubescent medially but glabrous laterally .... 294. *griseus* (Panzer)
- Hind-angle acute or right (in dwarf specimens sometimes a little obtuse), not rounded at tip. Three last abdominal sternites punctate and pubescent laterally but almost glabrous medially .... 295. *rufipes* (Degeer)

15(1) Elytra densely punctate and pubescent on the outermost intervals and at apex (may be difficult to see in *calceatus*) 16
- Elytra with single rows of coarse, setiferous punctures at most on third, fifth and the outermost intervals, but without any dense punctuation and pubescence 17

16(15) Upper surface without trace of metallic colour. Third elytral interval without dorsal puncture. Tarsi pubescent above .... 296. *calceatus* (Duftschmid)
- Upper surface (with rare individual exceptions) more or less metallic. Third interval with one to 3 dorsal punctures. Tarsi glabrous above .... 297. *affinis* (Schrank)

17(15) Elytra without dorsal puncture on third interval 18
- Third elytral interval with one or more dorsal punctures 20

18(17) 11-13 mm. Entirely rufo-testaceous . 319. *flavescens* (Piller & Mitterpacher)
- Less than 8 mm. Piceous to black 19

19(18) Shoulder angulate and denticulate (Fig. 400). Pro-tibia usually with 5 apical spines (Fig. 402) .... 315. *picipennis* (Duftschmid)

– Shoulder rounded without evident tooth (Fig. 401). Pro-tibia usually with 3 apical spines (Fig. 403) .... 316. *pumilus* (Sturm)

20(17) Seventh and eighth (sometimes also third and/or fifth) elytral intervals with a short row of punctures near apex (Figs 404, 405)... 21

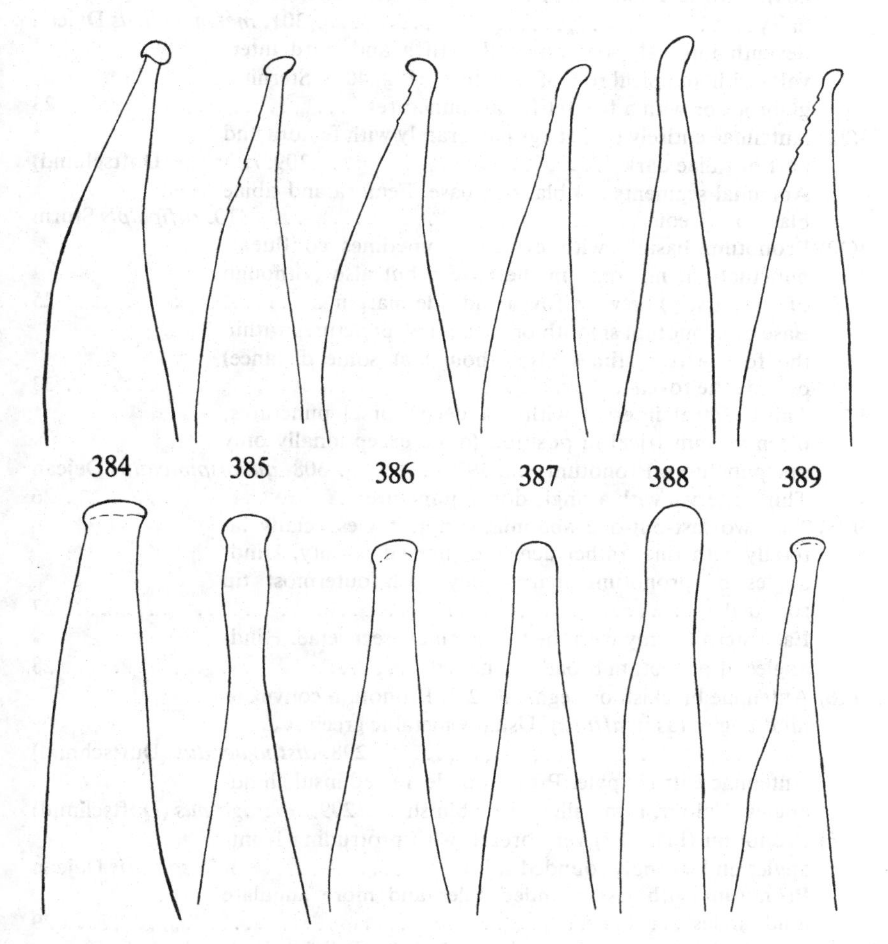

Figs 384-389. Penis of *Harpalus* subg. *Ophonus,* upper row in lateral view, lower row in dorsal view. – 384: *rupicola* Sturm; 385: *puncticollis* (Payk.); 386: *melletii* Heer; 387: *rufibarbis* (F.); 388: *schaubergerianus* (Puel) and 389: *puncticeps* (Stph.).

- Only ninth elytral interval with a row of coarse punctures; a single punctures adjoining apex of seventh stria 24

21(20) Third elytral interval with at least 2-4 dorsal punctures. Wings quite reduced .... 302. *autumnalis* (Duftschmid)

- Third interval only with the usual single dorsal puncture. Wings full 22

22(21) Eighth elytral interval with an apical row of punctures (Fig. 404). Two last abdominal sternites densely punctate and hairy .... 301. *melancholicus* Dejean

- Seventh interval (sometimes also fifth and third intervals) with an apical row of punctures (Fig. 405). Sternites glabrous or with a few setiferous punctures 23

23(22) Antennae entirely pale. Legs pale, rarely with femora and part of tibiae dark .... 309. *rubripes* (Duftschmid)

- Antennal segments 2-4 black at base. Femora and tibiae black or piceous .... 310. *rufipalpis* Sturm

24(20) Pronotum basally with evident, sometimes confluent, punctuation, not only in the foveae but also (although often sparsely) between fovea and side-margin 25

- Base of pronotum smooth or with a few punctures within the foveae (sometimes also, though at some distance) outside the fovea 33

25(24) Third elytral interval with 2-4 deep dorsal punctures, often asymmetrical in position (quite exceptionally only one puncture). Pronotum: fig. 393 .... 308. *quadripunctatus* Dejean

- Third interval with a single dorsal puncture 26

26(25) The two last-but-one abdominal sternites especially laterally with fine, rather dense, depressed pilosity. Hind-angles of pronotum sharp, only with outermost tip rounded 27

- Each sternite only with the two normal erect setae. Hind-angles of pronotum broadly rounded 28

27(26) Antennae blackish on segments 2-4. Pronotum convex to hind-angles (as in *affinis*). Usually metallic green .... 298. *distinguendus* (Duftschmid)

- Antennae entirely pale. Pronotum depressed inside hind-angles. Colour, if metallic, more bluish ... 299. *smaragdinus* (Duftschmid)

28(26) Pronotum (Fig. 383) very broad, with protruding front-angles and strongly rounded sides................. 303. *solitaris* Dejean

- Pronotum with less rounded sides and more angulate hind-angles 29

29(28) Tarsi and antennae from segment 3 onwards infuscated. Northern species .... 305. *nigritarsis* Sahlberg

- Legs and antennae entirely pale 30

30(29) Pronotum (Fig. 391) with sides clearly rounded in anterior half, inside hind-angles somewhat depressed (as in *smaragdinus,* though less produced). Elytra with evident reticulate microsculpture in both sexes . . . . 306. *luteicornis* (Duftschmid)

– Pronotum more rectangular in outline, not depressed laterally. Elytra virtually without microsculpture in the

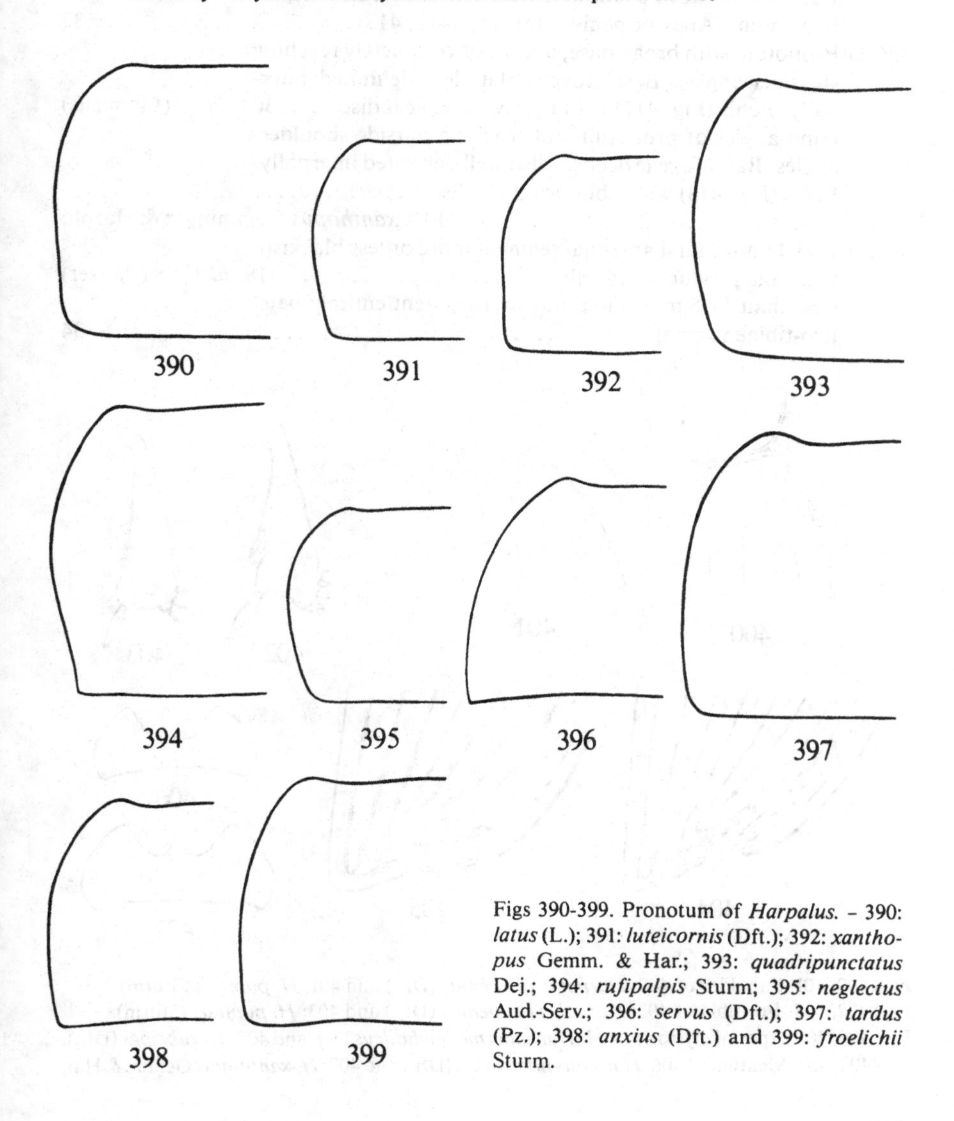

Figs 390-399. Pronotum of *Harpalus.* – 390: *latus* (L.); 391: *luteicornis* (Dft.); 392: *xanthopus* Gemm. & Har.; 393: *quadripunctatus* Dej.; 394: *rufipalpis* Sturm; 395: *neglectus* Aud.-Serv.; 396: *servus* (Dft.); 397: *tardus* (Pz.); 398: *anxius* (Dft.) and 399: *froelichii* Sturm.

male 31

31(30) Lateral margin of pronotum not translucent. At least the male with a bluish hue. Apex of penis very long, with terminal disc more oblique than in *xanthopus*.......... .... 309. *rubripes* (Duftschmid) var.

– Lateral margin of pronotum clearly translucent. Metallic hue absent. Apex of penis short (Figs 411, 413) 32

32(31) Pronotum with broad base, almost or completely reaching shoulder-angles. Basal foveae flat, less delimited internally. Penis (Fig. 411) with transverse apical disc .... 304. *latus* (Linnaeus)

– Hind-angles of pronotum not reaching outside shoulder-angles. Basal foveae deeper, also well delimited internally. Penis (Fig. 413) with oblique apical disc............... .... 307. *xanthopus* Gemminger & Harold

33(24) 13.5-15 mm. First antennal segment more or less blackish. Pro-tibia produced apically .... 318. *hirtipes* (Panzer)

– Less than 11.5 mm. First antennal segment entirely pale. Pro-tibiae normal 34

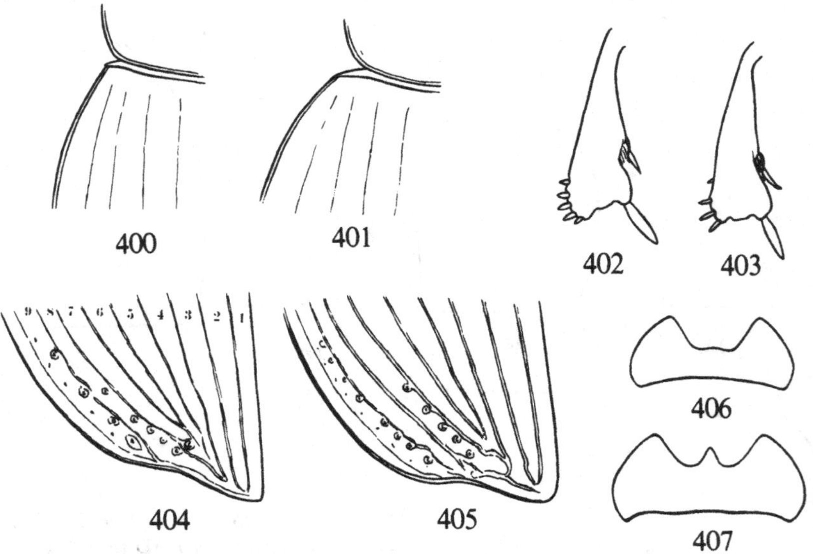

Figs 400, 401. Shoulder of 400: *Harpalus picipennis* (Dft.) and 401: *H. pumilus* (Sturm).
Figs 402, 403. Pro-tibia of 402: *Harpalus picipennis* (Dft.) and 403: *H. pumilus* (Sturm).
Figs 404, 405. Apex of elytron of 404: *Harpalus melancholicus* Dej. and 405: *H. rubripes* (Dft.).
Figs 406, 407. Mentum of 406: *Harpalus luteicornis* (Dft.) and 407: *H. xanthopus* Gemm. & Har.

34(33) The two last-but-one sternites with numerous punctures, each setiferous 35
– Sternites glabrous, each with two setae 37
35(34) Antennae entirely pale. Pronotum: fig. 399 .... 317. *froelichii* Sturm
– Antennae strongly infuscated from segment 2 36
36(35) Pro-tibia with 3 preapical spines (as in fig. 403). Abdomen with long, rather dense hairs. Pronotum: fig. 395 .... 311. *neglectus* Audinet-Serville
– Pro-tibia with 4-6 preapical spines (as in fig. 402). Abdomen only with very few, short extra hairs ..... 300. *serripes* (Quensel) var.
37(34) Antennae entirely pale or very faintly infuscated from segment 2 38
– Antennae strongly infuscated from segment 2 40
38(37) Pro-tibia with 4-6 preapical spines (as in fig. 402). Base of pronotum approximately straight, hind-angles broadly rounded (Fig. 397) .... 313. *tardus* (Panzer)
– Pro-tibia with 3 preapical spines (as in fig. 403). Base of pronotum with lateral backward sinuation, hind-angles less rounded 39
39(38) Piceous to brown, antennae pale. Pronotum: fig. 396 .... 312. *servus* (Duftschmid)
– Black, antennae darkened from segment 2. Pronotum: fig. 398 .... 314. *anxius* (Duftschmid)
40(37) Pro-tibia with 4-6 preapical spines (as in fig 402). Antennal segments 2-4 almost black. More than 9 mm.... 300. *serripes* (Quensel)
– Pro-tibia with 3 spines (as in fig. 403). Antennae evenly infuscated from segment 2. Less than 8.2 mm ... 314. *anxius* (Duftschmid)

Subgenus *Ophonus* Stephens, 1828

Ophonus Stephens, 1828, Ill. Brit. Ent. Mand. 1: 67.
Type-species: *Carabus obscurus* Fabricius, 1792.
Metophonus Bedel, 1897, Cat. Col. N.Afr. 1: 111.
Type-species: *Harpalus syriacus* Dejean, 1829.
Semiophonus Schauberger, 1933, Kol. Rdsch. 19: 131.
Type-species: *Carabus signaticornis* Duftschmid, 1812.

At once recognized on the entirely punctate and pubescent upper surface, including frons and temples. Appendages entirely pale, except in *signaticornis*. Wings full (dimorphic in *azureus*).

285. *Harpalus rupicola* Sturm, 1818

Figs 372, 384.

Harpalus rupicola Sturm, 1818, Deutschl. Fauna Ins. 5 (4): 105.

7-9 mm. A narrow, stretched species with markedly parallel-sided elytra. Underside brown, upper surface piceous, first elytral interval and forebody more or less pale; elytra usually with bluish or greenish hue. Appendages pale. Pronotum (Fig. 372) narrow, lateral sinuation short but deep, base not margined. Shoulder completely rounded. Punctuation of pronotum and elytra sparse but very strong. Microsculpture weaker than in the following members of the subgenus, virtually absent on the elytra of the male, obsolete in the female. Penis (Fig. 384) stout, transverse apical disc strong.

Distribution. Denmark: very rare; can only be recorded from a few localities: LFM: Gedser, 1 specimen 1887; Høje Møn, repeatedly collected, often in numbers, since 1861; SZ: Klinteby Klint, in numbers 1978; NWZ: Overby, 3 specimens 1979; B: Sose, repeatedly collected in numbers. - Sweden: a very limited distribution. Öl.: rare and recently established; Gtl.: throughout the island and not rare; Sdm.: Trosa, repeatedly collected in the last years; Upl.: Lovön, after 1959 in numbers. - Not in Norway or East Fennoscandia. - From W.Europe to Asia Minor, Iran and the Urals.

Biology. In open, fairly dry and warm habitats, usually on chalky or clayey gravel with sparse but often tall vegetation, e.g. *Daucus carota* and *Centaurea scabiosa.* Often on southern hill-sides and in gravel pits. The species is less thermophilous and xerophilous than most other typical "limestone species" (Lindroth, 1949). It is sometimes found in company with *H. azureus, melletii* and *Brachinus.* Its peak activity is in May-June, when reproduction takes place; newly emerged beetles occur in late July and August.

286. ***Harpalus nitidulus*** (Stephens, 1828)

Ophonus nitidulus Stephens, 1828, Ill. Brit. Ent. Mand. 1: 161.
Carabus punctatulus Duftschmid, 1812, Fauna Austriae 2: 89; *nec* Fabricius, 1792.

8.5-11 mm. Largest species of the subgenus and almost constantly with strong metallic reflection. Black, paler underneath, elytra dull. Entire upper surface metallic blue or green. Pronotum broad, sides sinuate posteriorly, with sharp, rectangular hind-angles and base entirely margined. Elytra with strong shoulder-tooth; punctuation fine and dense.

Distribution. Denmark: in Jutland very rare; 7 localities in eastern SJ and EJ. Scattered distributed on the islands, but usually only single individuals are recorded; only in number on Nekselø (NWZ) 1979. Also B: Sose, 1 specimen 1977. - Sweden: Öl., Gtl., G. Sand.; rather distributed, at least on Öl., but uncommon; also Sk., Kivik, 1986 (R. Baranowski leg.). - Unrecorded from Norway. - Finland: a recent addition to the fauna, found first time in 1931; now rare but recorded from many localities along the southern coast. In Vib in the USSR records from the beginning of this century. - From W.Europe to the Caucasus and W.Siberia.

Biology. A heat-preferent species, living on moderately dry, notably gravelly and chalky soil, often rich in humus. The species usually occurs in slightly shaded habitats, e.g. in light deciduous forests, forest edges and thickets, also in open country with rather tall vegetation. Adult beetles, which are partly phytophagous, have been encountered in the umbels of *Daucus carota.* The species is most numerous in May-June. According to Barndt (1976) breeding occurs in spring, but some females probably lay eggs in autumn (Larsson, 1939).

287. ***Harpalus puncticollis*** (Paykull, 1798)
Figs 373, 385.

Carabus puncticollis Paykull, 1798, Fauna Suec. Ins. 1: 120.

7-10 mm. Earlier often confused with *puncticeps* and *melletii* but can be separated on the form of the pronotum and the penis. Almost pure black, underside more piceous, appendages more clear rufous than in all following species. Pronotum (Fig. 373) compared to elytra strikingly broad, cordiform, sides strongly rounded anteriorly and with a long, deep posterior sinuation. Punctuation on disc sparse (as in *rufibarbis*) but very coarse; basal bead strong and continuous. Basal elytral margin, inside shoulders, arcuate. Male with penis (Fig. 385) constricted before apex, which has a strong disc. Internal sac with a tooth at middle and a single field of strong, slender spines in apical two-thirds.

Distribution. In Denmark only few records: SJ: near Haderslev, 1 specimen 1900; NEJ: near Ålborg, 3 specimens 1936; Klæstrup at Nibe, 2 specimens 1916 and 4 in 1982; Kongerslev, 2 specimens 1982, 4 in 1983, and several in 1984. - Sweden: very rare to rare with scattered distribution in the south. Sk.: Löderups strand, repeatedly but rare; Öl.: Borgholm; a few records also from Ög., Vg., Nrk., Sdm., Upl. and Vstm. - Norway: rare and local in the south-eastern part, from TEy to On. - Finland: rare in the south. Also Vib and southern Kr in the USSR. - From W.Europe to Asia Minor and C.Siberia.

Biology. In dry, open country, usually occurring on chalky gravel-soil with sparse vegetation, e.g. in chalk pits and on southern hill-sides. The beetles may climb umbelliferous plants; they have been observed to feed on the fruits of *Pimpinella saxifraga.* Mainly in spring.

288. ***Harpalus melletii*** Heer, 1837
Figs 374, 375, 386.

Harpalus Melletii Heer, 1837, Käf. Schweiz 2: 11.
Ophonus rectangulus Thomson, 1870, Opusc. Ent. 3: 323.
Harpalus brevicollis s. Jeannel, 1942; *nec* Serville, 1821.

5.5-8.6 mm. A variable species and therefore often confused with any of the related species. The form of the penis is sometimes the only reliable character. Piceous, forebody usually paler. Sides of pronotum (Fig. 374) approximately parallel before hindangles, which are about right, though not as sharp as in *puncticollis;* basal bead often quite obsolete. Shoulders with small but sharp denticle; intervals somewhat more densely and equally punctate than in *rufibarbis.* Microsculpture weaker than in related species, almost absent in the more shiny male. The penis (Fig. 386) short and stout, apical disc protruding dorsally; internal sac with a large central tooth and two small separate groups of spines.

Distribution. Denmark: very scattered and rare. SJ: 4 localities; EJ: Århus, Glatved (4 specimens 1970). Several localities in F, LFM, SZ, NEZ and B. - Sweden: uncommon but rather distributed in Sk., Öl. and Gtl.; also one record at the east coast of Sm. - Norway: unrecorded. - Also unknown in East Fennoscandia. - W. and C.Europe, east to W.Poland and Hungary.

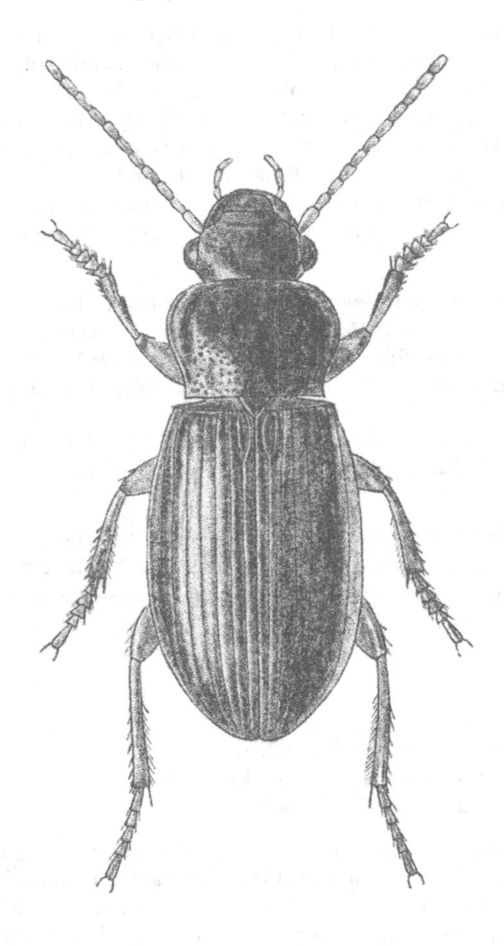

Fig. 408. *Harpalus rufibarbis* (F.), length 6.2-9.5 mm. (After Victor Hansen).

Biology. A pronounced "limestone species", living in warm open country, usually on rather dry, chalky gravel with sparse but often tall vegetation. It may climb Umbelliferae to feed on their fruits. Locally occurring in company with *H. azureus.* It is most numerous in May-August, probably reproducing in spring; newly emerged adults have been encountered in late summer.

289. ***Harpalus rufibarbis*** (Fabricius, 1792)
Figs 376, 387, 408.

Carabus rufibarbis Fabricius, 1792, Ent. Syst. 1: 159.
Ophonus subpunctatus Stephens, 1828, Ill. Brit. Ins. Mand. 1: 163.
Harpalus seladon Schauberger, 1926, Coleopt. Cbl. 1: 154.
Harpalus brevicollis auct., *nec* Serville, 1821.

6.2-9.5 mm. The commonest species of the subgenus and best characterized by the sparse punctuation on the upper surface. Piceous black, under surface paler, forebody rarely paler, appendages usually brownish. Pronotum short, clearly rounded anteriorly and with distinct lateral sinuation. Hind-angles sharp but varying from slightly acute to slightly obtuse; never with a regular basal bead. Punctuation of disc sparser than in other members of the subgenus; also sparse on elytra, with at most 3 punctures in width on each interval; punctures sometimes faint. Penis (Fig. 387) without apical disc (as in *schaubergerianus*).

Distribution. Denmark: distributed and common, more sparse in western Jutland. - Sweden: generally distributed and common in the south, north to S. Vrm., SE. Dlr. and Gstr.; isolated in central Sweden: one locality in Med. - Norway: local but common in the south-east, mainly in coastal areas. Isolated records in Ry and HOy. - In Finland fairly common in the southern provinces. Also in Vib and southern Kr of the USSR. - From W.Europe to the Caucasus. Introduced in North America.

Biology. The least xerophilous *Ophonus,* occurring in more or less shady habitats with not too dry, notably clayey soil rich in humus. It is rather eurytopic, preferring meadows, parks, gardens and arable fields, also in open woodland. The beetle in partly phytophagous; it has been found in the umbels of *Anthriscus silvestris.* Predominantly in May-August, reproducing in spring.

290. ***Harpalus schaubergerianus*** (Puel, 1937)
Figs 377, 388.

Metophonus schaubergerianus Puel, 1937, Misc. ent. 38: 91.
Harpalus brevicollis auct., *nec* Serville, 1821.

7.5-10 mm. Similar to *H. rufibarbis,* but pronotum much narrower and punctation denser. Piceous, underside somewhat paler, legs brownish to piceous. Pronotum (Fig. 377) comparatively narrow, strongly sinuate. Penis (Fig. 388) slender, sides parallel at apex.

Distribution. A Central European species. In our area only one occurrence reported from Finland, N: Helsinki in 1946 (E. Kangas).

Biology. Reportedly preferring limestone areas. The Finnish record was from a gravelly courtyard among remains from an old wood-pile (Kangas, 1978).

291. ***Harpalus puncticeps*** (Stephens, 1828)
Figs 378, 389.

Ophonus puncticeps Stephens, 1828, Ill. Brit. Ent. Mand. 1: 163.
Ophonus angusticollis J. Müller, 1921, Wien. ent. Ztg. 38: 137.

6.5-9 mm. With a stretched, parallel-sided stature (almost as in *rupicola*). Piceous, pronotum with somewhat paler margins, appendages brownish testaceous. Pronotum (Fig. 378) very narrow, widest part narrower than elytral width at the shoulders, inside hind-angles with a faint oblique depression. Basal bead very fine, often incomplete or almost lacking, base more or less oblique laterally. Elytra comparatively longer than in any other *Ophonus*, intervals densely punctate, basal margin moderately arcuate. Penis in side view (Fig. 389) more clearly serrate than in related species. Internal sac without tooth but with two widely separated fields of very dense and slender spines.

Distribution. Denmark: fairly distributed but rare in eastern Jutland north to Mols; also some records from central Jutland; absent in western and northern Jutland. Rather distributed on the inlands. - Sweden: rare and very local. Sk.: several localities, all from this century; Bl. and Hall.: each with one locality; Öl.: a few localities, after 1944; Gtl.: a few localities, after 1954. - Unrecorded from Norway. - Not in Finland, but one record from Vib: Muolaa (= Ogonki) in 1938. - From W.Europe to Syria and the Caucasus. Introduced in North America.

Biology. A thermophilous species living in open country on rather dry, clayey or chalky gravel with sparse but often tall vegetation of Umbelliferae, etc. For instance occurring in gravel and chalk pits, also on cultivated soil. The adult beetles are sometimes found in the umbels of *Daucus carota*, the seeds of which form the major part of the diet of the larva (Brandmayr & Brandmayr, 1975). The species is most numerous in autumn, mainly in August, when oviposition takes place; newly emerged beetles have been found in June-July.

Note. The species is a recent immigrant to Scandinavia and is still expanding.

292. ***Harpalus azureus*** (Fabricius, 1775)
Fig. 379.

Carabus azureus Fabricius, 1775, Syst. Ent.: 244.

6.2-9.2 mm. A small species, characterized both by coloration and form of pronotum. Piceous black, upper surface almost constantly with a strong blue, green or violaceous, reflection. A variety ("*similis* Dej.") can be separated from *melletii* by the pronotum and the coarse punctuation. Pronotum (Fig. 379) with well marked obtuse hind-angles, rounded at tip, and the absence (or only suggestion) of a lateral sinuation; basal bead raised. Shoulders forming an obtuse angle. Wings dimorphic.

Distribution. Denmark: Very rare. SJ: Jels, 1 specimen 1897; F: Ærø, 1 specimen 1964; LFM: Høje Møn, repeatedly collected in numbers, latest capture in 1941; B: Sose and Arnager, several records, often in numbers, latest in 1979. - Sweden: distributed and rather common on Öl. and Gtl. Elsewhere only a single locality at the coast of Sm. (Oskarshamn). - Not recorded from Norway or East Fennoscandia. - From W.Europe to Syria, Iran and Turkestan.

Biology. A heat-loving "limestone species" preferring temperatures at about 30°C (Lindroth, 1949). It occurs in dry, open habitats, usually on gravelly, clay-mixed chalky soil with short, sparse vegetation, e.g. in dry meadows and grassland, often on southern hill-sides; it is a characteristic member of the alvar fauna of Gtl. and Öl. The species is regularly found in company with *H. melletii, rupicola, Brachinus,* etc. It is usually considered a typical spring breeder, but finds of newly emerged beetles in mid and late June suggest that autumn breeding may also occur.

293. ***Harpalus signaticornis*** (Duftschmid, 1812)

Carabus signaticornis Duftschmid, 1812, Fauna Austriae 2: 91.

6-7 mm. Easily separated on coloration and pilosity; perhaps merited of forming a subgenus of its own. Piceous black, unmetallic; appendages rufo-testaceous, except that base of tarsal segments, femora and middle antennal segments are piceous. Pronotum most like that of *azureus,* but no basal bead. Head, most of disc of pronotum, and upper surface of tarsi are devoid of extra pubescence.

Distribution. Denmark: very rare and singly occurring. LFM: 4 records along the southern coast; B: several records from the coast; a total of 31 specimens were taken in the period 1919-77. All records certainly represent stray individuals. - Sweden: very rare, only stray individuals have been recorded from the southern coast of Sk.: Yngsjö, Vitemölle, Sandhammaren, Falsterbo). Usually singly but in numbers in 1972. - Unrecorded from Norway or East Fennoscandia. - C. and S.Europe, Asia Minor and the Caucasus.

Biology. On open, sandy or chalky ground. All Scandinavian specimens have been found on beaches, undoubtedly washed ashore (c.f. Baranowski & Gärdenfors, 1974), and the species has apparently not established permanent populations in our area.

Subgenus ***Pseudoophonus*** Motschulsky, 1844

Pseudoophonus Motschulsky, 1844, Ins. Sibér.: 223.
Type-species: *Carabus ruficornis* Fabricius, 1775 (= *rufipes* Degeer, 1774).
Pseudophonus auct.

Base of prothorax, all elytral intervals, and upper surface of tarsi densely pubescent.

Many North American species have, however, reduced vestiture of upper surface. Two large species in our fauna.

294. ***Harpalus griseus*** (Panzer, 1797)
Fig. 380.

Carabus griseus Panzer, 1797, Fauna Ins. Germ. 38: 1.

9-11.8 mm. Very closely allied to the following species and distinguished only in the following points: pronotum always with obtuse hind-angles, rounded at apex (Fig. 380); its disc still more indistinctly punctate. Elytra somewhat more strongly punctate and with fainter subapical sinuation of side margin. The last three sternites evidently punctate and hairy medially but glabrous laterally.

Distribution. Denmark: very rare. A few records from EJ, F, LFM, SZ and NWZ; more records from NEZ and B. Most records represent immigrating individuals found on beaches. However, in 3 localities (EJ: Mols, 3 records 1944-77; SZ: Mogenstrup, 9 records 1921-38 in a gravel pit; NEZ: Ørholm, several records 1860-1887 in a gravel pit) the species has certainly maintained a population for a period. - Sweden: rare and local from Sk. north to Vg. and G. Sand.; only in Sk. more generally distributed. Most records seem to be occasional but breeding populations have been observed at Degeberga and Löderup in Sk. - Norway: very rare, only two records from AK and Bø. - In Finland only two records, Ab: Turku, Ruissalo 1960 and N: Sjundeå 1966. Also in Vib and at the river Svir in Kr (USSR). - Transpalaearctic, from France to China and Japan, south to North Africa and Iran. Also the Azores.

Biology. A xerophilous species which occurs on dry, sandy meadows and grassland (e.g. Corynephoretum) with sparse vegetation, also in fallow fields. It is a nocturnal species, regularly coming to light; during daytime under stones or among plant roots, e.g. of *Artemisia campestris*. The species has a good dispersal power, and the northernmost finds in Scandinavia are undoubtedly due to accidental stragglers; it is sometimes found in sea drift (Baranowski & Gärdenfors, 1974). *H. griseus* is an autumn breeder, mainly occurring in August; newly emerged beetles have been found in July.

295. ***Harpalus rufipes*** (Degeer, 1774)
Pl. 3: 12.

Carabus rufipes Degeer, 1774, Mém. Hist. Ins. 4: 96.
Carabus ruficornis Fabricius, 1775, Syst. Ent.: 241.
Carabus pubescens Müller, 1776, Zool. Dan. Prodr.: 77.

10-16.7 mm. Piceous to almost black, appendages rufous or testaceous, legs exceptionally brown; elytral vestiture dense, yellowish. Pronotum with dense confluent punctuation at base but almost smooth on disc; hind-angles usually sharp, rectangular, sides sinuate posteriorly; in dwarf specimens the sinuation is obsolete and the angles obtuse though the apex of the latter is never rounded. The last three sternites of abdomen punctate and hairy laterally but almost glabrous medially.

Distribution. In Denmark very distributed and very common. - Sweden: very distributed and common in the south, north to southern Dlr., Gstr., Hls., Med., Jmt., Ång., and Vb. (one locality). - Norway: fairly common in all provinces north to NT. - Also in Finland very common, found north to Ks. Also Vib and Kr. in the USSR. - Transpalaearctic, from Britain to Japan, south to North Africa and Iran. Also the Azores and Madeira. Introduced in North America. Once found on Iceland.

Biology. A eurytopic species which occurs on almost every kind of open ground, notably on clayey, mull-rich soil. It is especially typical of cultivated fields, meadows and gardens, also on waste land and ruderal places. Less frequent in open woodland. The adults are omnivorous, feeding for instance on insect larvae, aphids and seeds of plants, e.g. of *Chenopodium album, Polygonum aviculare,* cereals and strawberries, which are often damaged by the beetles. *H. rufipes* is a nocturnal autumn breeder having its main period of reproduction in July-September. During this period the beetles may swarm in warm evenings, when the air temperature exceeds 18°C; they are attracted to light. According to Briggs (1965) most females do not lay eggs until a year after emergence, and some lay again in the following year. Hibernating adults survive buried deeply in the soil. The species is often found in shore drift.

Subgenus *Pardileus* Gozis, 1882

Pardileus Gozis, 1882, Mitt. schweiz. ent. Ges. 6: 289.
Type-species: *Carabus calceatus* Duftschmid, 1812.

296. ***Harpalus calceatus*** (Duftschmid, 1812)
Fig. 381.

Carabus calceatus Duftschmid, 1812, Fauna Austriae 2: 81.

10.5-14 mm. Piceous, almost black above, extreme sides of pronotum, antennae, and of legs at least tarsi, paler. Pronotum without evident basal foveae, but punctuated along entire base and sides, which are obliquely depressed posteriorly; hind-angles about right, sides almost straight in basal half. Elytra without dorsal puncture on third interval; ninth interval and part of eighth with very fine punctuation and pubescence. Can possibly be confused with *Anisodactylus signatus* (see that species).

Distribution. Denmark: very rare and usually singly occurring. In Jutland only in EJ: several localities in the Århus-Ebeltoft-Samsø area, which probably is the only Danish area with breeding populations in these years. In NEZ breeding populations occurred *ca* 1860-70. Other records represent stray individuals. - Sweden: rare and local, only in Sk. more repeatedly recorded. One record in each of the following districts: Bl., Sm., Öl., Gtl., G. Sand., Vg., Boh. and Upl., and two records from Hall, all representing stray individuals. Breeding populations have been observed in Sk.: Degeberga and Löderup. - In Norway only one record from Ak: Asker. - In Finland recorded from Ab: Mynämäki and N: Tusby. Also Vib and at the river Svir in Kr of the USSR. - Europe, the Caucasus, Siberia and China.

Biology. A xerophilous species, occurring on open, sandy ground with sparse vegetation of grasses, *Artemisia campestris*, etc.; also on agricultural land, e.g. in fallow fields. It is a nocturnal species, often flying at night and readily coming to light. *H. calceatus* is an autumn breeder, having its main period of activity in August.

Subgenus ***Harpalus*** s.str.

Actephilus Stephens, 1839, Man. Brit. Col.: 46.
Type-species: *Carabus vernalis* Duftschmid, 1812.

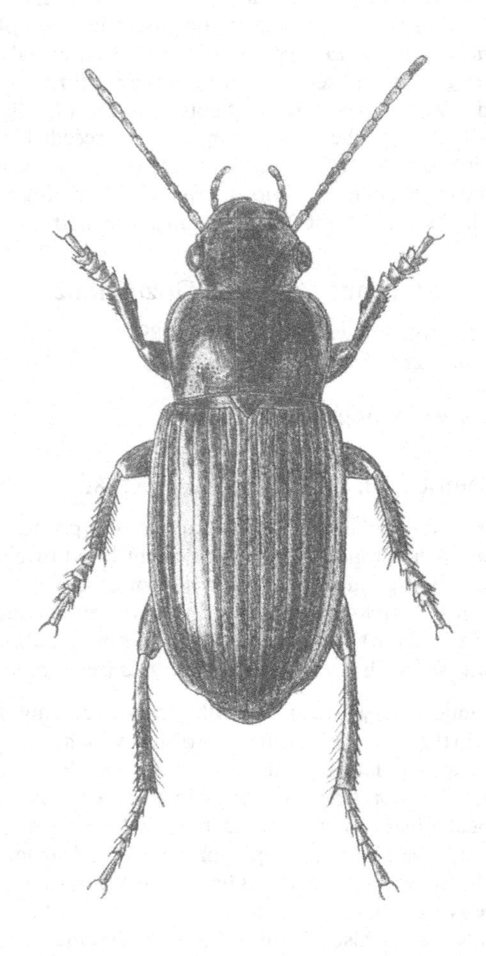

Fig. 409. *Harpalus affinis* (Schrk.), length 8.5-12 mm. (After Victor Hansen).

Acardystus Reitter, 1908, Fauna Germ. 1: 173.
Type-species: ***Harpalus rufus*** Brüggemann, 1873.
Haploharpalus Schauberger, 1926, Col. Centralbl. 1: 44.
Type-species: ***Harpalus froelichii*** Sturm, 1818.

A very large and diverse subgenus, often divided into several subgenera.

297. ***Harpalus affinis*** (Schrank, 1781)
Fig. 409; pl. 7: 1.

Carabus affinis Schrank, 1781, Enum. Ins. Austr.: 212.
Carabus aeneus Fabricius, 1775, Syst. Ent.: 245; *nec* Degeer, 1774.

8.5-12 mm. With regard to coloration the most variable of all *Harpalus,* but can be safely identified on the elytral pubescence. Underside black, upper surface most often metallic green, but often brass, coppery, bluish to almost black, rarely more or less rufinistic. Appendages pale, legs from clear rufous to piceous ("*confusus* Dej."), middle segments of antennae sometimes infuscated. Elytra with 1-3 dorsal punctures. Their outer invervals and the apical part punctate and pubescent, rarely also on middle intervals ("*semipunctatus* Dej."). Side-margin with deep subapical sinuation in the female. Abdominal sternites punctate and pubescent.

Distribution. Denmark: very distributed and common. - Sweden: very distributed and common except for the western mountains in the north. Recorded from all districts from Sk. to Lu. Lpm. - Norway: fairly common in all provinces north to Ns, mainly in coastal areas. - In Finland very common except for the northernmost parts. Also Vib and Kr of the USSR. - Europe and Siberia, south to Asia Minor and Iran. Introduced in North America.

Biology. One of the most eurytopic *Harpalus*-species, occurring on all kinds of open ground, notably on sandy, grassy or weedy soil. Often on agricultural land, especially in fallow fields, also in gardens, ruderal sites, etc. The adults are predominantly phytophagous and mainly nocturnal. Their main period of activity is in early spring, when most beetles reproduce. Another, but smaller part of the population are autumn breeders having winter larvae. Some beetles hibernate twice and reproduce for a second time (Schjøtz-Christensen, 1965). The species has been found in shore drift in spring.

298. ***Harpalus distinguendus*** (Duftschmid, 1812)

Carabus distinguendus Duftschmid, 1812, Fauna Austriae 2: 76.

8.8-11.2 mm. Very similar to the green and brass forms of *affinis,* but without punctuation and pubescence of the outer elytral intervals. Antennal segments 2 and 3 always with blackish base. Pronotum posteriorly with sides more sinuate than in *affinis,* therefore more parallel-sided, and hind-angles sharper. Shoulder-tooth more evident; only one dorsal puncture, sides with faint subapical sinuation in the female. Abdominal sternites punctate and pubescent, but less so than in *affinis*.

Distribution. Denmark: very rare; only LFM: 3 records from Falster and Møn; B: Hammeren. - In Sweden very locally distributed over the southern half of the country, lacking in large areas; north to Hls. - Norway: rare and local in the south-east: AK: HEs and Bø. - Finland: only three records, Ab: Turku, Ab: Lojo and Sa: Punkaharju; also once in Vib. - From W.Europe to North Africa, Syria, Iran and the Caucasus; also the Azores and Madeira.

Biology. On dry, sun-exposed ground, notably on clay-mixed sandy or gravelly soil with scattered vegetation of grasses or weeds. In our area usually in the vicinity of human habitations, for instance occurring on building sites and waste places in cities. It as a diurnal species, often seen running about and flying in warm sunshine. Mainly in spring.

299. ***Harpalus smaragdinus*** (Duftschmid, 1812)

Carabus smaragdinus Duftschmid, 1812, Fauna Austriae 2: 78.
Harpalus discoideus s. Erichson, 1837; *nec* (Fabricius, 1801).

9-11.4 mm. Best recognized on the oblique depression along the side margin of pronotum. Piceous brown, upper surface darker, but margins of pronotum and usually elytral suture more or less pale. Only elytra of male with strong metallic (bluish or greenish) lustre, female faintly metallic. All appendages pale. Hind-angles of pronotum almost rectangular, sharp. Shoulder tooth protruding, sinuation of elytra evident. One dorsal puncture. Abdominal punctuation and pubescence in median position.

Distribution. Denmark: rather distributed but rare in C.Jutland, southern Djursland, northern Zealand and Bornholm. Otherwise very scattered or absent. - Sweden: somewhat scattered but not rare from Sk. to southern Gstr. - Norway: local in the south-eastern parts, isolated finds in On and VAy. - Finland: rare but found north to Kb. - Also in Vib and Kr of the USSR. - Europe, W. & C. Siberia, and the Caucasus.

Biology. A pronounced xerophilous and rather heat-preferent carabid which occurs in grassland and heaths, on sandy soil with scattered vegetation of grasses (e.g. *Corynephorus*), *Calluna, Artemisia campestris,* etc. It is a night-active species; during daytime usually buried in the sand at the roots of plants. Schjøtz-Christensen (1965) showed that a population of *H. smaragdinus* consists of a spring breeding group having summer larvae, and an autumn breeding sub-population with winter larvae. The adults may live for several years and reproduce more than once.

300. ***Harpalus serripes*** (Quensel, 1806)
Figs 382, 410.

Carabus serripes Quensel, 1806, *in* Schönherr, Syn. Ins. 1: 199.
Harpalus tardoides Hansen, 1940, Ent. Meddr 20: 577.

9.3-11.5 mm. A stout, convex species with dark antennae. Black, rarely with faint bluish hue; palpi somewhat infuscated. First antennal segment rufo-testaceous, segments 2-4 black, the following segments subsequently paler. Tarsi and sometimes also tibiae dark brown. Pronotum with base straight (Fig. 382) and sides evenly rounded. Protibia with 4-6 preapical spines, the large ordinary spine bigger and less arcuate than in *tardus*. Abdominal sternites rarely with a few extra, irregular, setigerous punctures.

Distribution. In Denmark very scattered and rare, nearly exclusively associated with coastal slopes. Only six localities in Jutland, mainly in the Ebeltoft-Samsø area. The main occurrence on the islands is along the coasts of district NWZ, but also records

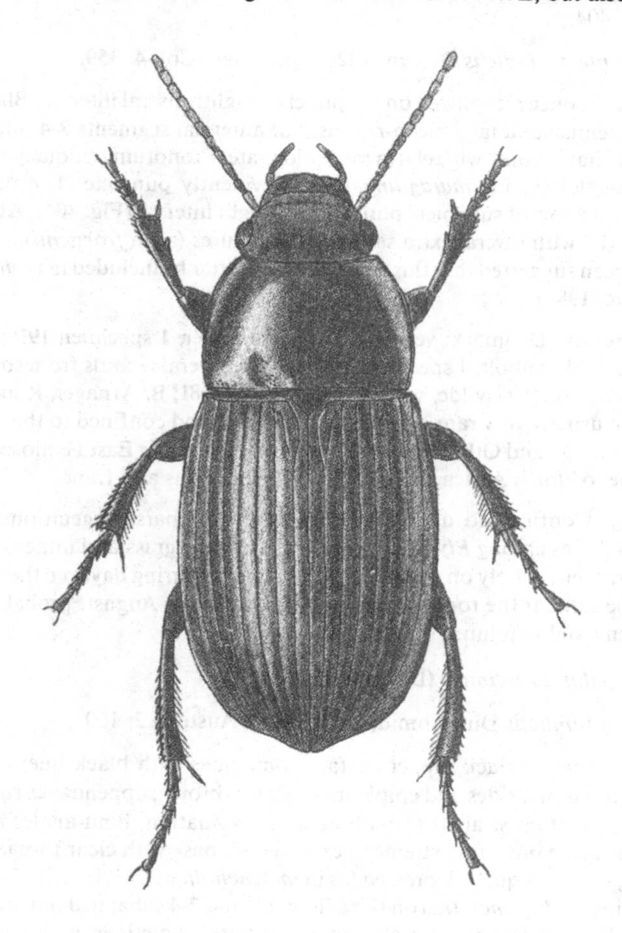

Fig. 410. *Harpalus serripes* (Quensel), length 9.3-11.5 mm. (After Victor Hansen).

from F, LFM, SZ, NEZ and B. - Also in Sweden a rare species, distributed in the coastal areas of the south-east: in Sk., Öl. and Gtl. several lovalities, in Bl. and Upl. only single records. - Not recorded from Norway or East Fennoscandia. - From W.Europe to North Africa, Asia Minor and the Caucasus.

Biology. Very xerophilous and rather heat-preferent. On dry, sandy or gravelly soil with sparse and short vegetation, usually near the sea coast on dry, grassy, southern slopes. It is most numerous in spring when propagation takes place.

301. ***Harpalus melancholicus*** Dejean, 1829
Fig. 404.

Harpalus melancholicus Dejean, 1829, Spec. Gén. Col. 4: 359.

10-11 mm. At once recognized on the punctate eighth elytral interval. Black or dark piceous, antennae and tarsi piceo-rufous, but antennal segments 2-4 often infuscated. Narrower than *tardus,* with elytra more elongate. Pronotum obliquely depressed inside hind-angles (as in *smaragdinus*); base evidently punctate. The only *Harpalus*-species with a row of subapical punctures on eight interval (Fig. 404). Abdominal sternites 4 and 5 with several extra setiferous punctures (as in *froelichii*).

It has been suggested that this species might better be included in *Ophonus.* (Brandmayr *et al.,* 1980).

Distribution. Denmark: very rare. EJ: Femmøller, 1 specimen 1919; Strandkær, 1 specimen 1944; Anholt, 1 specimen 1974; LFM: several records from southern Falster before 1900; NEZ: Tisvilde, several records 1891-1981; B: Arnager, Rønne, 1860-1921. - Sweden: usually very rare, especially after 1950, and confined to the south and east coast of Sk., Öl. and Gtl. - Not recorded from Norway or East Fennoscandia. - From W.Europe to North Africa, Asia Minor, the Caucasus and Iran.

Biology. Confined to dry, sandy habitats with sparse vegetation. Usually near coasts: in dunes among *Elymus* vegetation, on sandy grassland under *Artemisia, Corynephorus,* etc. Rarely on similar inland habitats. During daytime the beetles are buried in the sand at the roots of plants. Mainly in May-August, probably breeding in both spring and autumn.

302. ***Harpalus autumnalis*** (Duftschmid, 1812)

Carabus autumnalis Duftschmid, 1812, Fauna Austriae 2: 100.

9-10 mm. Piceous black, upper surface sometimes with black hue; entire pronotal margin, the suture, sides and epipleura of elytra, brown; appendages rufo-testaceous. Sides of pronotum straight or with posterior sinuation, hind-angles rectangular or somewhat acute but with extreme apex rounded, base with clear foveae but these impunctate, sides obliquely depressed (as in *melancholicus*). Elytra with 2-5 dorsal punctures (as in *quadripunctatus*) on third interval, and 2-4 subapical punctures in seventh interval. The two last-but-one abdominal sternites sometimes with a few small extra punctures. Wings reduced into scalelike rudiments.

Distribution. In our area only one locality is known in Sweden: Sk.: Löderup, 5 specimens 1951-52, 3 specimens 1979. - C.Europe, south to Yugoslavia and east to the Urals.

Biology. On dry, open, sandy fields. According to Barndt (1976), it is a characteristic inhabitant of grassland with xerophytic vegetation, e.g. *Corynephorus canescens, Armeria maritima* and *Hieracium pilosella,* living in association with *H. froelichii* and *smaragdinus, Mesoreus wetterhali,* etc. Also on sandy *Calluna*-heaths. It is an autumn breeder.

303. ***Harpalus solitaris*** Dejean, 1829
Fig. 383.

Harpalus solitaris Dejean, 1829, Spec. Gén. Col. 4: 337.
Carabus fuliginosus Duftschmid, 1812, Fauna Austriae 2: 83; *nec* Panzer, 1809.
Harpalus nigritarsis s. Thomson, 1859; *nec* Sahlberg, 1827.

8.8-10.4 mm. At once recognized on the form of the pronotum. Black, without any metallic hue, outermost margin of pronotum usually transparent; antennae, tarsi and tibiae (except apex) rufo-testaceous ("*germanicus*" with entirely red legs). Pronotum (Fig. 383) very broad with strongly rounded sides and hindangles; front-angles protruding; basal foveae shallow, with surrounding punctuation very dense and confluent. Elytra short with rounded sides; one dorsal puncture. Microsculpture of elytra well developed in both sexes, though stronger in the dull female. Penis strongly serrate along ventral side. Internal sac with a clutch of strong spines about middle.

Distribution. Denmark: known from all districts in Jutland, but only fairly distributed and absent from many areas. F: Fåborg, before 1900; LFM: several records from eastern Falster; SZ: Feddet v. Præstø; NWZ, NEZ and B: several localities, but only few records after 1950. - Sweden: distributed over the entire country, but still unrecorded from several districts. - Norway: recorded from several provinces in the southern part, but scattered and local. Also in Fø. - Finland: rare but found over most of the country. Also a few finds from Vib, Kr and Lr in the USSR. - From France to the Caucasus and through Siberia to Japan; North America.

Biology. On sandy or gravelly soil, notably on open heaths with sparse vegetation of *Calluna, Cladonia* and grasses; also on dry grassland and in light pine forest, etc. Regularly occurring in the birch region of the mountains, occasionally above the timber line. The species is sometimes found together with *Miscodera* and *Cymindis vaporariorum.* It is most numerous in May-August. Reproduction takes place in spring, probably also in autumn (Schjøtz-Christensen, 1966a).

304. ***Harpalus latus*** (Linnaeus, 1758)
Figs 390, 411.

Carabus latus Linnaeus, 1758, Syst. Nat. ed. 10: 415.

8.2-11 mm. The commonest of the red-legged unmetallic species, of stout stature. Black, margin of pronotum pale. All appendages rufo-testaceous. Elytral epipleura rarely black. Head remarkably large. Mentum with a tooth (though sometimes fainter than in fig. 407). Sides of pronotum almost parallel-sided in basal half (Fig. 390), hind-angles about rectangular but with broadly rounded apex; the base usually reaching outside humeral tooth, which is denticulate. Microsculpture of elytra lacking in the male, at least anteriorly, generally distributed in the dull female. Penis (Fig. 411) with transverse apical disc; internal sac with 2 fields of enormous spines.

Distribution. In Denmark very distributed and common. - Sweden: known from all districts except G. Sand., common in the south, much more local in the north. - Norway: widespread and obviously common north to NT; only a few records from N and TR. - Very common in Finland, lacking only in the subarctic-arctic zone. Also in Vib, Kr and Lr of the USSR. - Europe including Iceland, to the Caucasus and Siberia.

Biology. Very eurytopic, occurring on almost every kind of soil, most abundantly on not too dry, clay-mixed gravelly soil with humus, less frequently on dry, sandy ground. Both in open country, e.g. grassland, heaths, meadows, arable land, and in light deciduous forest. Rarely living in the subarctic or arctic regions of the mountains. The beetles are found almost throughout the year, reproduction takes place in both spring and autumn.

305. ***Harpalus nigritarsis*** Sahlberg, 1827

Harpalus nigritarsis "var. b" Sahlberg, 1827, Ins. Fenn. 1: 237.
Harpalus proximus Leconte, 1848, Ann. Lyceum Nat. Hist. 4: 298.

7-9.2 mm. Like small *latus* but darker. Black, also elytral epipleura at least piceous, margins of pronotum hardly translucent. Two last segments of maxillary palpi (almost constantly), sometimes also antennae from third segment, infuscated. Legs bright rufous but never entirely so; tarsi, or at least last tarsal segment, piceous, in some individuals also apex of femora. Sides of pronotum more rounded (almost as in *solitaris).* Shoulder-tooth less pronounced. Penis with oblique disc (as fig. 411); internal sac with three separated groups of well developed teeth.

Distribution. Not in Denmark. - In Sweden extremely rare, only two specimens recorded; one labelled "Lapponia borealis", more than one hundred years old, and probably originating in Lu. Lpm., and one found near Kalix in Nb. in 1977. - Unrecorded from Norway. - Finland: a few specimens labelled "Lapponia" and collected in the 19th century. - Circumpolar, but very rare in Eurasia.

Biology. Unknown in Scandinavia. In North America on open, rather dry, firm soil, mostly gravel, with dense and short vegetation of grasses.

Note. Since the "typical form" of *nigritarsis* Sahlberg is a synonym of *solitaris* Dejean, the "var. b" should rightly be named *proximus* Leconte. The usage followed here was introduced by Thomson (1871) and Reitter (1900). Also, the name "*nigritarsis*" fits the "var. b" only.

306. ***Harpalus luteicornis*** (Duftschmid, 1812)
Figs 391, 406, 412.

Carabus luteicornis Duftschmid, 1812, Fauna Austriae 2: 86.

6-7.8 mm. Separated from *latus* and *xanthopus* by the pronotum (Fig. 391) being narrowing posteriad, and therefore with more obtuse hind-angles having a slight obtuse depression inside. Mentum (Fig. 406) without or with a just suggested tooth. Also in the male the elytra have microsculpture all over. Penis (Fig. 412) narrower with apical disc more oblique.

Distribution. Not in Denmark. – In Sweden a very rare species, recorded from Sk., Bl., Hall., Öl., Gtl., Vg., Boh., Dlsl. and Vstm., but only a few localities are known from

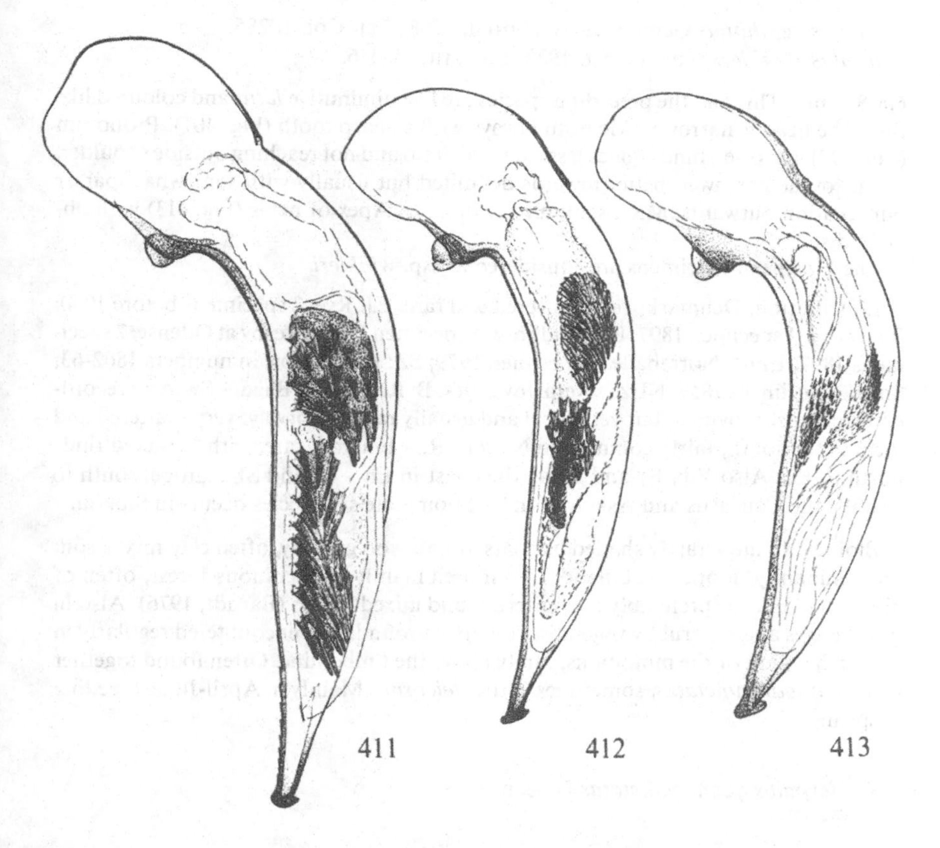

Figs 411-413. Penis of *Harpalus*. – 411: *latus* (L.); 412: *luteicornis* (Dft.) and 413: *xanthopus* Gemm. & Har.

that area. Most recorded specimens may be regarded as stray individuals, breeding populations probably only existing along the west coast. - Norway: very rare and local, only found in AK and B. - Finland: rather rare, found only in the southern provinces. Also Vib and southern Kr in the USSR. - Europe, the Caucasus and Siberia.

Biology. The habitat preference of this species is very imperfectly known, also because it was first very recently separated from *H. xanthopus*, with which it is sometimes associated. It seems to occur mainly in light deciduous forest. Also in open country. Mainly in spring.

307. ***Harpalus xanthopus*** Gemminger & Harold, 1868
Figs 392, 407, 413.

Harpalus xanthopus Gemminger & Harold, 1868, Cat. Col. 1: 285.
Harpalus Winkleri Schauberger, 1923, Ent. Anz. 3: 116.

6.8-8.5 mm. This and the preceding species are like diminutive *latus* and coloured like this. The head is narrower. Mentum always with a sharp tooth (Fig. 407). Pronotum (Fig. 392) narrower, hind-angles less rounded at tip and not reaching outside shoulder. Basal foveae narrower, better inwards delimited but usually with somewhat sparser punctuation outwards. Microsculpture as in *latus*. Apex of penis (Fig. 413) with oblique disc.
The European specimens are considered as ssp. *winkleri*.

Distribution. Denmark: very scattered and rare. EJ: Rye, 2 specimens before 1900; Silkeborg, 1 specimen 1897; F: Svendborg, 1 specimen 1932; Næsby at Odense, 7 specimens 1897; LFM: Nørreballe, 1 specimen 1979; SZ: Knudsskov, in numbers 1862-63; Sorø, 3 specimens 1859; NEZ: several lovalities; B: Rønne and Bagå. - Sweden: recorded from most provinces but very local and usually rare. - Norway: very scattered and local in the south, fairly common in N and TR. - Finland: rare, with scattered finds north to Kb. Also Vib, Kr and along the coast in Lr. - C. and SE.Europe, south to N.Italy, the Caucasus and Asia Minor. The nominate subspecies occurs in Siberia.

Biology. In moderately shaded habitats, notably on gravelly, often clay-mixed soil; predominantly in open woodland, in our area mainly in deciduous forest, often of birch, in C.Europe preferably in coniferous and mixed stands (Barndt, 1976). Also in forest edges and in scrubby vegetation on open ground. It is encountered regularly in the birch region of the mountains, rarely above the timber line. Often found together with *H. quadripunctatus*, sometimes with *luteicornis*. Mainly in April-June, breeding in spring.

308. ***Harpalus quadripunctatus*** Dejean, 1829
Fig. 393.

Harpalus quadripunctatus Dejean, 1829, Spec. Gén. Col. 4: 326.
Harpalus serie-punctatus Gyllenhal, 1827, Ins. Suec. 1 (4): 434; *nec* Sturm, 1818.

9.5-12 mm. Very characteristic through the presence of more than one dorsal puncture on third interval. Black, appendages pale; coloured as in *latus*, except that the side-margin of pronotum is rarely translucent. The male elytra often with slight steel-blue reflection. Pronotum (Fig. 393) broader, with more rounded sides, protruding front-angles, somewhat obtuse hind-angles, and more pronounced basal foveae. Elytra with weak shoulder tooth, third interval with 2-4 large dorsal punctures, usually asymmetrically placed (in abnormal cases also adjacent intervals or reduced to a single puncture).

Distribution. Denmark: fairly distributed but uncommon in eastern Jutland north to the Limfjord, and on the islands. Very scattered or missing in large areas of southern, western and northern Jutland. - Sweden: very distributed and common over the entire country. - Norway: widespread and fairly common in all districts. - Finland: fairly common all over the country. Also in Vib, Kr and Lr of the USSR. - Europe, Asia Minor, the Caucasus and Siberia.

Biology. The most prominent forest-dwelling *Harpalus*. It lives mainly in fairly shaded deciduous or mixed forests, on moderately humid, notably gravelly or stoney ground with *Rubus* etc., occurring among leaves and moss. Also in open country under bushes, in forest edges and in tall meadow vegetation; in the mountains reaching the birch region. The species is most numerous in spring, found in shore drift in June; probably breeding in both spring and autumn.

309. ***Harpalus rubripes*** (Duftschmid, 1812)
Fig. 405.

Carabus rubripes Duftschmid, 1812, Fauna Austriae 2: 77.

8.5-12.2 mm. Similar to *latus* but head and pronotum narrower. Normal specimens are easily recognized on the punctures of the seventh elytral interval. Black, female dull, margin of pronotum not so clearly translucent as in *latus*, upper surface, notably in the male, with blue or green reflection. Appendages rufo-testaceous; femora, rarely also tibiae, sometimes darkened. Base of pronotum densely and confluently punctate, at least between foveae and hind-angles, which are about right and hardly rounded; sides straight or faintly sinuate posteriorly. Seventh elytral interval with a row of 2-5 (in dwarf specimens only one) punctures (Fig. 405). Punctures may also occur on fifth interval. Pro-tibiae with at least 4 preapical spines. Apex of penis more slender than in *luteicornis*. The last-but-one abdominal sternites may exceptionally carry a few small setiferous punctures (in addition to the 2 normal large ones).

Distribution. In Denmark rather distributed and rather common. - Sweden: distributed in the south, north to Vrm. Rather common in eastern Sk. and on Öl and Gtl. In the rest of the area more local and less abundant. No records from the highlands of Sm. and Vg. - Norway: scattered in the south-eastern part, but local and uncommon. In the north one isolated record, Nsy: Bodø. - Finland: a few records only, Al: Finström, Ab: Lojo and Karislojo, Sa: Punkaharju and Sa: Parikkala. Also along the river Svir in Kr. - Europe, the Caucasus and W.Siberia.

Biology. A rather heat-preferent and xerophilous species which occurs on gravelly or sandy, sometimes clay-mixed or chalky soil with sparse and short vegetation, e.g. *Thymus, Armeria maritima* and grasses. For instance on sandy commons and fields, but scarsely in dry Corynephoretum; also on the alvar of Öl and Gtl., and in habitats influenced by human activity, e.g. gravel pits, fallow fields and ruderal places. It is active chiefly in May-August, breeding in both spring and autumn.

310. ***Harpalus rufipalpis*** Sturm, 1818
Fig. 394.

Harpalus rufipalpis Sturm, 1818, Deutschl. Fauna Ins. 5 (4): 70.
Carabus rufitarsis Duftschmid, 1812, Fauna Austriae 2: 82; *nec* Illiger, 1802.
Harpalus decipiens Dejean, 1829, Spec. Gén. Col. 4: 313.

8.7-11 mm. Like *rubripes* with a row of subapical punctures on seventh elytral interval but with infuscated antennae. Black, female dull, upper surface (especially in the male) often with faint steel-blue hue. Appendages piceous, except antennae, which have segment 1 rufo-testaceous but are infuscated from second or third segment; terminal segments again paler. Sides of pronotum (Fig. 394) narrowing towards base and usually sinuate, hind-angles slightly obtuse; sparsely punctuated in the foveae and near hind-angles. A small form with more obtuse hind-angles and non-sinuate sides has been termed *"decipiens* Dej.", but does not seem to have subspecific character. Seventh elytral interval with at least 2 subapical punctures. The two last-but-one abdominal sternites often with a few small setiferous punctures (besides the 2 normal large ones). Penis with apex bent ventral and spatulate in dorsal view.

Distribution. Denmark: rare and scattered but with increasing frequency of records since 1900, and especially since 1950. - Sweden: usually very rare and recorded from a few localities in Sk., Hall., Öl., Gtl. and Ög. Only on the eastern coast of Sk. less rare and with increasing frequency since 1950 (as in Denmark). Stray specimens are often found on seashores in SE.Skåne, sometimes in great numbers. - Not recorded from Norway or East Fennoscandia. - Europe, Asia Minor and the Caucasus.

Biology. In warm and dry, sandy habitats with scattered vegetation of xerophilous plants, e.g. *Corynephorus canescens, Hieracium pilosella* and *Armeria maritima.* Usually in open, sun-exposed country, but also in light pine forest, especially on clear-felled areas; notably in coastal regions, rarely in similar inland localities. Sometimes occurring in drift material on seashores. It is predominantly diurnal. Its peak activity is in spring when breeding takes place.

311. ***Harpalus neglectus*** Audinet-Serville, 1821
Fig. 395.

Harpalus neglectus Audinet-Serville, 1821, Faune Franc. 1: 26.

7-9 mm. With characteristic pronotum. Black, mouth-parts and main part of antennae piceous, but first segment rufo-testaceous, segments 2-4 almost black. Pronotum (Fig. 395) strongly constricted in basal half, hind-angles quite rounded, sparsely punctate in the fovea as well as near side-margin. The two last-but-one abdominal sternites with sparse punctures and long hairs. Dimorphic: wings either full or strongly reduced.

Distribution. Denmark: in Jutland rather distributed along the western coast from Skallingen northwards, and along the eastern coast from the Skaw to Horsens; also a number of inland records. Also rather distributed on the northern coast of Zealand and on Bornholm. Otherwise very scattered and rare. - Sweden: rare and very local, confined to coastal areas in the south: Sk., Bl., Hall., Öl., Gtl., G. Sand. and Vg. - Not known from Norway or East Fennoscandia. - From W.Europe to N.Italy and the Caucasus.

Biology. A very xerophilous species which is confined to open habitats with loose sand and sparse vegetation, usually near the coast, for instance in dune and shifting sand regions. Also on dry, virgin grassland with scattered tussocks of *Corynephorus canescens;* as such a habitat changes into a more dense grass or *Calluna* heath the number of beetles declines (Schjøtz-Christensen, 1965). It is often found in company with *H. smaragdinus, anxius, servus,* etc. During daytime buried in the sand at plant roots. The main period of activity is in May-September. Most animals reproduce in spring, but some have autumn reproduction and winter larvae (Schjøtz-Christensen, op. cit.).

312. ***Harpalus servus*** (Duftschmid, 1812)
Fig. 396; pl. 7: 2.

Carabus servus Duftschmid, 1812, Fauna Austriae 2: 101.

7.5-8.5 mm. General form very *Amara*-like. Upper surface piceous or brown; margins of pronotum and often also elytra reddish; if so, the latter sometimes with darker suture; palpi, antennae, and of legs at least tarsi, rufo-testaceous. Pronotum (Fig. 396) with base strongly produced laterally and hind-angles acute. Pro-tibiae with preapical spines.

Distribution. Denmark: very rare. In Jutland only found in a narrow strip of land on the west coast from Fanø to Ringkøbing fjord. Also repeatedly found in NWZ: Røsnæs and Rørvig; NEZ: Lynæs, Tisvilde, Asserbo and Melby; and B: several localities. Very scattered and accidental in other parts of the islands. - Sweden: rather distributed but rare in Sk., only more abundant, sometimes numerous, in a few localities (Kåseberga, Kivik, Ripa) in the south-eastern part. A few records of single specimens from Hall., Gtl., and Öl. - No records from Norway or East Fennoscandia. - From W.Europe to Siberia, Mongolia and N.China.

Biology. On fine, loose sand with sparse, xerophilous vegetation, usually in dunes on the coast, rarer on sandy inland habitats. During daytime it is mostly buried in the sand at the roots of *Corynephorus canescens* or under *Artemisia campestris,* etc.

Often found together with *H. neglectus* and *anxius*, sometimes with *melancholicus*. It is a spring breeder, most numerous in May-June.

313. ***Harpalus tardus*** (Panzer, 1797)
Fig. 397.

Carabus tardus Panzer, 1797, Fauna Ins. Germ. 37: 24.

8.4-11 mm. Black without any metallic hue; sides of pronotum more or less translucent, antennae and palpi rufo-testaceous, tarsi and at least base of tibiae brown. Pronotum (Fig. 397) almost rectangular with slightly rounded sides and about rectangular hind-angles; base impunctate or with a few punctures in basal fovea and at hind-angle. Pro-tibia with 4-6 preapical spines.

Distribution. Denmark: rather distributed and common, but apparently more scattered in WJ. - Sweden: generally distributed and common in the south, north to Med. - Norway: fairly common along the coast in the south-east, local in the inner parts; northernmost in Os: Ringebu. - Finland: common in the south, found north to Ks: Kuusamo. Also in Vib and Kr in the USSR. - Europe, Asia Minor, the Caucasus and Siberia.

Biology. Our commonest dark *Harpalus;* rather xerophilous, usually occurring in open country on sandy, sometimes clay-mixed soil with more or less dense vegetation, for instance on dunes, grassland (e.g. Corynephoretum) and sandy heaths; also on cultivated, light soil. Less abundant in moderately shady habitats like forest edges, open pine forest, etc. It is most predominant in May-June, which is the breeding period.

314. ***Harpalus anxius*** (Duftschmid, 1812)
Fig. 398.

Carabus anxius Duftschmid, 1812, Fauna Austriae 2: 101.

6.6-8.2 mm. Large specimens may be confused with small specimens of *tardus,* but the body is flatter and the pronotal form as well as the colour of the antennae are distinctive. Coloured as *tardus,* except that the tarsi usually are darker and first antennal segment evidently paler than the following segments. Base of pronotum (Fig. 398) somewhat concave, sides less rounded in basal half, so that the hind-angles become less rounded. Pro-tibiae with 3 preapical spines.

Distribution. Denmark: rather distributed in Jutland, but seemingly absent from large areas of western and northern parts. Scattered records in F, LFM and SZ. Well distributed along the north coast of Zealand (in NWZ and NEZ) and on Bornholm. - Sweden: rather distributed and common in coastal areas of Sk., southern Hall., western Bl., and on Öl and southern Gtl. Elsewhere very rare and very local, recorded from Ög. (2 localities), Nrk. (1 locality) and Upl. (1 locality). - Unrecorded from Norway. - Finland: known from two localities in the east, Sa: Rautjärvi and Kb: Kitee. Also a few records from Vib. - From W.Europe to North Africa, Asia Minor, Mongolia and E.Siberia.

Biology. Decidedly xerophilous. In dry, open sandy ground, notably in coastal habitats with scattered, short vegetation, for instance in dry *Corynephorus canescens* vegetation, where it is often the most dominant *Harpalus*. Also on dunes, etc. During daytime occurring under stones or buried at the roots of grasses, *Artemisia campestris*, etc. Less often on gravel, e.g. on the alvar of Öl. It is frequently found in company with *H. smaragdinus* and *neglectus*. The species is most numerous in spring, when the majority of beetles reproduce. A smaller part of the population are autumn breeders having winter larvae. The adults may survive for several years and may reproduce more than once (Schjøtz-Christensen, 1965).

315. ***Harpalus picipennis*** (Duftschmid, 1812)
Figs 400, 402, 414.

Carabus picipennis Duftschmid, 1812, Fauna Austriae 2: 102.
Harpalus multisetosus Thomson, 1884, Annls. Soc. Ent. France (6) 3: cxxi.

6-7.2 mm. Often confused with *pumilus* and separated almost only on relative characters. It is larger, more convex, and somewhat broader. Tibiae darker, the foremost pair carries 5 (exceptionally 4) pre-apical spines (Fig. 402). Pronotal foveae larger and more punctate. Shoulder-angle and tooth evident (Fig. 400). The upper surface is more dull through the stronger reticulate microsculpture; this is especielly marked on the head. Dimorphic: long- and short-winged individuals may occur in the same population. Penis: fig. 414.

Distribution. Denmark: only 3 accidental records in SZ: Knudshoved, 1926 & 1928, Svinø strand *ca* 1860-70. - Sweden: only found in a few localities in eastern Sk. and in one on Öl.; usually very rare. - Not recorded from Norway or East Fennoscandia. - From W.Europe to the Caucasus and E.Siberia.

Biology. A xerophilous species, living exclusively on dry, sandy soil with scattered plants, for instance on dunes and in dry grassland. It is sometimes found together with the following species, but *picipennis* seems to have a stronger preference for loose, almost sterile sand than *pumilus*. In Germany, Barndt (1976) recorded the species numerously in *Corynephorus canescens* vegetation, while *pumilus* occurred in adjacent dry meadows with e.g. *Dianthus deltoides, Armeria maritima* and *Festuca ovina*. It is a pronounced spring animal having summer larvae.

316. ***Harpalus pumilus*** (Sturm, 1818)
Figs 401, 403, 415.

Carabus pumilus Sturm, 1818, Deutschl. Fauna Ins. 5 (4): 77.
Carabus vernalis Fabricius, 1801, Syst. Eleuth. 1: 207; *nec* Panzer, 1796.
Harpalus picipennis auct.; *nec* (Duftschmid, 1812).

5.3-6.2. The smallest of our *Harpalus*-species. Piceous black, margin of pronotum and usually elytral suture paler, appendages rufo-testaceous, except that femora and apex of tibiae are infuscated. Pronotum with hind-angles entirely rounded and basal foveae very small. Elytra without dorsal puncture; shoulder rounded and only with suggested tooth (Fig. 401). Pro-tibia with 3 (exceptionally 4) preapical spines (Fig. 403). Apparently constantly with reduced wings. Penis: fig. 415.

Distribution. Denmark: very rare and scattered, almost exclusively on sun-exposed coastal slopes, SJ: Halk; EJ: Mols and Samsø; F: Ærø, Ristinge and Thurø; LFM: Tillitze, Gedser, Bøtø and Møns klint; SZ: Vejlø; NWZ and NEZ: several localities from Asnæs to Tisvilde; B: several localities. - Sweden: rare and local in areas at or near the coast in Sk., Öl., Gtl. and G. Sand.; also Ög.: Alvastra, 1967. - Not in Norway or East Fennoscandia. - From W.Europe to the Caucasus and W.Siberia.

Biology. In open, sun-exposed country on dry, usually sandy soil with sparse vegetation, for instance on dry hill-sides and coastal slopes. Also on gravelly soil, e.g. on the alvar of Öl. During daytime at the roots of plants. Mainly in spring and early summer, predominantly breeding in spring.

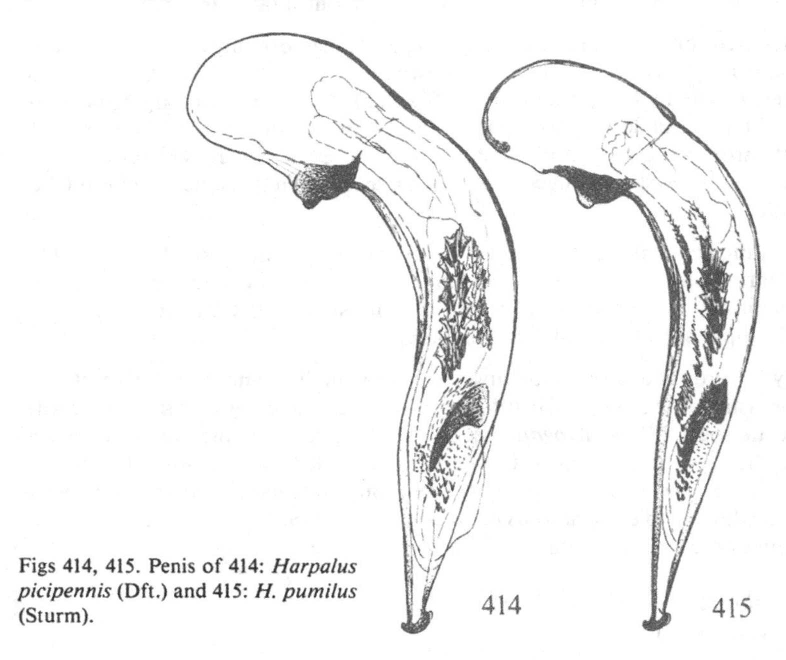

Figs 414, 415. Penis of 414: *Harpalus picipennis* (Dft.) and 415: *H. pumilus* (Sturm).

317. ***Harpalus froelichii*** Sturm, 1818
Fig. 399.

Harpalus Frölichii Sturm, 1818, Deutschl. Fauna Ins. 5 (4): 117.

8.5-10.4 mm. A stout species, very like *serripes.* Piceous black, mouth-parts and antennae rufo-testaceous, also tarsi pale. Pronotum (Fig. 399) narrow in comparison with elytra. The three last-but-one abdominal sternites strongly punctate and with long hairs. Hind femora (as in *melancholicus*) with more than 10 long setae along hind margin. The preapical spines of pro-tibiae are not isolated but only constituting the distal part of the row on the lower surface.

Distribution: Denmark: very scattered and rare, especially after 1950. - Sweden: usually very rare and only found in a few localities in Sk., one in southern Hall., and two in southern Öl. In three localities in Sk. (Haväng, Degeberga and Ripa) observed in great numbers. Not in Norway or East Fennoscandia. - From W.Europe to the Caucasus, E.Siberia and N.China.

Biology. Confined to dry, sandy areas on open ground, notably grassland and heaths, with scattered vegetation of *Corynephorus canescens, Calluna, Artemisia campestris,* etc. Also in dunes. During the day it is usually burrowed in the sand at plant roots. The main periods of activity is May-June and August-September; a *froelichii* population apparently comprises a spring as well as an autumn breeding group of animals (Schjøtz-Christensen, 1966a).

318. ***Harpalus hirtipes*** (Panzer, 1797)

Carabus hirtipes Panzer, 1797, Fauna Ins. Germ. 38: 5.

12.5-15 mm. The largest species of *Harpalus* s.str. in our fauna, easily rcognized on the pro-tibiae. Black with paler palpi and tarsi; antennae paler towards apex but first segments black or almost so. Flat and broad, especially the pronotum, which is strongly depressed along the side-margin. The two last-but-one abdominal sternites punctate and hairy. Pro-tibia with produced outer corner, preapical spines as in *froelichii.*

Distribution. Denmark: very rare and scattered, since 1950 only found in several localities in southern Djursland. Older records from all districts except WJ and NWJ. - Sweden: rare to very rare, found in Sk. and along the coast of Hall., and also known from Bl. (1 locality), Öl. (3 localities) and Gtl. (one locality). At Ripa in eastern Sk. it has been common. - Unrecorded from Norway and East Fennoscandia. - From C.Europe to Rumania, the Caucasus and C.Siberia.

Biology. This large *Harpalus* is probably the most xerophilous member of the genus. It is confined to sandy soil in dry, open country, occurring in habitats with sparse vegetation of grasses (e.g. *Corynephorus canescens*), *Calluna, Artemisia campestris,* etc., and is most often encountered in coastal regions, rarer on inland sands. During daytime the beetles are usually buried in the sand under *Artemisia,* among grass roots, etc. The species is sometimes found together with *H. froelichii.* It occurs mainly in May-June and in August-September. A *hirtipes* population is composed of two subpopulations, one breeding in spring having summer larvae, the other reproducing in autumn giving rise to winter larvae (Schjøtz-Christensen, 1966a).

319. ***Harpalus flavescens*** (Piller & Mitterpacher, 1783)

Carabus flavescens Piller & Mitterpacher, 1783, Iter. Per Pos.: 98.
Carabus ferrugineus Fabricius, 1775, Syst. Ent.: 244; *nec* Linnaeus, 1758.
Harpalus rufus Brüggemann, 1873, Abh. Naturw. Ver. Bremen 3: 459.

11-13 mm. A large, entirely rufo-ferrugineous species. The elytra without basal puncture, and confusion with *Amara fulva* is therefore possible, but the front-angles of pronotum are not produced, the head has a single supra-orbital puncture, and the body entirely devoid of metallic hue. Abdominal sternites sparsely punctate and hairy.

Distribution. Denmark: only one record from Bornholm, 1843. - Sweden: extremely rare. In Sk. one old record from Ystad and three recent records: Vomb 1971, Haväng 1966, and Ravlunda abundant since about 1965. Also one record from Öl: Stora Rör, 1910. - Not recorded from Norway. - In East Fennoscandia one record from Ab: Turku in 1916. - C.Europe to N.Italy, Bosnia and the Caucasus.

Biology. In dry open country on loose sand with scattered vegetation, for instance in dunes and on dry grassland. During the day it is buried in the sand at the roots of plants (e.g. *Corynephorus* and *Ammophila*) and therefore difficult to discover. Mainly in spring and autumn.

Genus *Diachromus* Erichson, 1837

Diachromus Erichson, 1837, Käf. Mark Brandenburg 1 (1): 43.
Type-species: *Carabus germanus* Linnaeus, 1758.

Contains a single medium-sized species with entire upper surface densely punctate and with erect pubescence. At once recognized on the striking colour pattern.

Pronotum with a long seta at hind-angle. Tarsi pubescent on upper side. Fore tarsus with four, and middle tarsus with two, dilated segments in the male, and also with adhesive hairs of the *Anisodactylus*-type underneath. Terminal spur of fore tibia very large. Hind-angles of pronotum sharply right-angled. Wings fully developed.

320. ***Diachromus germanus*** (Linnaeus, 1758)
Pl. 7: 3.

Carabus germanus Linnaeus, 1758, Syst. Nat. ed. 10: 415.

8-10 mm. Black, pronotum with blue or green reflection and sides narrowly pale. Head rufo-testaceous, elytra ferrugineous, with a common bluish black spot near apex. Appendages pale.

Distribution. In our area now probably an extinct species. Only recorded from Denmark: some 19th century records from the southern parts of the country. - From W.Europe to N.Africa, Asia Minor, Syria, Iran and NW.China.

Biology. On moderately dry meadow ground, often in somewhat shaded sites, e.g.

sparsely tree growth. In Central Europe frequently found in cereal fields. It dwells among grass roots and under stones.

Genus *Anisodactylus* Dejean, 1829

Anisodactylus Dejean, 1829, Spec. Gén. Col. 4: 132.
Type-species: *Carabus binotatus* Fabricius, 1787.

Species of moderate size, superficially similar to *Harpalus* (Fig. 421). Most important diagnostic characters are the more stretched basal segment of the hind tarsus (Fig. 420), especially when compared with the apical tibial spine, and the multiserially arranged adhesive hairs of the underside of the fore tarsus of the male (Fig. 417).

Frons with a pair of small, sometimes indistinct, reddish spots. Elytra with humeral tooth. Third interval (except in *signatus)* with at least one dorsal puncture; outer intervals punctate and pubescent (as in *Harpalus aeneus*). Wings full. Fore and mid tarsi of male with 4 dilated segments.

Key to species of *Anisodactylus*

1 Upper surface with metallic, usually green lustre. Apical spur of fore tibia trifid (as in *Amara* subg. *Zezea,* fig. 328) .... 321. *poeciloides* (Stephens)
– Upper surface black, usually without metallic lustre. Apical spur of fore tibia simple 2
2(1) Third elytral interval without a dorsal puncture. Hind-angles of pronotum blunt .... 324. *signatus* (Panzer)

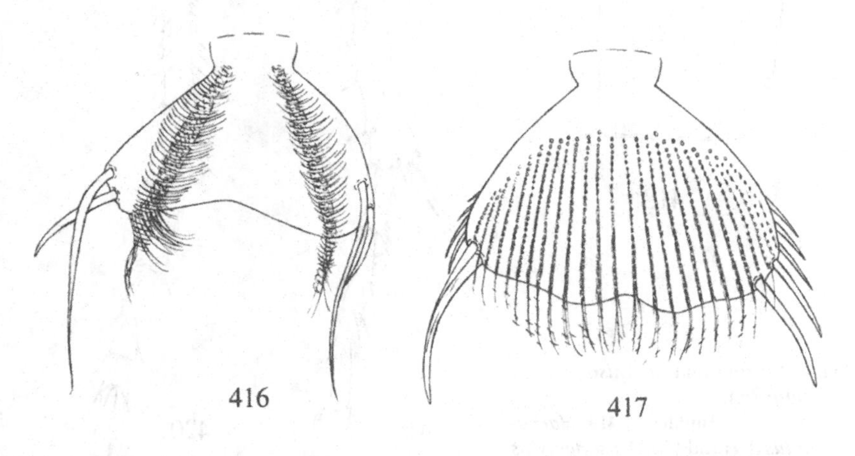

Figs 416, 417. Underside of dilated protarsal segment of 416: *Harpalus* and 417: *Anisodactylus.*

- Third elytral interval with a setiferous dorsal puncture. Hind-angles of pronotum denticulate (Fig. 418) 3

3(2) Elytra punctate and pubescent only on the two or three outermost intervals. Shoulders angulate ..... 323. *nemorivagus* (Duftschmid)

- Elytral punctuation and pubescence expanding apically over all intervals. Shoulders rounded .... 322. *binotatus* (Fabricius)

321. ***Anisodactylus poeciloides*** (Stephens, 1828)

Harpalus poeciloides Stephens, 1828, Ill. Brit. Ent. Mand. 1: 154
Anisodactylus pseudoaeneus auct.; *nec* Dejean, 1829.

10-13.5 mm. Black, underside faintly, upper surface strongly metallic: green or brassy, rarely bluish; appendages dark, except that first antennal segment is rufous, at least underneath. The specific name refers to the superficial similarity to *Poecilus* (sg. of *Pterostichus)* but among other differences, the protibial terminal spur is trifid, and the outer elytral intervals are pubescent. Fore femora incrassate in male. Pronotum with completely rounded hind-angles. Elytra with 1-3 dorsal punctures on 3rd interval.

Distribution. Denmark: very rare; after 1900 only a few records from Amager

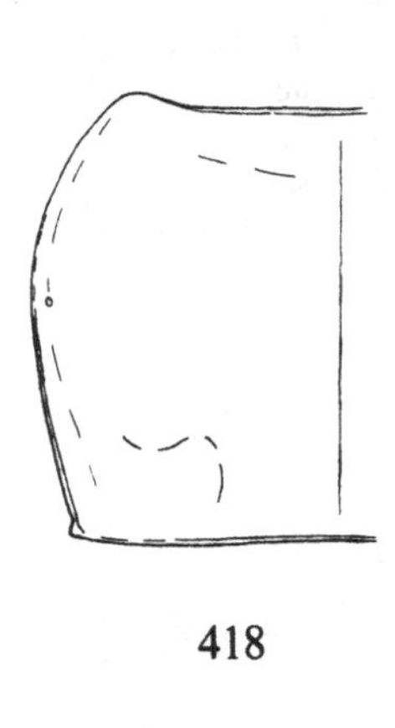

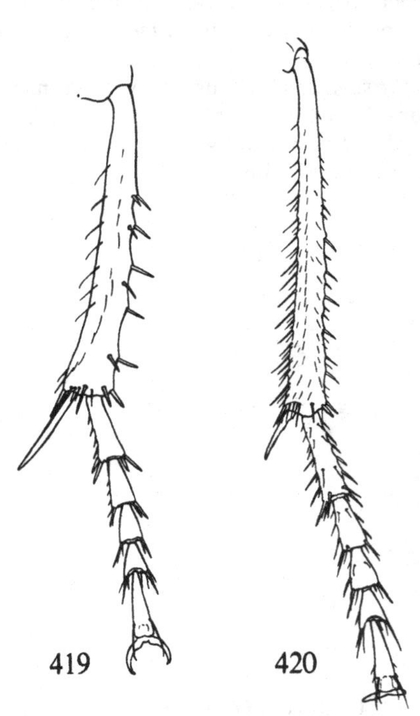

Fig. 418. Pronotum of *Anisodactylus binotatus* (F.).
Figs 419, 420. Hind leg of 419: *Harpalus tardus* (Pz) and 420: *Anisodactylus binotatus* (F.).

(NEZ), latest record 1972. Several records between 1860 and 1882 from LFM, SZ and NEZ. – Sweden: extremely rare, only a few records in the 19th century (latest record 1888) at the coast of southwestern Skåne; now apparently extinct. – Along the coast from S. England and N. France to Schleswig-Holstein; inland records from France to DDR and Rumania.

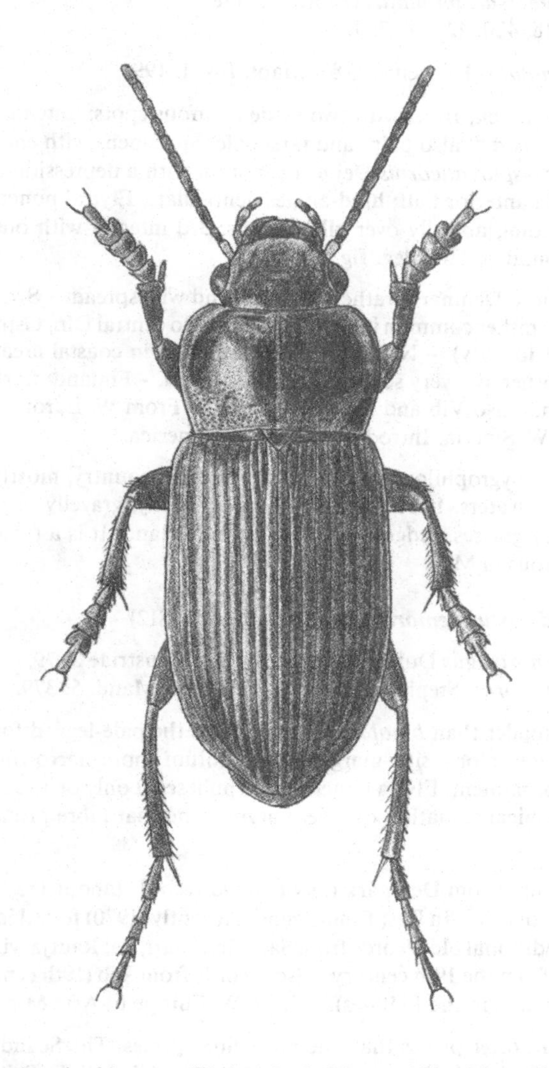

Fig. 421. *Anisodactylus binotatus* (F.), length 10-12.8 mm. (After Victor Hansen).

Biology. In Scandinavia confined to salt-marshes near the sea, occurring on clay-soil with halophytes, e.g. *Salicornia*. In C. Europe also on saline inland localities. The beetles are usually found among roots of plants or under stones, mostly in spring and autumn. The larva develops in summer.

322. ***Anisodactylus binotatus*** (Fabricius, 1787)
Figs 418, 420, 421; pl. 7: 4.

Carabus 2-notatus Fabricius, 1787, Mant. Ins. 1: 199.

10-12.8 mm. Black, frons with two evident rufous spots; antennae with one or two basal segments red, also palpi and tarsi pale. Specimens with entirely pale legs have been named »*spurcaticornis* Dej.«. Pronotum with a depression along side-margin, narrowing in anterior half; hind-angles denticulate. Elytral punctuation and pubescence expanding apically over all intervals; 3rd interval with one dorsal puncture. Shoulder rounded. Hing leg: fig. 420.

Distribution. Denmark: rather common and widespread. - Sweden: generally distributed and rather common in the south, north to central Dlr., Gstr., Hls. (3 localities) and Ång. (1 locality). - Norway: locally common in coastal areas in the south and southeast; otherwise very scattered north to 62°N. - Finland: fairly common, found north to Om. Also Vib and Kr in the USSR. - From W. Europe to N. Africa, Asia Minor and W. Siberia. Introduced in North America.

Biology. A hygrophilous species, living in open country, mostly near standing or slow-running waters. It occurs on clay-mixed sandy, gravelly or peaty soil with tall vegetation of grasses, sedges, etc. Also in arable land. It is a typical spring breeder, most numerous in May.

323. ***Anisodactylus nemorivagus*** (Duftschmid, 1812)

Carabus nemorivagus Duftschmid, 1812, Fauna Austriae 2: 79.
Harpalus atricornis Stephens, 1835, Ill. Brit. Ent. Mand. 5: 379.

8-10 mm. Smaller than *binotatus*. Coloured as the pale-legged form of this species. Depressed area along side-margin of pronotum more narrowing forwards; hind-angles less prominent. Elytra punctate and pubescent only on the 2 or 3 outermost intervals; subapical sinuation of side-margin somewhat more pronounced; shoulders angulate.

Distribution. From Denmark only two old records (about 1850) in NEZ. - Not in Sweden or Norway. - In East Fennoscandia recently (1970) found in Finland: Ka, Kotka. Some additional old records from Sa: Taipalsaari, Sa: Rautjärvi, Kivijärvi, and Sb: Kuopio, all from the 19th century. Also records from Vib (19th century) and Kr (both 19th century and in the 1940-ies). - From W. Europe to Asia Minor and N. Iran.

Biology. In drier places than the preceding species. The Scandinavian specimens have been found on high ground, on sandy or gravelly soil. In Britain the species has been encountered on sandy heaths; in Germany it shows a preference for heather - covered peaty soil, but also occurs on cultivated land.

324. ***Anisodactylus signatus*** (Panzer, 1797)

Carabus signatus Panzer, 1797, Fauna Ins. Germ. 38: 4.

11.5-14 mm. In habitus more like a *Harpalus*, eg. *calceatus*, but separated from this i.a. by the dark antennae. Black, in the male often with a faint metallic hue, elytra sometimes piceous, elytral epipleura brown, antennae piceous (even the first segment), brownish apicad; also palpi and tarsi somewhat paler. Pronotum broad, little constricted, basal hind-angles about rectangular with tip more or less rounded, base with dense rugulose punctuation. Elytra with deep stria, 3rd interval devoid of a dorsal puncture; 8th stria removed from side-margin at middle.

Distribution. Denmark: only 2 accidental stragglers found about 1860 in NEZ and LFM. - Not in Fennoscandia. - From W. Europe to E. Siberia, Japan and China, north to Latvia.

Biology. In C. Europe recorded from moist soil near shores.

Genus *Dicheirotrichus* Jacquelin du Val, 1857

Dicheirotrichus Jacquelin du Val, 1857, Gen. Col. Eur. 1: 35.
Type-species: *Harpalus obsoletus* Dejean, 1829.
Dichirotrichus auct.

This, the following genus, and also *Diachromus*, are separated from all other harpalin genera by the presence of an erect seta at the hind angle of pronotum. A further characteristic of this and the 4 following genera is the »clypo-ocular line«, running from the eye to the base of clypeus. This line has also been called »frontal furrow« but cannot be homologous with e.g. that of the Bembidiini and related tribes. Entire upper surface, incl. the eyes, densely punctate and pubescent. Tarsi pubescent above. Pronotum with sharp hind-angles. Elytra with an abbreviated scutellar stria. Separate dorsal puncture on 3rd interval absent. Wings full. Fore tarsus of male with 4 dilated segments; they are provided with adhesive hairs of the *Anisodactylus*-type (Fig. 417).

Key to species of *Dicheirotrichus*

1 Larger, 5.2-7.5 mm. Head of same colour as pronotum .. 325. *gustavii* Crotch
- Smaller, 4-4.5 mm. Head darker than pronotum, generally darkened in the middle .... 326. *rufithorax* (Sahlberg)

325. ***Dicheirotrichus gustavii*** Crotch, 1871
Fig. 422; pl. 7: 5

Dicheirotrichus Gustavii Crotch, 1871, List Descr. Col.: 11.

Carabus pubescens Paykull, 1790, Monogr. Carab. Suec.: 61; *nec* O.F. Müller, 1776.

5.2-7.5 mm. Sexes usually differently coulored: female entirely testaceous or with a dark spot on head, pronotum and each elytron. These spots are sometimes more expanded, but extreme margin of pronotum and a broad border along sides and apex of elytra are always pale; legs pale. Male entirely black or, usually with two spots on head; margins of pronotum, shoulder, side-margins and suture of elytra rufous; legs more or less infuscated. The palest males are similar in colour to the darkest females. Punctuation and pubescence of upper surface coarse.

Distribution. Denmark: scattered occurrence along the coasts, but apparently absent from the NW coast of Jutland between Thyborøn and the Skaw. Not recorded from Bornholm. - Sweden: widespread and often common along the western coast (Sk.-Boh.); locally distributed at the Baltic Sea: Sm. (Kalmar), Öl., southern Gtl., Ög. (Bråviken). Doubtfully recorded from Sdm. (Mörkö). - Norway: locally common in coastal areas, except for some districts in the west, between 60° and 63°N. - Not in Finland. - In the USSR common in Kr and Lr, along the coasts of the White Sea and the Arctic Ocean. - The Atlantic coast of Europe, at the S. Baltic to Oder; inland records from C. Europe to W. Siberia and NW. China.

Biology. A halobiontic species, living in marine salt-marshes on soft clay-soil, preferably at salt-concentrations at about 20‰ (Heydemann, 1968) and accordingly not occurring on the Baltic Sea coasts. In C. Europe also in saline inland localities. In coastal hatitats the species lives in the upper intertidal zone, preferably where a dense vegetation of e.g. *Puccinellia maritima, Obione* sp. and *Plantago maritima* is present; it is less abundant in sparse *Salicornia*-vegetation. It experiences regular tidal submergence, during which most beetles avoid direct contact with the water by staying in water-free crevices in the clay (Evans *et al.,* 1971), some by climbing the vegetation (Treherne & Foster, 1977). Those individuals that are directly exposed to sea water show a pronounced reduction in oxygen consumption and a subsequent oxygen dept (Evans *et al.,* op. cit.). *D. gustavii* is a night-active, carnivorous ground beetle. Propagation time is in summer; the larvae and some of the adults overwinter.

326. ***Dicheirotrichus rufithorax*** (Sahlberg, 1827)
Fig. 423.

Harpalus rufithorax Sahlberg, 1827, Ins. Fenn. 1: 260.

4-4.5 mm. Distinctly smaller than *gustavii.* Same colour pattern in both sexes, but male often somewhat darker than female. Underside (except prothorax) piceous; upper surface reddish or brownish yellow, head more or less infuscated; elytra usually with a faint, dark, long spot behind middle. Appendages rufo-testaceous. Pronotum as in fig. 423. Elytral striae impunctate. Terminal segment of palpi incrassate.

Distribution. Not in Denmark or Norway. - Sweden: rather distributed in the east (Ög.-Dlr.), most records after 1920. Spreading towards the west and south has recently taken place: Gtl. (Stånga 1975), Sm. (Hornsö 1980), Vg. (Borås 1975), Sk. (Sandhammaren 1969, accidental). – Finland: fairly common in the southernmost part, occasionally found north to Kb.: Kesälahti. Also Vib and Kr in the USSR. – From C. Europe to C. Siberia, south to Austria.

Biology. In Scandinavia predominantly synanthropic, yet in Finland less so, mostly occurring in or in the immediate vicinity of cities, for instance on construction sites and waste lands. Predominantly on clayey, in Finland also on sandy, soil sparsely covered with tall vegetation of ruderal plants, e.g. *Artemisia vulgaris.* In C. Europe the species prefers sandy soil, frequently occurring on river banks. It is most numerous in early spring and autumn.

Genus *Trichocellus* Ganglbauer, 1892

Trichocellus Ganglbauer, 1892, Käfer Mitteleur. 1: 366.
Type-species: *Harpalus placidus* Gyllenhal, 1827.
Oreoxenus Tschitschérine, 1899, Hor. Soc. Ent. Ross. 32: 445.
Type-species: *Bradycellus mannerheimii* F. Sahlberg, 1844.

Similar to *Dicheirotrichus* but of smaller size. Upper surface hairy, at least along margins of elytra. Scutellar stria lacking. The seta at hind-angle of pronotum is sometimes

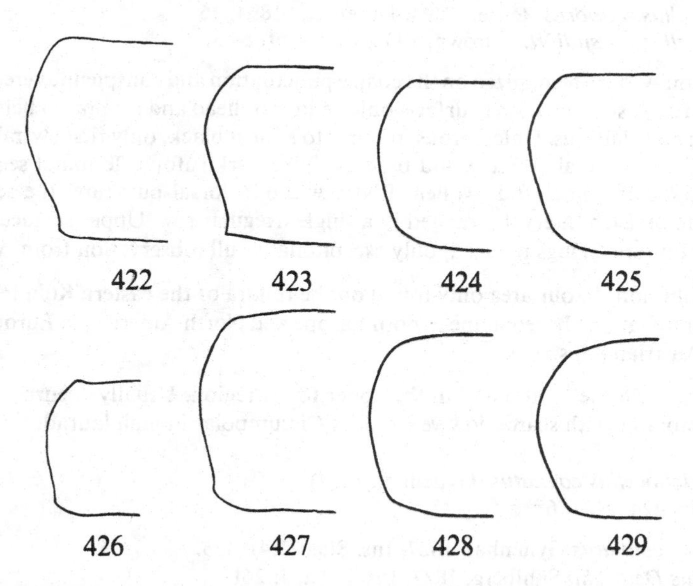

Figs 422-429. Pronotum of 422: *Dicheirotrichus gustavii* Crotch; 423: *D. rufithorax* (Sahlbg.); 424: *Trichocellus cognatus* (Gyll.); 425: *T. placidus* (Gyll.); 426: *Bradycellus ruficollis* (Stph.); 427: *B. verbasci* (Dft.); 428: *B. harpalinus* (Aud.-Serv.); 429: *B. caucasicus* (Chaud.).

difficult to observe. The main character is the completely rounded hind-angles of pronotum, and above all, the 4 dilated protarsal segments of the male. The adhesive hairs of these segments are of the *Harpalus*-type (Fig. 416), not of the *Anisodactylus*-type. Wings full, however dimorphic in *mannerheimii.*

Note. Despite the difference in structure of the adhesive protarsal hairs the genera *Dicheirotrichus* and *Trichocellus* are closely related taxa.

Key to species of *Trichocellus*

1 All elytral intervals with hairs .... 327. *mannerheimii* (F. Sahlberg)
- Only outer intervals with hairs 2
2 (1) Pronotum (Fig. 424) with rounded sides, the dark discal colour reaches margins .... 328. *cognatus* (Gyllenhal)
- Sides of pronotum (Fig. 425) straight behind the middle, disc with dark spot that does not reach margins .... 329. *placidus* (Gyllenhal)

327. ***Trichocellus mannerheimii*** (F. Sahlberg, 1844)

Bradycellus Mannerheimii F. Sahlberg, 1844, Faun. Ins. Ross. Symb.: 51.
Bradycellus Ponojensis J. Sahlberg, 1875, Notis. Sällsk. Faun. Fl. Fenn. Förh. 14: 131.
Tachycellus oreophilus K. & J. Daniel, 1890, Dt. ent. Z. 1890: 209.
Trichocellus setiporus Reitter, 1894, Dt. ent. Z. 1894: 36.
Trichocellus porsildi W.J. Brown, 1932, Can. Ent. 64:3.

3.2-4 mm. At once recognized on the coarse punctuation and conspicuous erect pubescence of almost entire upper surface, only centres of head and pronotum being partly smooth and glabrous. Unicolorous, piceous to almost black, only first elytral interval paler. First antennal segment and base of tibiae dark rufous. Terminal segment of maxillary palpi short and swollen. Elytra without dorsal puncture; the setigerous puncture on each interval arranged in a single irregular row. Upper surface without microsculpture. Wings reduced, only exceptionally full (observation from Alaska).

Distribution. In our area only found on the tundra of the eastern Kola Peninsula. - Circumpolar and boreoalpine in both Europe and North America; in Europe found in the Austrian Alps.

Biology. On the tundra and in the upper forest region. Usually occurring on dry, sandy moraine with scarce, low vegetation. Circumpolar in high latitudes.

328. ***Trichocellus cognatus*** (Gyllenhal, 1827)
Fig. 424; pl. 7: 6.

Harpalus cognatus Gyllenhal, 1827, Ins. Suec. 1(4): 455.
Harpalus Deutschii Sahlberg, 1827, Ins. Fenn. 1: 261.
Several N. American synonyms.

3.5-4.2 mm. Pubescence restricted, fine and depressed. Piceous to black, side-margins and base of pronotum dark rufous. Elytra rufo-testaceous, each elytron with a large longitudinal macula, which may expand so to leave only 1st interval, base and outer margin, pale. Antennae piceous with segment 2 rufous; legs more or less infuscated but at least base of tibiae pale. Anterior part of body with sparse punctures. The setiferous punctures of elytra small, most evident on the outer intervals, rarely sparsely distributed over all of them. Third interval with dorsal puncture. Pronotum : fig.424.

Distribution. Denmark: very rare; several records from Central Jutland, but otherwise very scattered. Only a few records from the islands: LFM (Horreby lyng 1978), SZ (Ølby lyng 1977), NWZ (Yderby lyng 1966), NEZ (Ishøj strand 1963; also older records), and B (Sose 1933, 1979, 1980). - Sweden: distributed over the entire country and common in the north. In the south more scattered and usually rare; unrecorded from a few districts. - Norway: scattered in the south, common and widespread north of 67°N. - Finland: rare and scattered in the south, more common in the north. - Also Kr and Lr in the USSR. - Circumpolar, including Iceland and Greenland, in Europe south to England and Austria.

Biology. In southern Scandinavia and in C. Europe this is a stenotopic species, almost confined to oligotrophic bogs, usually occurring on moist, heather-covered peaty soil. In northern Fennoscandia it is most predominant in the conifer and birch regions, preferring open or slightly shaded, dry, sandy heaths, usually with sparse *Calluna* or *Empetrum* vegetation. The species is at least in C.Europe a winter breeder. Here reproduction takes place from late autumn to early spring (Den Boer, 1980).

329. ***Trichocellus placidus*** (Gyllenhal, 1827)
Fig. 425.

Harpalus placidus Gyllenhal, 1827, Ins. Suec. 1(4): 453.

4-5.5 mm. Pubescence still more restricted than in *cognatus*, and also paler than that species. Piceous brown, pronotum rufous or yellowish with a central dark spot of varying size, though not reaching any of the margins. Elytra brownish yellow, longitudinal spot usually only occupying 3rd interval . Appendages pale, of antennae at least 2 or 3 basal segments pale. Pronotum as in fig. 425, flatter and somewhat more widened anteriorly. Punctuation and pubescence usually visible only near the side-margin and before apex.

Distribution. Denmark: common and very distributed, - Sweden: very common and distributed in the south to 61°N, in the north mainly along the coast and not common ; almost completely lacking in the west from northern Dlr. to T. Lpm. - Norway: common in most districts north to 70°N. - Finland: very common in the southern parts, found north to LkW. Also in Vib and Kr in the USSR. - From W. Europe to Italy, Asia Minor and the Caucasus.

Biology. Rather eurytopic, occurring in moderately shaded and moist sites, dwelling among moss and litter under deciduous trees and bushes in thin forest, forest edges, thickets or fens. It avoids very wet and dark forest swamps. Found almost throughout the year. In C.Europe it breeds in the winter time (Den Boer, 1980).

Genus *Bradycellus* Erichson, 1837

Bradycellus Erichson, 1837, Käf. Mark Brandenburg 1 (1): 64.
Type-species: *Carabus collaris* Paykull, 1798.
Tetraplatypus Tschitschérine, 1897, L'Abeille 29: 62.
Type-species: *Acupalpus similis* Dejean, 1829.
Tachycellus auct.; *nec* Morawitz, 1862.

Small beetles (Fig. 430), more convex than *Acupalpus* (cf. Figs 441, 442) and with entirely pale pubescence. Separated from other small harpalines by the presence of a mentum tooth (as in Fig. 407). Colour brown or piceous, to almost black with pale suture, but otherwise without paler marks. Microsculpture absent or rudimentary; elytra not iridescent. The presence of a clypeo-ocular line (cf. fig. 436) separates the genus from small species outside the harpalines. Anterior margin of prosternum elevated. Wings often reduced. Male with at least fore tarsus dilated. The genus is also distinguished by an oval, punctate and pubescent, fovea on 3rd abdominal sternite.

Key to species of *Bradycellus*

1 Pronotum with evident hind-angles; sides in front of them at least faintly sinuate (Figs 426, 427) 2
- Hind-angles of pronotum only suggested or quite indistinct; sides not sinuate (Figs 428, 429) 4

2 (1) Abbreviated scutellar stria lacking. Abdominal sternites (except for the 2 normal long setae) glabrous ...... 331. *ponderosus* Lindroth
- Abbreviated scutellar stria developed between 1st and 2nd normal striae. All abdominal sternites with numerous setigerous punctures 3

3 (2) 4.5-5.2 mm. Rufo-testaceous, often with paler suture. Pronotum: fig. 427 .... 332. *verbasci* (Duftschmid)
- 2.5-3.4 mm. Elytra piceous to black, suture pale. Pronotum: fig. 426 .... 330. *ruficollis* (Stephens)

4 (1) Pronotum unicolorous reddish brown or rufo-testaceous, at most with a faint shadow on centre or basally 5
- Pronotum piceous to black, with all margins more or less pale 6

5 (4) Pronotum (Fig. 428) smaller and narrower as compared with elytra, sides less rounded, with marginal bead prolonged upon outermost part of base (outside fovea) .... 333. *harpalinus* (Audinet-Serville)
- Pronotum (Fig. 429) larger, elytra narrower, entire body there-

fore more cylindrical. Basal bead of pronotum (outside fovea) feebly developed or almost obsolete . . . . 335. *caucasicus* Chaudoir

6 (4) Eyes strongly protrudent, virtually hemispherical. Basal foveae of pronotum rather deep. Penis: fig. 431 . . . . 333. *harpalinus* (Audinet-Serville) var.

– Eyes somewhat flattened. Basal foveae shallower. Penis: fig. 432 . . . . 334. *csikii* Laczó

330. ***Bradycellus ruficollis*** (Stephens, 1828)
Fig. 426.

Trechus ruficollis Stephens, 1828, Ill. Brit. Ent. Mand. 1 (1): 168.
Acupalpus similis Dejean, 1829, Spec. Gén. Col. 4: 474.
Acupalpus circumcinctus F. Sahlberg, 1834, Bull. Soc. Nat. Moscou 7: 268.

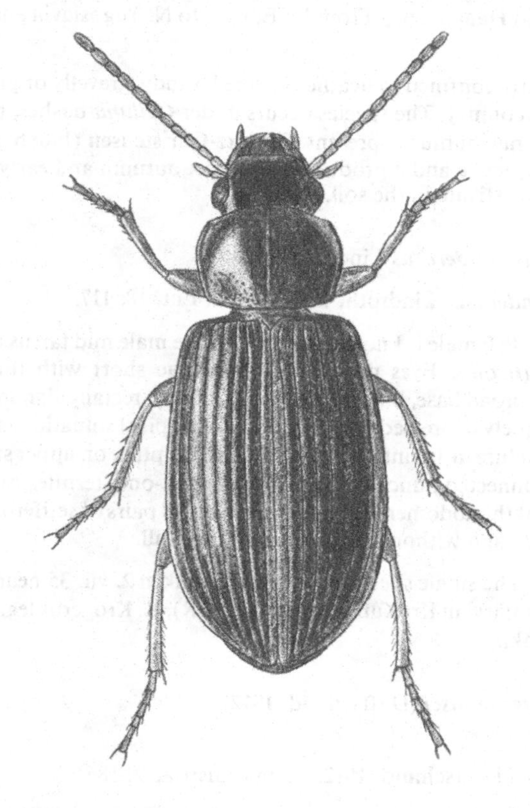

Fig. 430. *Bradycellus harpalinus* (Aud.-Serv.), length 3.8-4.2 mm.

2.5-3.4 mm. The smallest species of the genus. The male has also the mid tarsus dilated. Therefore sometimes referred to a subgenus of its own: *Tetraplatypus* Tschitschérine.

Piceous black, pronotum often paler, elytral suture and extreme margin rufous; appendages rufo-testaceous. Eyes rather flat. Pronotum (Fig. 426) with deep, more or less punctate basal foveae. Elytra with complete scutellar stria, subapical sinuation of side-margin more pronounced than in *caucasicus*. Wings full, at least in specimens from our area.

Distribution. Denmark: uncommon and rather widespread, but apparently sparse or missing in the same areas as described for *caucasicus*. - Sweden: in the south rather common and widespread, north to 60°N, but unrecorded from Gtl. Further north scattered finds: S.Dlr., Gstr., Hls., Ång., and Vb. - Norway: restricted to the southern coastal districts. - Finland: rather rare, found in the southern parts, with isolated finds in Om: Veteli and Haapavesi. - From W. Europe to N. Yugoslavia and the Gorki area in the USSR.

Biology. Strictly confined to heather-covered, sandy, gravelly or peaty soil, usually in entirely open country. The species occurs under *Calluna*-bushes, preferably where a thick layer of raw humus is present. Schjøtz-Christensen (1966b) showed that the adult beetles are active and reproduce in very late autumn and early spring; in May-September they aestivate in the soil.

331. ***Bradycellus ponderosus*** Lindroth, 1939

Bradycellus ponderosus Lindroth, 1939, Notul. Ent. 18: 117.

Since only a single female is known, the state of the male mid tarsus can not be given. Coloured as *ruficollis*. Eyes prominent. Antennae short with thick 1st segment. Pronotum with broad base, hind-angles sharp, almost rectangular, inside hind-angles pronotum obliquely depressed. Elytra without subapical sinuation, scutellar stria absent, dorsal puncture in 3rd interval small. Microsculpture on upper surface consisting of minute, unconnected punctures, on the 2 last-but-one sternites of transverse lines. The underside of the abdomen, except for the normal pairs of setiferous punctures, entirely impunctate and without pubescence. Wings full.

Distribution. The single specimen known was taken 2. vii. 35 near the western end of the lake Paanajärvi in Kr: Kuusamo (now USSR), R. Krogerus leg., on the banks of river Oulankajoki.

332. ***Bradycellus verbasci*** (Duftschmid, 1812)
Fig. 427.

Carabus verbasci Duftschmid, 1812, Fauna Austriae 2: 186.

4.5-5.2 mm. Our largest species, best recognized on the pronotal hind-angles. Ground-colour rufo-testaceous, elytra often clouded apically, except along suture. Pronotum

(Fig. 427) with marked hind-angles; lateral bead prolonged inside hind-angles. Elytra with evident subapical sinuation. Wings full.

Distribution. Denmark: rather common and widespread, but only a few records from districts WJ, NWJ and NEJ. - Sweden: rapidly spreading and now locally common at least in Skåne; before 1950 only known from a few localities in SW.Skåne, the first record in 1934. Also recorded from Bl., Hall., Sm., Öl. and Boh. - Not in Norway or East Fennoscandia. - From W. Europe to Asia Minor and N. Iran.

Biology. In open or moderately shaded ground, usually on sandy or gravelly soil. Is is expecially typical of woodland glades, forest edges, and of clear-felled areas, where it may occur together with *Pterostichus quadrifoveolatus*. Also on sandy heaths under *Calluna*-bushes, etc. The species often enters the vegetation at dusk and regularly comes to light. It is most numerous in August-September and is probably an autumn breeder.

333. ***Bradycellus harpalinus*** (Audinet-Serville, 1821)
Figs 428, 430, 431, 434; pl. 7:7.
Acupalpus harpalinus Audinet-Serville, 1821, Faune Franc. 1: 84.

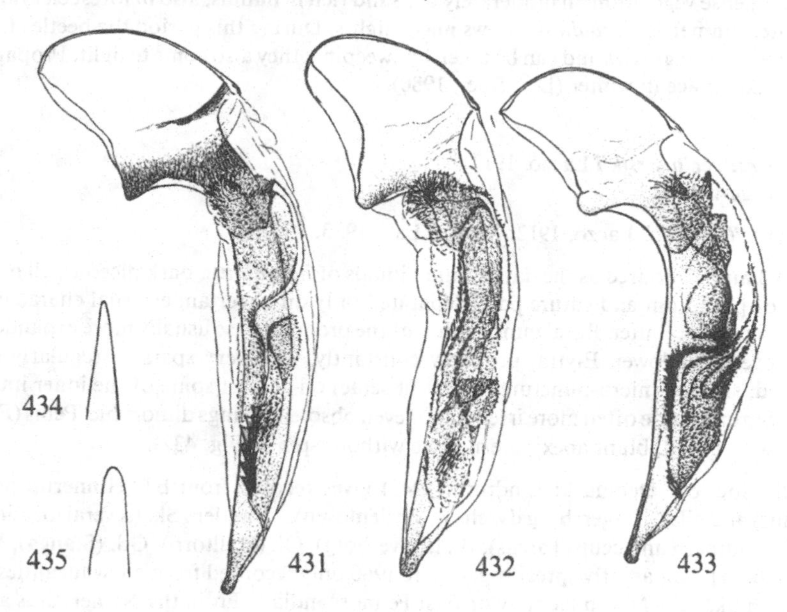

Figs 431-433. Penis of *Bradycellus*. - 431: *harpalinus* (Aud.-Serv.); 432: *czikii* Laczó; 433: *caucasicus* Chaud.
Figs 434, 435. Apex of penis in dorsal view of 434: *Bradycellus harpalinus* (Aud.-Serv.) and 435: *B. czikii* Laczó.

3.8-4.2 mm. Much varying in colour. In our area normally almost as pale as *verbasci*, rarely and especially in British specimens often as dark as *csikii*. Eyes strongly protruding, virtually hemispherical. Basal foveae of pronotum rather deep; hind-angles of pronotum (Fig. 428) as a rule not entirely absent; punctuation usually restricted to the basal foveae and their nearest surroundings. The abbreviated scutellar stria of elytra always developed, though sometimes irregular. Wings full, except in very rare individuals. Penis (Fig. 434) with narrow, pointed apex; internal sac ventrally near apex with a group of slender spines (Fig. 431).

Distribution. Denmark: rather common and distributed. - Sweden: widely distributed and common in Sk.; elsewhere uncommon but probably spreading, recorded from the western coastal districts and from Bl., Sm., Öl. and Gtl. - Norway: very rare, a few records from VAy and Ry. - Finland: this species has been listed from Ok: Suomussalmi (leg. Sorsakoski), but the record must be considered doubtful (cf. pt. 1, p. 145). - In the USSR it has been recorded from Vib: Koivisto (=Primorsk) in 1935. - From W. Europe to the Caucasus and Estonia. Introduced in North America.

Biology. In open country on sandy or peaty soil, especially prominent on heaths, for instance under *Calluna* together with *B. caucasicus*. Further in coastal duneland, in rather dense vegetation on moderately dry sand rich in humus; also in forest clearings. In late summer *B. harpalinus* shows mass flights. During this period the beetles feed on the seeds of grasses and can be taken by sweeping; they also come to light. Propagation takes place in winter (Den Boer, 1980).

334. ***Bradycellus csikii*** Laczó, 1912
Figs 432, 435.

Bradycellus Csikii Laczó, 1912, Rovart. Lap. 19: 3.

3.5-4.3 mm. Coloured as the darkest individuals of *harpalinus:* dark piceous, all margins of pronotum and suture pale. Separated only on uncertain external characters: eyes somewhat flatter. Basal punctuation of the pronotal base usually more expanded, the foveae shallower. Elytra, probably constantly, with very sparse, irregularly arranged, shallow micro-punctures (without setae) on at least some of the inner intervals. Sutural striae often more irregular or even obsolete. Wings dimorphic. Penis (Fig. 435) with broad, blunt apex; internal sac without spines (Fig. 432).

Distribution. Denmark: Lindroth (1943) gives records from EJ (Hinnerup near Århus) and NEZ (Jægersborg dyrehave, Bøllemosen). - Sweden: Sk. (several localities in the southern and central areas), Bl. (Sölvesborg), Öl. (Halltorp), Gtl. (Stånga), Vg. (Göteborg). Apparently spreading, before 1960 only recorded from a few localities in western Skåne. - Not in Norway or East Fennoscandia. - From the Netherlands and England (1 specimen) to Poland and Hungary, south to N. Italy.

Biology. Little is known about the biology of this species. It seems to prefer more or less clayey soil in open country.

335. ***Bradycellus caucasicus*** (Chaudoir, 1846)
Fig. 433.

Acupalpus caucasicus Chaudoir, 1846, Enum. Car. Cauc.: 187.
Carabus collaris Paykull, 1798, Fauna Suec. Ins. 1: 146; *nec* Herbst, 1784.

3-3.9 mm. The commonest and most widespread species. More or less pale reddish brown, underside of abdomen often almost black; pronotum brownish or yellowish rufous, sometimes with a faint central shadow; appendages pale. Concerning form of pronotum, separating the species from *harpalinus,* see the key. Scutellar stria of elytra usually irregular or absent. Dimorphic; both wing-forms often occurring together. Internal sac of penis (Fig. 433) without spines but with a characteristic folding pattern.

Distribution. Denmark: rather common and widespread, but apparently absent from SJ and southern F; in LFM only one record (Rødby); unrecorded from central and western Zealand. - Sweden: common and widespread, only unrecorded from Ås. Lpm. - Norway: common and widespread, except for the extreme north. - In East Fennoscandia common and widespread. - From W. Europe to the Caucasus and C. Siberia.

Biology. In open country, in southern Scandinavia notably in *Calluna*-vegetation on sand, gravel or peat, often in company with *B. ruficollis,* but according to Schjøtz-Christensen (1966b) it prefers soil poor in raw humus. Also in dry grassland, especially in the north. In the mountains it reaches the birch region, rare in the arctic zone. Schjøtz-Christensen (op. cit.) found that *B. caucasicus* is a winter-active species, having its peak of activity in October-December, with additional activity in early spring. Females with mature eggs are found in September-November, and egg-laying is probably resumed in spring. In May-September the adult beetles experience an aestivation period in which they are inactive.

Genus *Stenolophus* Dejean, 1821

Stenolophus Dejean, 1821, Cat. Coll. Col. B. Dejean: 15.
Type-species: *Carabus vaporariorum* Fabricius, 1787 (= *Carabus teutonus* Schrank, 1781).

Related to *Acupalpus* but of larger size, our species (Fig. 440) being over 5 mm long. The clypeo-ocular line is well developed (Fig. 436). Last segment of maxillary palpi blunt (Fig. 438). Pronotum with sides rounded to hind-angles which are virtually obsolete. Marginal punctures of elytra with a pronounced gab posteriorly (Fig. 437); scutellar stria present. Microsculpture of elytra transverse and dense, causing iridescence. Wings full. Legs pale. First segment of hind tarsus with thin keel externally. Fore tarsus of male dilated, 4th segment strongly bilobed.

Key to species of *Stenolophus*

1 Pronotum entirely rufous, basal foveae impunctate or with

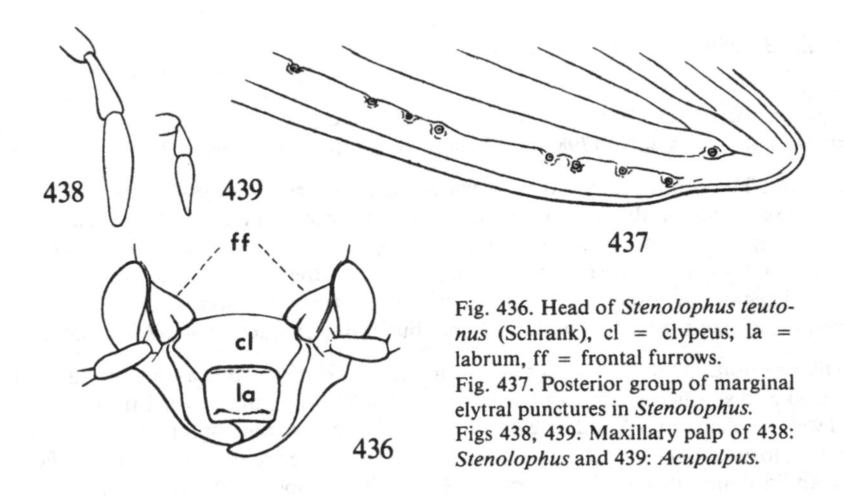

Fig. 436. Head of *Stenolophus teutonus* (Schrank), cl = clypeus; la = labrum, ff = frontal furrows.
Fig. 437. Posterior group of marginal elytral punctures in *Stenolophus*.
Figs 438, 439. Maxillary palp of 438: *Stenolophus* and 439: *Acupalpus*.

a few scattered punctures. Antennae with 2 pale basal segments 2
- Pronotum dark, only extreme margins pale; basal foveae more or less punctuate. Only first antennal segment pale 338. *mixtus* (Herbst)

2 (1) Lateral bead of pronotum prolonged upon base. Elytra entirely pale or indistinctly darker in apical third . . . . . 337. *skrimshiranus* Stephens
- Lateral bead of pronotum ceasing at hind-angle. Elytra with well-defined black macula, extending from apex to before middle . . . . 336. *teutonus* (Schrank)

336. ***Stenolophus teutonus*** (Schrank, 1781)
Figs 436, 440; pl. 7: 9.

Carabus teutonus Schrank, 1781, Enum. Ins. Austr.: 214.
Carabus vaporariorum Fabricius, 1787, Mant. Ins. 1: 205.
Stenolophus anglicus Schiødte, 1861, Naturh. Tidsskr. (3) 1: 185.

5.5-6.2 mm. Stouter and with the pale coloured parts more clear rufous than in *skrimshiranus*. Reminding of *Badister bipustulatus* in colour, but with pronotum much broader posteriorly (cf. Fig. 461). Black, entire pronotum and elytra across base and along side-margins rufo-testaceous; first 2 antennal segments, palpi and legs pale. Elytra strongly iridescent, their striae very deep towards base with very convex intervals.

Distribution. Denmark: very rare and probably not a regular breeder. After 1900 the following records: LFM, St. Musse, 2 specimens 1978 in a gravel pit; B: several localities. - Sweden: very rare and never established. The following records have been made in southern Skåne: Kämpinge 1879, Trälleborg 1862, Sandhammaren 1972 a few speci-

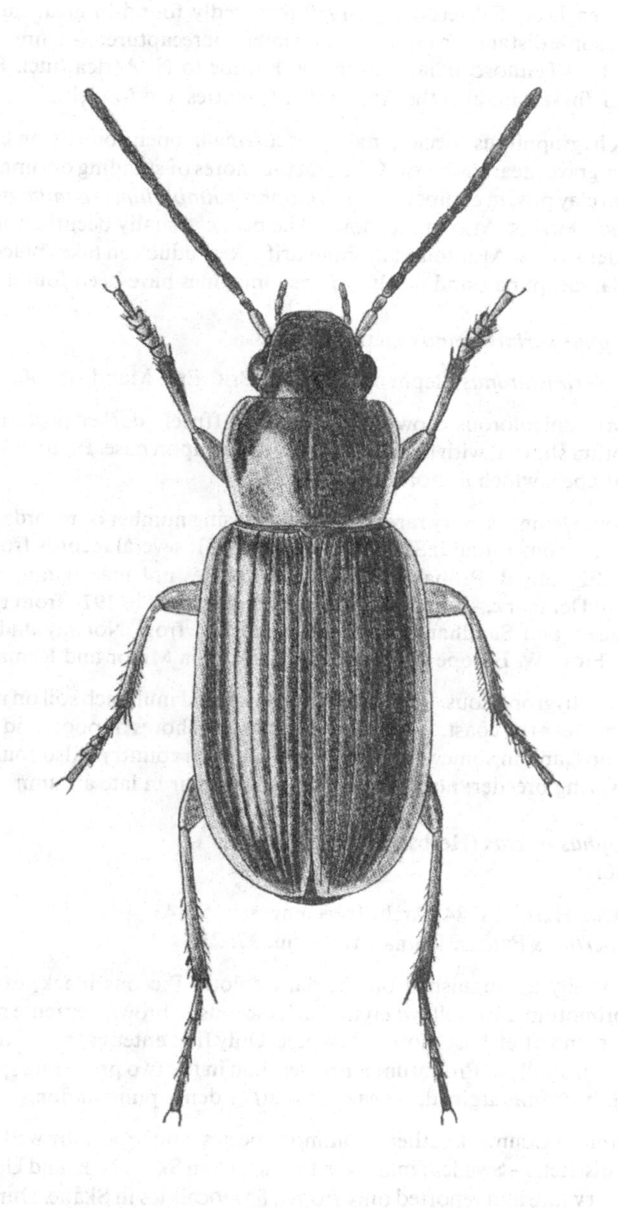

Fig. 440. *Stenolophus teutonus* (Schrank), length 5.5-6.2 mm. (After Victor Hansen).

mens on the seashore, Falsterbo 1977-1980 repeatedly found in great numbers at a small pond at some distance from the sea and later not recaptured. - Unrecorded from Norway and East Fennoscandia. - From W. Europe to N. Africa (incl. Egypt), the Caucasus and Turkestan; also the Azores, the Canaries and Madeira.

Biology. A hygrophilous species, mainly occurring in open country on clay or clay-mixed sand or gravel near the coast. Often on the shores of standing or running waters, for instance in clay pits, in company with *Omophron limbatum, Agonum marginatum* and *Chlaenius vestitus.* Also in meadows. The beetles usually occur among roots of plants or under stones. Also found in shore drift. Reproduction takes place in spring; fully grown larvae, pupae, and newly emerged imagines have been found in August.

337. ***Stenolophus skrimshiranus*** Stephens, 1828.

Stenolophus Skrimshiranus Stephens, 1828, Ill. Brit. Ent. Mand. 1: 166.

5-6 mm. Elytra unicolorous brownish yellow or diffusely darker posteriorly. Head black. Pronotum shorter, with lateral bead prolonged upon base. Elytra with intervals less convex at apex, which is more produced.

Distribution. Denmark: very rare but with increasing number of records since 1950. Only one record from Jutland: SJ, Als Nørreskov 1981; several records from F, LFM, SZ, NWZ, NEZ and B. Probably a migrating species not maintaining permanent populations in Denmark. - Sweden: two accidental records in 1972 from the coast of Sk.: Ljunghusen and Sandhammaren. - Unrecorded from Norway and East Fennoscandia. - From W. Europe to N. Africa, Syria, Asia Minor and Rumania.

Biology. Very hygrophilous, preferring wet, clayey and mull-rich soil on more or less shaded ground near the coast. It usually occurs on the shores of pools and ponds, e.g. in alder-swamps, and in somewhat shaded sites in open country. Also found in shore drift. It is a spring breeder; newly emerged beetles occur in late autumn.

338. ***Stenolophus mixtus*** (Herbst, 1784)
Pl. 7: 10.

Carabus mixtus Herbst, 1784, Arch. Insectengesch. 5: 143.
Carabus vespertinus Panzer, Fauna Ins. Germ. 37: 21.

5.1-5.6 mm. Easily distinguished on the dark colour. Piceous black, extreme side-margins of pronotum dark yellow; elytra dark piceous to brown, extreme margins, including suture, and often base more or less pale. Only first antennal segment pale. Legs and most of palps yellow. Pronotum narrower than in the two preceding species, sides less rounded, base unmargined, foveae with rather dense punctuation.

Distribution. In Denmark rather a common species, and expecially well distributed in the island districts. - Sweden: mainly in the east from Sk. to Nrk. and Upl.; also Vg. Before 1945 very rare and reported only from a few localities in Skåne. During the last decades strongly increasing in numbers and spreading north. - Norway: very rare, only a few records from AAy, VE and AK. A recent immigrant, first record dating from 1977. - Also in Finland recently established, recorded for the first time in 1955 in N: Ekenäs, now found in many localities in the south and locally abundant. In the adjacent parts of the USSR found in Vib already in the 1920-ies, and at river Svir in Kr in the 1940-ies. - From W. Europe to N. Africa, the Caucasus, N. Iran and W. Siberia.

Biology. In wet habitats, for instance in swamps and at the margin of standing waters like small ponds and pools, less often along rivers. The species prefers muddy soil with rich vegetation of sedges etc., occurring both in woods and in wet meadows. It flies at night and is attracted to light. It is most numerous in spring, when breeding takes place.

Note. The species has increased its range during this century (Lindroth, 1972) and is still expanding in Scandinavia (Kvamme, 1978; Palm, 1982).

Genus *Acupalpus* Dejean, 1829

Acupalpus Dejean, 1829, Spec. Gén. Col. 4: 391.
Type-species: *Carabus meridianus* Linnaeus, 1761.
Anthracus Motschulsky, 1850, Käf. Russl.: T. 7, p. 21.
Type-species: *Carabus consputus* Duftschmid, 1812.
Balius Schiødte, 1861, Naturh. Tidsskr. (3) 1: 184.
Type-species: *Carabus consputus* Duftschmid, 1812.

Small species (Fig. 451) under 4.5 mm long. Separated from *Stenolophus* by the continuous posterior row of setiferous punctures on 9th elytral interval. Mentum tooth absent. Body metallic, but with pale pattern in some species; elytra (except in *meridianus)* iridescent from dense, transverse microsculpture. Last segment of maxillary palp acuminate (Fig. 439). Abdomen without central fovea, which separates the genus from *Bradycellus,* from which it is also separated by the generally more flattened body (Fig. 441). Hind tarsus not carinated externally. Abdomen more or less pubescent.

All species of this genus are hygrophilous.

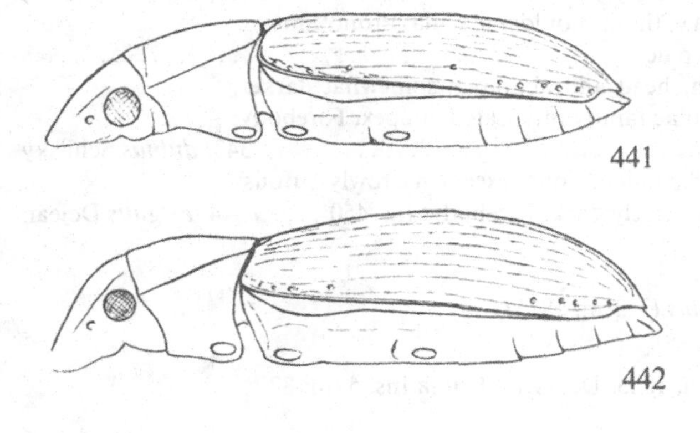

Figs 441, 442. Body profile of 441: *Acupalpus* and 442: *Bradycellus.*

Key to species of *Acupalpus*

1 Pronotum (Fig. 445) with sides sinuate behind; hind-angles sharp, rectangular .... 345. *consputus* (Duftschmid)

- Pronotum with sides not or barely sinuate; hind-angles rounded (Figs 443, 444, 446) 2

2 (1) Very shiny, elytra without microsculpture. Pronotum with entire base punctate .... 341. *meridianus* (Linnaeus)

- Elytra more or less iridescent due to very dense transverse microsculpture. Base of pronotum punctate only laterally or quite impunctate 3

3 (2) Elytra without dorsal puncture 4

- Elytra behind middle with a dorsal puncture on third interval, adjoining second stria 5

4 (3) Head black, pronotum rufo-testaceous (Fig. 451). .... 339. *flavicollis* (Sturm)

- Head and pronotum both dark, piceous to black. .... 340. *brunnipes* (Sturm)

5 (3) Pronotum bright rufous, sometimes with a central dark spot, and strongly contrasting against the black head and the dark markings of the bicoloured elytra 6

- Pronotum brown to black, often with paler margins, little contrasting against head and ground colour of elytra; the latter often immaculate 7

6 (5) Pronotum (Fig. 446) convex with strongly rounded sides and obsolete hind-angles. Fore tibia stout (Fig. 447), fourth tarsal segment of male strongly bilobed .... 346. *elegans* (Dejean)

- Pronotum (Fig. 444) with less rounded sides and at least suggested hind-angles. Fore tibia as in fig. 448; fourth tarsal segment of male only emarginate .... 342. *parvulus* (Sturm) var.

7 (5) At least 3 mm. Elytra with well developed pale shoulder macula. Pronotum (Fig. 444) much broader than head, with margins more or less narrowly pale .... 342. *parvulus* (Sturm)

- Below 3 mm. Elytra without shoulder macula. Pronotum narrower, margins not pale 8

8 (7) Entire body brown, head and abdomen somewhat darker; legs pale or with tibiae faintly infuscated at apex. Forebody: fig. 449 .... 343. *dubius* Schilsky

- Body piceous black, unicolorous, except narrowly rufous along suture; tibiae largely dark. Forebody: fig. 450 ..... 344. *exiguus* Dejean

339. ***Acupalpus flavicollis*** (Sturm, 1825)
Fig. 451.

Trechus flavicollis Sturm, 1825, Deutschl. Fauna Ins. 5 (6): 87.

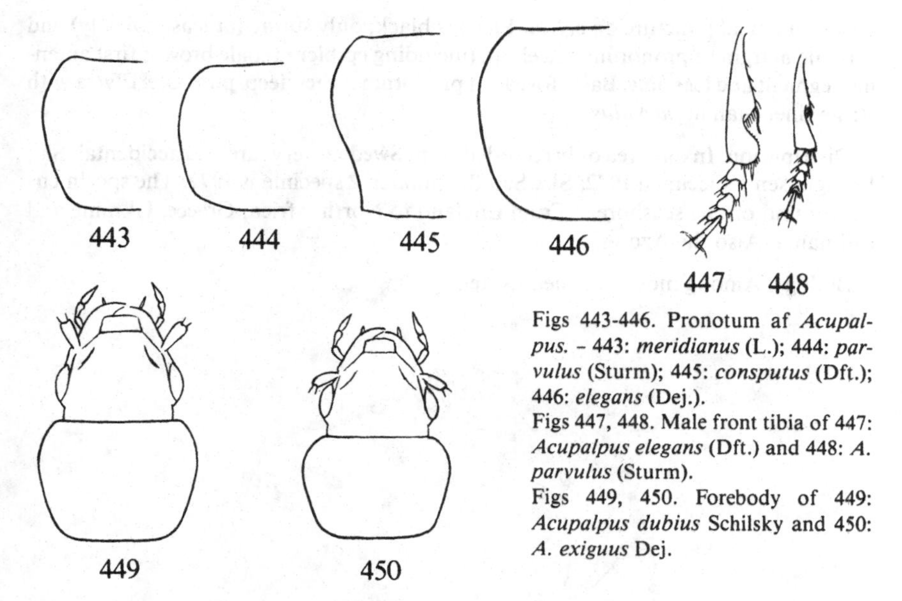

Figs 443-446. Pronotum af *Acupalpus.* – 443: *meridianus* (L.); 444: *parvulus* (Sturm); 445: *consputus* (Dft.); 446: *elegans* (Dej.).
Figs 447, 448. Male front tibia of 447: *Acupalpus elegans* (Dft.) and 448: *A. parvulus* (Sturm).
Figs 449, 450. Forebody of 449: *Acupalpus dubius* Schilsky and 450: *A. exiguus* Dej.

2.6-3.5 mm. Black, pronotum rufo-testaceous; elytra piceous with base and suture rufous; antennae with two pale basal segments. Basal foveae of pronotum shallow and impunctate. Separated from the palest form of *parvulus* by shorter and more convex body, and by the lack of a dorsal puncture on elytra.

Distribution. Denmark: rare. In Jutland some records in SJ, WJ and EJ, especially since 1950. Also recorded from F, LFM, NWZ, NEZ and B. – Sweden: widespread and rather common in the south, north to southern Dlr., Gstr. and Hls. (Bergvik). – Norway: restricted to a few districts in the south-east. – Finland: rare but rather widespread in the southern parts, recorded as far north as Kb. Also Vib and Kr in the USSR. – From W. Europe to Asia Minor and the Caucasus.

Biology. The preferred habitat of this species seems to be river banks in open country, especially such with fine, moist sand covered by a thin layer of silt, and with sparse vegetation of *Juncus, Carex,* mosses, etc. In this type of habitat *flavicollis* often occurs together with *A. parvulus.* Also found at the margin of lakes and ponds, both in open country and in forest; less often in bogs, occurring on bare spots between pillows of *Sphagnum.* The species is often encountered in sea-drift. It is a typical spring breeder, most abundant in June.

340. ***Acupalpus brunnipes*** (Sturm, 1825)

Trechus brunnipes Sturm, 1825, Deutschl. Fauna Ins. 5 (6): 88.

3-3.5 mm. Easily recognized on the combination of uniform dark coloration and the lack of a dorsal puncture on elytra. Piceous black, only suture (at least apically) and extreme margin of pronotum and elytra (including epipleura) pale brown; first antennal segment and legs pale. Basal foveae of pronotum rather deep, punctate. Elytra with striae finer than in *parvulus*.

Distribution. In our area only recorded from Sweden: very rare and accidental; Sk., Ljunghusen 1 specimen 1972; Sk., Sandhammaren 2 specimens 1972. The specimens were found on the seashore. - From England to North Africa, Greece, Ukraine and Lithuania. Also the Azores.

Biology. Among moss, etc., near water.

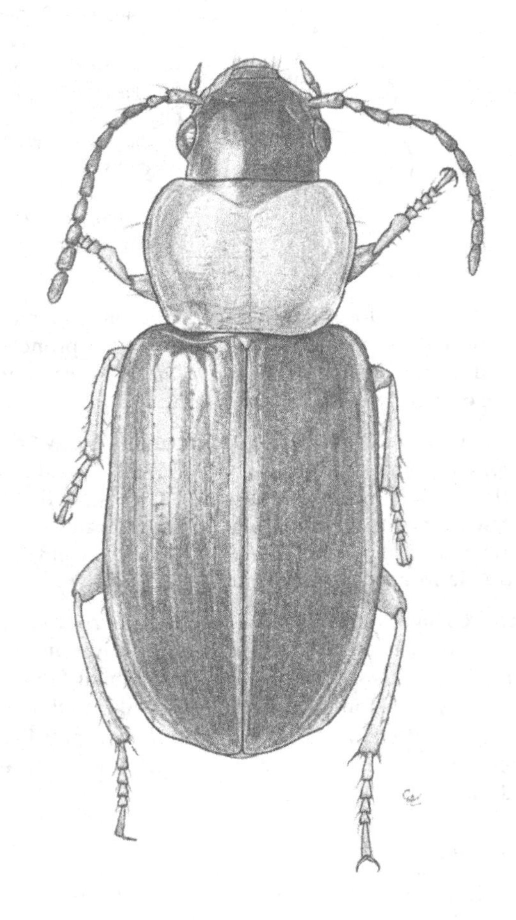

Fig. 451. *Acupalpus flavicollis* (Sturm), length 2.6-3.5 mm.

341. ***Acupalpus meridianus*** (Linnaeus, 1761)
Fig. 443; pl. 7:11.

Carabus meridianus Linnaeus, 1761, Fauna Suec. ed. 2: 221.
Carabus dorsalis Fabricius, sensu Thomson, 1859.

3.3-3.8 mm. Easily distinguished on the lacking microsculpture and the consequently very shiny upper surface. Sharply bicoloured: black, pronotum often dark rufous, elytra with long, oblique shoulder-macula and suture rufo-testaceous; mouth parts and first two antennal segments pale; femora and apex of tibiae often slightly infuscated. Pronotum (Fig. 443) more constricted posteriorly than in all other species, sides sinuate before the rounded hind-angles; entire pronotal base punctate. Elytral striae punctate anteriorly.

Distribution. Denmark: uncommon. In Jutland expecially in the eastern parts, north to Limfjorden. More widespread on the islands. - Sweden: scattered in the southern part, north to Upl. and S. Vrm., obviously completely lacking in the southern highlands. Usually rare but sometimes found in numbers, for instance in western Sk., on Öl. and Gtl., and around Göteborg and Stockholm. - Norway: rare, a few records in the south-east. - In Finland a comparatively recent addition to the fauna, found since 1914 in the Helsinki area, since 1949 in the Turku area; also recorded in Ka: Virolahti in the 1970-ies. Also in Vib in the USSR. - From W. Europe to the Urals, the Caucasus and Asia Minor. Introduced in North America.

Biology. Less hygrophilous than the other species of the genus, occurring on open, sun-exposed ground, notably moderately humid, clayey or clay-mixed sandy soil. The species usually occurs on bare spots in short vegetation of grasses, etc. It is frequently encountered in arable land; also in waste areas in cities, e.g. in vegetable refuse. Baranowski & Gärdenfors (1974) found the species in great numbers in sea drift in June. It is a pronounced spring breeder.

342. ***Acupalpus parvulus*** (Sturm, 1825)
Figs 444, 448.

Carabus dorsalis Fabricius, 1787, Mant. Ins. 1: 205; *nec* Pontoppidan, 1763.
Trechus parvulus Sturm, 1825, Deutschl. Fauna Ins. 5 (6): 77.
Stenolophus Gyllenhali Thomson, 1859, Skand. Col. 1: 288.
Trechus brunnipes Sturm, sensu Thomson, 1859.
Stenolophus Thomsoni Roth, 1898, Ent. Tidskr. 18: 133.

3-4 mm. A much variable species in regard to coloration. The palest form resembles *flavicollis* or *elegans,* the darkest *brunnipes.* Piceous, pronotum rarely entirely rufo-ferruginous, usually black with narrowly pale margins, elytra with a well delimited pale shoulder macula. Antennae with first, rarely also second, segments pale; palps and legs likewise more or less extensively pale. Pronotum (Fig. 444) much wider than head, with hind-angles usually suggested, basal foveae at most very sparsely punctate. Elytra with a dorsal puncture. Penis with less than 10 teeth in internal sac.

Distribution. Denmark: rather common and distributed. - Sweden: rather common and distributed in the south, north to S. Vrm., S. Dlr. and Gstr. In the north very rare, only a few records from near the coast in Hls., Ång., Vb. and Nb. - Norway: restricted to the south-eastern districts, but here locally common. - Finland: common in the southern parts, found north to LkW: Muonio. Also Vib and Kr in the USSR. - From W. Europe to E. Siberia, Turkestan and the Caucasus.

Biology. On open, sun-exposed or slightly shaded ground, at the margin of standing or running waters. The species usually occurs on clay-mixed sandy soil, on bare spots in rich vegetation of *Juncus, Carex, Equisetum,* etc., sometimes together with *A. flavicollis.* Also on peaty soil. It has a high power of dispersal and is frequently found in abundance in sea-drift. A spring breeder.

343. ***Acupalpus dubius*** Schilsky, 1888
Fig. 449.

Acupalpus exiguus v. *dubius* Schilsky, 1888, Dt. ent. Z. 1888: 189.
Acupalpus luridus auct., *nec* Dejean.

2.5-2.7 mm. This and the following species are the smallest in the genus; both are almost unicolorous. Entire body brown, head and abdomen somewhat darker; legs pale or with tibiae faintly infuscated apically. Pronotum often indistinctly darker at middle and base, and elytra usually darker apically, except pale along suture. Fore-body: fig. 449. Pronotum with rather deep basal foveae. Elytra stretched, with a dorsal puncture.

Distribution. Denmark: very rare, only known from 8 localities: WJ (Fanø, Esbjerg, Søndervig), EJ (Svejbæk), NWJ (Hundborg), NEJ (Skagen), F (Svendborg) and NEZ (Tisvilde). - Sweden: very rare and local, only a few records: Sk. (Kämpinge 1886), Sk. (Skanörs ljung and Falsterbo, repeatedly found (sometimes in numbers) after 1968), Bl. (Sturkö 1947-48 in numbers). Also a specimen in sea drift at Sandhammaren (Sk.) in 1972. - Unrecorded from Norway and East Fennoscandia. - From W. Europe to the Caucasus and Estonia.

Biology. In moist, more or less shaded sites, both on clayey, sandy and peaty soil. It is usually found at the margin of ponds and pools, both in open country and in forest, dwelling among moss and dead leaves. Also in *Sphagnum*-bogs.

344. ***Acupalpus exiguus*** Dejean, 1829
Fig. 450; pl. 7: 8 (as *Bradycellus*).

Acupalpus exiguus Dejean, 1829, Spec. Gén. Col. 4: 456.

2.2-2.8 mm. Body piceous black, unicolorous, but narrowly rufous along suture; tibiae largely dark, other parts of legs and also palpi and first antennal segment indistinctly pale. Head (Fig. 450) compared to pronotum wider than in any other species of the genus. Pronotal foveae shallow. Superficially somewhat similar to *Syntomus* but with complete striae and iridescent elytra.

Distribution. Denmark: in Jutland a scattered occurence in the eastern parts; in the island districts widespread on Lolland (LFM), and in NEZ and B, scattered in the rest. - Sweden: rare and very scattered in the south, north to S. Vrm. (Visnum-Kil) and Sdm. (Nacka, Dammtorpssjön). Absent from high elevated areas. - Unrecorded from Norway. - Finland: rather rare, but found widely in the south, north to Oa; especially along the south coast mainly found after 1930. Also in Vib and Kr of the USSR. - From W. Europe to the Caucasus and E. Siberia; also the Canaries and Madeira.

Biology. In marshy, somewhat shaded sites, usually occurring at the margin of ponds and pools, both in forest and in open country; in Denmark predominantly in forest swamps. It prefers clayey soil with rich vegetation of grasses, sedges, etc. The species has been found in abundance in sea drift in Skåne (Baranowski & Gärdenfors, 1974). It is most numerous in spring.

345. ***Acupalpus consputus*** (Duftschmid, 1812)
Fig. 445; pl. 7: 12.

Carabus consputus Duftschmid, 1812, Fauna Austriae 2: 148.

3.8-5 mm. Because of the form of the pronotum (Fig. 445) and the densely pubescent abdomen often referred to a subgenus of its own *(Anthracus* Motsch., *Balius* Schiødte). Dark brown, head almost black; pronotum often paler, rufous, or with pale margins; elytra yellowish, each with an oblong dark macula, which may expand so as to leave only shoulders and margins pale. Antennae very long. Head almost as broad as pronotum. Elytra long, parallel-sided; third interval with a dorsal puncture.

Distribution. Denmark: in Jutland only 6 records of single individuals after 1950; on the islands rather widespread on Lolland, Falster, and in NEZ; otherwise scattered; B: several records. - Sweden: distributed but uncommon in Sk., Öl. and Gtl. In the following districts very scarce: Bl., Hall., Sm., G. Sand., Ög., Vg., Sdm. and Upl. The published record from Nrk. can not be verified. - Not recorded from Norway. - Finland: only a few records - Ab: Lojo 1942 2 specimens, N: Hangö 1936, N: Tvärminne 1939, N: Snappertuna 1939. - From W. Europe to Asia Minor, the Causasus and Turkestan.

Biology. At the margin of small stagnant waters, often such which dry up in summer, usually living on mull-rich clay soil in moderately shaded sites, for instance in light deciduous forest and bushy meadows. The beetles dwell among grass and leaves on sparsely vegetated spots. It is sometimes found in company with *A. exiguus.* Most numerous in May-July.

346. ***Acupalpus elegans*** (Dejean, 1829)
Figs 446, 447.

Stenolophus elegans Dejean, 1829, Spec. Gén. Col. 4: 412.

3.5-4.5 mm. Very similar to *flavicollis* but separated on the following points: larger; pronotum exceptionally dark at middle, entire prosternum pale. Pronotum (Fig. 446) with more rounded sides, with a few punctures in basal foveae. Elytra with dorsal puncture in third interval. Fore tibia broader (Fig. 447), fourth segment of fore tarsus bilobed (not only emarginate as in related species). Internal sac of penis with about 15 large teeth.

Distribution. Two old records from NEZ (Copenhagen before 1840, Roskilde fjord before 1870); both may represent accidental stragglers. - Not found in Fennoscandia. - From W. Europe to the Caucasus and the Caspian Sea; a separate subspecies in N. Africa.

Biology. Restricted to saline habitats on the coast; in marshes and among refuse.

Tribe Perigonini

Genus *Perigona* Laporte de Castelnau, 1835

Perigona Laporte de Castelnau, 1835, Étud. Ent.: 151.
Type-species: *Perigona pallida* Laporte de Castelnau, 1835.
Trechicus LeConte, 1853, Trans. Amer. Phil. Soc. 10: 386.
Type-species: *Trechicus umbripennis* LeConte, 1853 (= *P. nigriceps* (Dejean, 1831)).

A small genus, characterized by the eighth elytral stria being deepened as a furrow in posterior half, and by the eighth and ninth intervals being finely pupescent. Our single species is small, in general outline reminding of a *Tachys* or a *Trechus*.

Antennae short, pubescent from second segment, outer segments squarish. Pronotum with sides and hind-angles rounded. Elytral striae, except for the eighth, obliterated and replaced by rows of feable punctures; three small dorsal punctures. Wings full; these beetles are good flyers. Microsculpture reticulate on head and pronotum, however composed of transverse lines on disc, as well as on elytra. Fore tarsus of male faintly dilated. Penis (Fig. 453) with a long, coiled flagellum in the internal sac.

347. ***Perigona nigriceps*** (Dejean, 1831)
Figs 452, 453; pl. 7:13.

Bembidium nigriceps Dejean, 1831, Spec. Gén. Col. 5: 44.

2-2.5 mm. Yellowish brown with black head. Elytra usually infuscated apically and/or along the suture.

Distribution. In Denmark very rare. First record is NEZ: Dyrehaven 1948 (and later). First record from Jutland is SJ: Haderslev 1953. After 1975 captured in several localities in WJ and EJ. - Sweden: rare and scattered, but still spreading. Records from Sk. to Ly. Lpm., with a large gab between Gstr. and Vb. The first record is Sk.: Alnarp

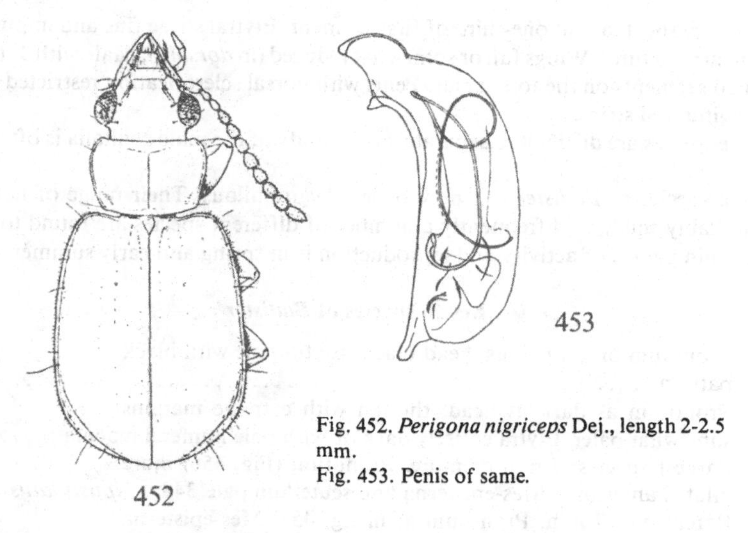

Fig. 452. *Perigona nigriceps* Dej., length 2-2.5 mm.
Fig. 453. Penis of same.

1948. – Norway: a single record from AK. – Finland: first recorded from Sa: Joutseno in 1960; subsequent records from Sb: Kuopio in 1964 and N: Helsinki in 1966. Later recorded from ObN: Keminmaa in 1978 and Ks: Kuusamo, Juuma in 1977. Now at least established in the Helsinki area from where it has spread to N: Hyvinkää (1978) and Ta: Loppi (1985). – Cosmopolitan, probably originating in South Asia. Found in W. Europe already in the 19th century, in C. Europe since 1902.

Biology. Among vegetable refuse, for instance in compost heaps in gardens, in heaps of straw in deer parks, etc. Often attracted to light.

Tribe Licinini

Genus *Badister* Clairville, 1806

Badister Clairville, 1806, Ent. Helv. 2: 90.
Type-species: *Carabus bipustulatus* Fabricius, 1792.
Trimorphus Stephens, 1828, Ill. Brit. Ent. Mand. 1: 180.
Type-species: *Trimorphus scapularis* Stephens, 1828 (=*Carabus sodalis* Duftschmid, 1812).
Baudia Ragusa, 1884, Nat. Sicil. 4: 3.
Type-species: *Carabus peltatus* Panzer, 1797.

Small species, either more or less uniformly dark, or with bright checkered pattern (Fig. 461). Elytra always more or less iridescent from dense transverse lines. Mouthparts peculiar (Figs. 454, 455), the labrum being almost cleft, and the mandibles asymmetric, one of them being provided with a large dorsal notch. Second antennal segment very short, about one-third of first segment. Elytral striae fine and impunctate; 2 dorsal punctures. Wings full or somewhat reduced (in *dorsalis).* Male with 3 strongly dilated segments on the fore tarsus. Penis with dorsal sclerotization restricted to 2 or 3 longitudinal strips.

The species are difficult to separate, and a study of the male genitalia is often decisive.

The species of *Badister* are more or less hygrophilous. Their range of habitat is remarkably small, and frequently a number of different species are found together. The main period of activity and reproduction is in spring and early summer.

Key to species of *Badister*

1 Pronotum bright rufous, head black. Elytra pale with black pattern 2
- Pronotum as dark as head, though with extreme margins somewhat paler. Elytra entirely dark or with pale humeral macula 5

2 (1) Largest species: 7 mm or more. Pronotum (Fig. 456) more dilated anteriorly. Mes-episterna and scutellum pale 348. *unipustulatus* Bonelli
- Rarely over 7 mm. Pronotum as in fig. 457. Mes-episterna black; scutellum almost constantly darker than surrounding parts of elytra . 3

3 (2) First antennal segment entirely rufous, exceptionally with faintest shadow apically. Microsculpture of elytra transverse but rather coarse, which causes only a moderate iridescence . . . . . 349. *bullatus* (Schrank)
- First antennal segment more or less darkened apically. Elytra with strong iridescence due to very dense microsculpture 4

4 (3) Dark elytra spot with rounded limit anteriorly. Microsculpture of pronotum also on centre with more or less transverse meshes . . . . . 350. *meridionalis* Puel
- Dark elytra spot (Fig. 462) anteriorly with transverse, rather irregular limit. Microsculpture of pronotum entirely isodiametric . . . . . 351. *lacertosus* Sturm

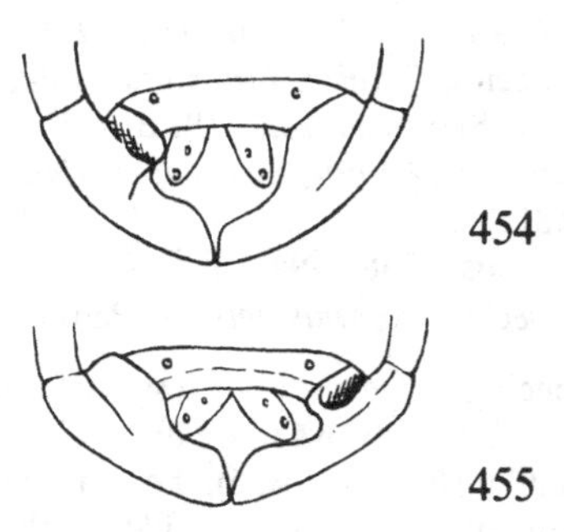

Figs 454, 455. Mouthparts (generalized) of 454: *Badister* s.str. and 455: *Badister* subgenus *Baudia.*

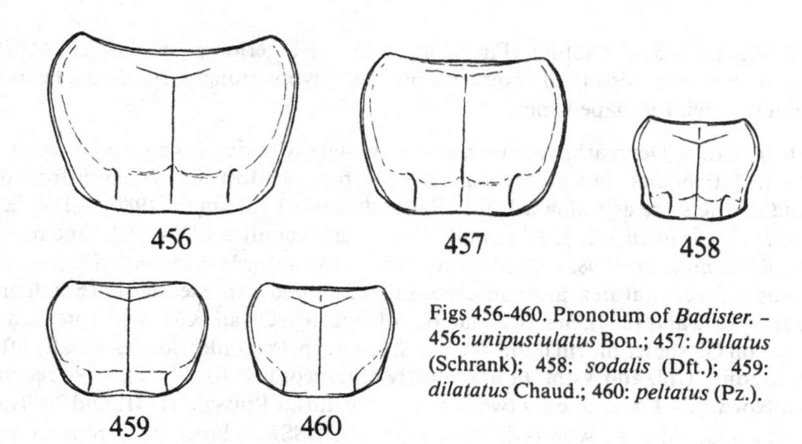

Figs 456-460. Pronotum of *Badister.* – 456: *unipustulatus* Bon.; 457: *bullatus* (Schrank); 458: *sodalis* (Dft.); 459: *dilatatus* Chaud.; 460: *peltatus* (Pz.).

5 (1) Elytra with well-defined yellow humeral spot. Legs entirely pale 6
– Elytra dark and without spots (diffusely paler anteriorly in teneral specimens). Legs (at least tibiae and tarsi) more or less infuscated 7
6 (5) Less than 5 mm. Head narrower than pronotum across front-angles . . . . 352. *sodalis* (Duftschmid)
– Above 5 mm. Head nearly as wide as pronotum across front-angles . . . . 353. *dorsiger* (Duftschmid)
7 (5) Pronotum (Fig. 460) with hind-angles small but evident, base inside them almost rectilinearly oblique 8
– Pronotum (Fig. 459) broader, sides more rounded, hind-angles indistinct, base laterally oblique but rounded . . 354. *dilatatus* Chaudoir
8 (7) Elytral striae exceedingly fine, intervals completely flat 356. *anomalus* (Perris)
– Elytral striae deeper, intervals slightly convex. . . . . 355. *peltatus* (Panzer)

Subgenus *Badister* s. str.

Elytra with large black spots. Penis hooked at apex. Right mandible notched (Fig 454).

348. *Badister unipustulatus* Bonelli, 1813
Fig. 456; pl. 7: 14.

Badister unipustulatus Bonelli, 1813; Mem. Acad. Sci. Turin 20:443.

7-9.1 mm. Black, pronotum and ground-colour of elytra bright rufous; each elytron with two large black spots, one at apex, the other just before middle, often they fuse along side-margin. Mes-episterna and scutellum pale as the base of elytra. Middle section of antennae and sometimes palpi infuscated, legs rufous. Head larger than in all following species. Pronotum (Fig. 456) widening anteriorly, on the centre with microsculpture consisting of transverse meshes. Elytra strongly iridescent. Penis in lateral view with S-shaped apex.

Distribution. Denmark: very rare, and seemingly with decreasing frequency after 1950. In Jutland only two 19th century records from SJ. Most recent records from the island districts are as follows: LFM (Radsted mose, 3 specimens 1980), LFM (Engestofte, 1 specimen 1950), SZ (Knudsskov, several captures 1975-1981). One record from B: Hammeren 1908. - Sweden: very rare, most records being old: Sk. (two old records in the central area, also some specimens captured on the seashore at Sandhammaren in 1972 and 1973), Sm. (Kalmar, one old record), Öl. and Gtl. (widespread, and at least on Öl. sometimes in numbers, e.g. at Halltorp, Petgärde träsk, especially after 1950), Sdm., Upl. and Vstm. (a few, mostly old, records before 1950). - No records from Norway. - Finland: only two records, Ab: Turku Ruissalo (1941) and N: Tvärminne (1976). Also a few records from Vib in the USSR. - From W. Europe to Asia Minor, the Caucasus and W. Siberia.

Biology. A stenotopic species, occurring on moist and warm, mull-rich clay soil, in rather shaded sites, usually near stagnant water. It is especially typical of forest swamps under deciduous trees and bushes, e.g. alder and birch. The beetles occur among moss and leaves; in winter under bark of trees. It is most numerous in May-July.

349. ***Badister bullatus*** (Schrank, 1798)
Figs 457, 461, 463; pl. 7: 15.

Carabus 2 pustulatus Fabricius, 1792, Ent. Syst. 1: 161; *nec* Fabricius, 1775.
Carabus bullatus Schrank, 1798, Fauna Boica 1: 623.

4.8-6.5 mm. First antennal segment entirely rufous (exceptionally with a faint shadow apically). General colour as in preceding species, but scutellum and mes-episterna dark; the black markings of elytra usually more expanded, the two spots often more broadly connected. Head narrower, antennae shorter. Pronotum: fig. 457. Microsculpture coarser than in the three other spotted species: on pronotum isodiametric on disc, on elytra transverse but coarser, resulting in a moderate iridescence only. Apex of penis (Fig. 463), in side view, hooked both dorsally and ventrally.

Distribution. Denmark: rather common and distributed. - Sweden: generally distributed and common in the south, north to about 60°N. The northernmost localities are in Upl. (Älvkarleö) and Dlr. (By og Nås). - Norway: scattered along the southern coast and in the eastern inland districts, north to STi. - Finland: fairly common in Al; otherwise a few records along the south coast. Also recorded in Vib and southern Kr in the USSR. - From E. Europe to W. Siberia, N. Iran, the Caucasus, and (in a separate ssp.) N. Africa. Records from North America should be referred to other, closely related species.

Biology. The most eurytopic *Badister*, occurring in dry as well as in rather moist

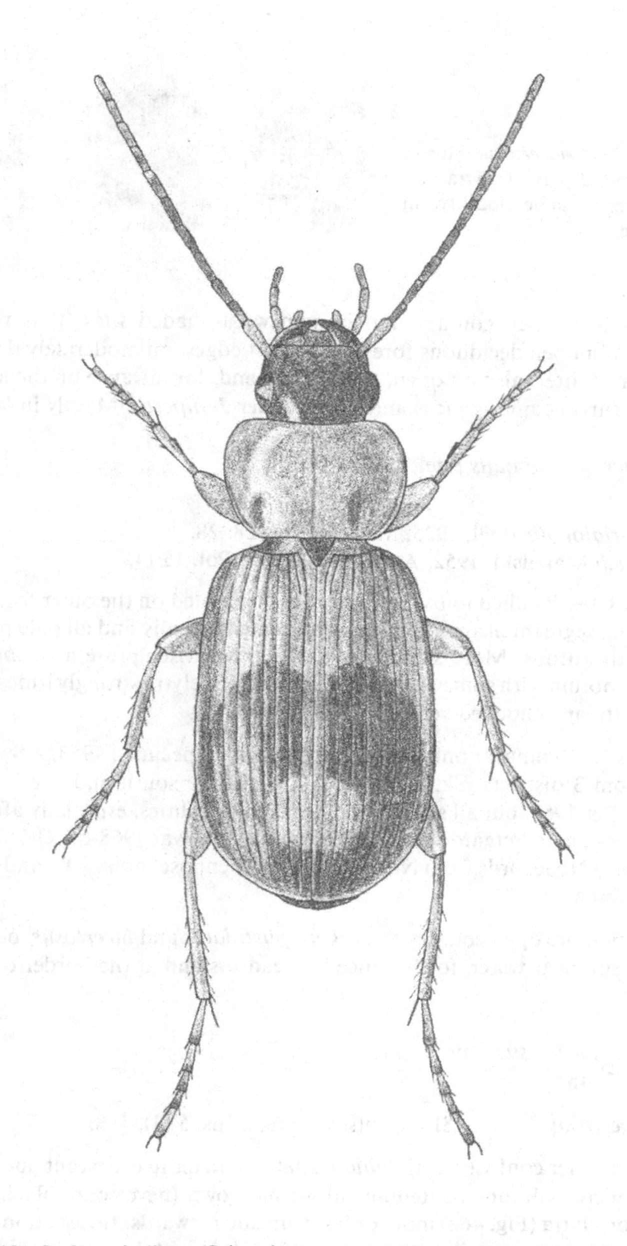

Fig. 461. *Badister bullatus* (Schrank), length 4.8-6.5 mm. (After Victor Hansen).

Fig. 462. *Badister lacertosus* Sturm, pattern of central part of elytra in 3 specimens from same locality in Skåne, Sweden.

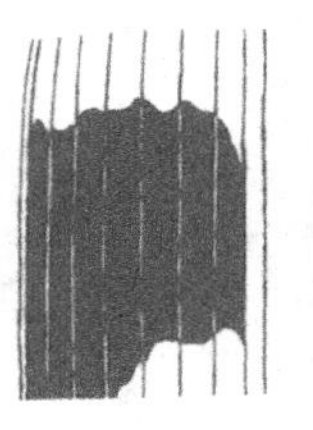
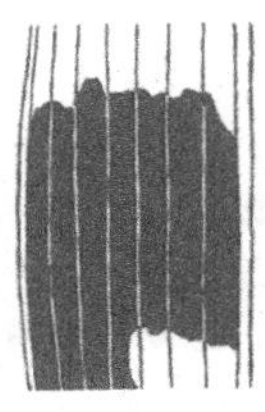

habitats, both in open country and in somewhat shaded sites. It is particularly predominant in open deciduous forest and forest edges, on moderately dry soil with a thick layer of litter; also on open, sandy grassland, for instance on the alvar of Öl. and Gtl., occurring among moss and leaves under *Juniperus.* Mainly in May-July.

350. ***Badister meridionalis*** Puel, 1925
Fig. 464.

Badister meridionalis Puel, 1925, Miscnea Ent. 28: 28.
Badister kineli Makólski, 1952, Annls Mus. Zool. Pol. 15:14.

6.2-7.2 mm. Closely allied to *bullatus* and best separated on the outer form of penis. First antennal segment also more or less infuscated apically and all pale parts a little more brightly rufous. More slender in habitus. Microsculpture as in *bullatus,* i.e. centre of pronotum with somewhat transverse meshes; elytra strongly iridescent. Penis (Fig. 464) with apex hooked ventrally only.

Distribution. Denmark: only one record, B: Rø, 1 specimen 1953. - Sweden: only recorded from 3 districts. Sk.: a few localities on the south and south-east coast, repeatedly after 1950, but all accidental; Öl.: a few localities, especially after 1959, in great numbers e.g. in Petgärde träsk 1970, and Resmo alvar 1968-69; Gtl.: 2 localities, När and Öja. - No records from Norway and East Fennoscandia. - W. and C. Europe, north to Estonia.

Biology. In more open country than *B. unipustulatus* and *lacertosus,* occurring on rather firm soil near water, for instance in meadows and at the border of eutrophic fens.

351. ***Badister lacertosus*** Sturm, 1815
Figs 462, 465.

Badister lacertosus Sturm, 1815, Deutschl. Fauna Ins. 5 (3): 188.

6.1-7.2 mm. Earlier confused with *bullatus,* but elytra more iridescent due to a denser transverse microsculpture. Scutellum rufous or brown (never pure black). Anterior black spot on elytra (Fig. 462) more or less truncate forwards. Infuscation of first antennal segment sometimes indistinct. Penis (Fig. 465) with long, straight apex.

Distribution. Denmark: rare, and with decreasing frequency after 1950. In Jutland only in the southern and eastern parts. More frequent in F, LFM, SZ, NWZ and NEZ. No records from B. – Sweden: rather distributed in the south, north to Hls., but uncommon. Seemingly absent from southern Sk. and the western coastal districts (Hall.-Boh.). – Norway: very rare, only SFi (1965) and MRi (1984). – Finland: fairly common in the southern and eastern districts, north to Sb and Kb; unrecorded from Al. Also in Vib and Kr of the USSR. – Central Europe; distribution not known in details.

Biology. In preference of habitat agreeing with, and sometimes occurring together with, *B. unipustulatus.* In moist, rather shaded sites, usually in mull-rich deciduous forest, for instance of ash, alder and beech, often in forest swamps. It dwells among moss and leaves. Mainly in May-June.

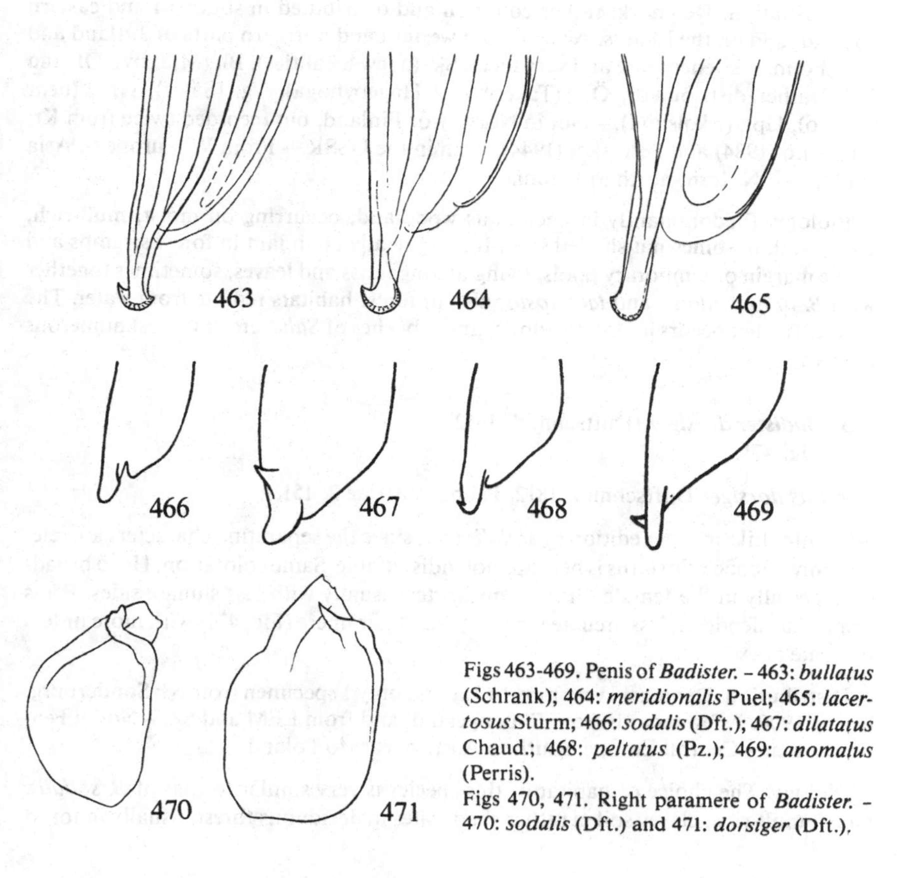

Figs 463-469. Penis of *Badister.* – 463: *bullatus* (Schrank); 464: *meridionalis* Puel; 465: *lacertosus* Sturm; 466: *sodalis* (Dft.); 467: *dilatatus* Chaud.; 468: *peltatus* (Pz.); 469: *anomalus* (Perris).

Figs 470, 471. Right paramere of *Badister.* – 470: *sodalis* (Dft.) and 471: *dorsiger* (Dft.).

Subgenus *Trimorphus* Stephens

Elytra with pale shoulder macula. Penis with simple apex.

352. ***Badister sodalis*** (Duftschmid, 1812)
Figs 458, 466, 470; pl. 7: 16.

Carabus sodalis Duftschmid, 1812, Fauna Austriae 2: 152.
Badister humeralis Bonelli, 1813, Mem. Acad. Sci. Turin 20: 443.

3.9-4.8 mm. Black or piceous, margins of pronotum and elytra, including suture, pale. Elytra with well defined, pale humeral spot. Legs testaceous, antennae dark medially. Right mandible notched. Pronotum (Fig. 458) without iridescence, microsculpture reticulate. Elytral striae impunctate. Wings, though with reflexed apex, somewhat reduced and probably not functionary. Penis: fig. 466. Right paramere: fig. 470.

Distribution. Denmark: rather common and distributed in southern and eastern Jutland, and on the islands. Absent from western and northern parts of Jutland and Bornholm. - Sweden: rare and very local, Sk. (a few localities), Bl. (Mjällby), Öl. and Gtl. (rather distributed), Ög. (Tåkern and Hagebyhöga), Vg. (Säter), Nrk. (near Örebro), Upl. (Ekolsund). - Not in Norway or Finland, but recorded twice from Kr: Suistamo (1934) and Petroskoi (1944), both in the USSR. - From W. Europe to Asia Minor and N. Iran, north to Estonia.

Biology. Predominantly in deciduous woodland, occurring on moist, mull-rich, clayey soil, in somewhat shaded sites. It is especially abundant in forest swamps and at the margin of temporary pools, living among moss and leaves, sometimes together with *B. unipustulatus* and *lacertosus;* also in forest habitats remote from water. The species further occurs in rich meadows, under bushes of *Salix,* etc. It is most numerous in May.

353. ***Badister dorsiger*** (Duftschmid, 1812)
Fig. 471.

Carabus dorsiger Duftschmid, 1812, Fauna Austriae 2: 151.

5-6.5 mm. Like a larger edition of *sodalis* and, since the separating characters are reletive only, the specific status is perhaps not indisputable. Same coloration. Head broader, especially in the female. Pronotum shorter, usually with less sinuate sides. Penis somewhat slenderer, less arcuate ventrally; right paramere (Fig. 471) with more or less truncate apex.

Distribution. Denmark: very rare. In Jutland only 1 specimen from SJ: Sønderborg, before 1900. A few records from F: Langeland, and from LFM and SZ. - Not in Fennoscandia. - Central Europe, south to Austria, east to Poland.

Biology. The choice of habitat of this species is very similar to that of *B. sodalis:* moist, mull-rich, clayey soil in rather shady sites in deciduous forest, usually in forest swamps and at the border of temporary pools. It occurs among moss and leaves, and is often found together with other species of *Badister,* e.g. *sodalis* and *unipustulatus.*

Subgenus *Baudia* Ragusa, 1884

Elytra dark. Penis with subapical tooth. Left mandible notched (Fig. 455).

354. ***Badister dilatatus*** Chaudoir, 1837
Figs 459, 467.

Carabus dilatatus Chaudoir, 1837, Bull. Soc. Nat. Mosc. 10 (3): 20.

5-5.9 mm. Distinguished on the form of the pronotum. Entire body stouter than in the two following species. Ground colour almost black, margins of pronotum and elytra, including suture, rufescent; appendages largely pale brownish but antennae, tarsi and apex of tibiae infuscated. Forebody broad, hind-angles of pronotum (Fig. 459) rounded, oblique lateral part of base slightly arcuate. Elytral apex more suddenly rounded. Ventral tooth of penis (Fig. 467) well removed from apex.

Distribution. Denmark: rare and with scattered occurence; most frequent in Jutland. - Sweden: rather rare and scattered in the south, north to Boh., Nrk., Upl., and an isolated record from Jmt. (Revsund 1961). - Norway: very rare, a single record from Ø near the Swedish border. - Finland: very rare, with scattered finds in the south. Also Vib and Kr in the USSR. - From W. Europe to the Caucasus.

Biology. Occurs on wet, muddy or clayey soil at the margin of eutrophic lakes and ponds, dwelling among leaves and dry reeds in tall vegetation of *Phragmites, Glyceria,* etc., or in the shade of bushes and scattered trees in forest meadows. Often found in company with *B. peltatus.* The species is regularly found in sea drift. It is most numerous in May-June.

355. ***Badister peltatus*** (Panzer, 1797)
Figs 460, 468.

Carabus peltatus Panzer, 1797, Fauna Ins. Germ. 37:20.

4.3-5.4 mm. Piceous, margins of body and partly appendages pale to the same extent as in *dilatatus,* but more ferrugineous than rufous. Hind-angles of pronotum (Fig. 460) better developed. Elytra with somewhat impressed striae and slightly convex intervals. Tooth of penis (Fig. 468) truly apical.

Distribution. Denmark: rare. In Jutland only EJ (Fussingø, in numbers 1977-78) and EJ (Samsø, 1 ♂1976). Scattered records from the islands and Bornholm. - Sweden: not common, distribution pattern and frequency nearly as in preceding species. Seemingly absent from higher elevations. North to Dlr. (Leksand) and Gstr. (Ovansjö). No records from Boh. and Dls. - No records from Norway. - Finland: rare and local in the southern part, northernmost find in Sb: Kuopio. Also in Vib and Kr of the USSR. - From W. Europe to the Caucasus and C. Siberia.

Biology. The habitat of *B. peltatus* is similar to that of *dilatatus*, and the two species are often found together. However, *peltatus* is more often encountered at the margin of small, temporary pools in shaded sites in deciduous forest. The species has been found in abundance in sea drift. It is most numerous in May-June.

356. ***Badister anomalus*** (Perris, 1866)
Fig. 469.

Olisthopus anomalus Perris, 1866, Annls Soc. Ent. Fr. (4) 6: 182.
Badister striatulus Hansen, 1944, Ent. Meddr 24: 93.

Almost impossible to separate decisively from *peltatus* except on male genitalia. The coloration of the two species are identical. Elytral striae extremely fine and intervals quite flat, which is virtually never the case in *peltatus*. The tooth of penis (Fig. 469) is subapical.

Distribution. Denmark: very rare. Unknown in Jutland; very scattered in the southern parts of the islands; two records from Bornholm. - Sweden: very rare and local, restricted to Sk., Öl. and Gtl. In Sk. first recorded from Ivö, Biskopskällaren (several specimens 1967-68). After 1969 8 records of accidental specimens in sea drift on the SE. coast of Sk. On Öl. recorded 1969 from Halltorp and Petgärde träsk (here in numbers, recaptured 1971 and 1977); also on »Stora alvaret« (a few localities, at least from 1970). Gtl.: Harudden (one specimen 1981). Not recorded from Norway or East Fennoscandia. - From W. Europe to the Middle East.

Biology. In the same habitats as described for *B. peltatus;* the two species often occur together.

Genus *Licinus* Latreille, 1802

Licinus Latreille, 1802, Hist. Nat. Crust. Ins. 3: 92.
Type-species: *Carabus cassideus* Fabricius, 1792.

Larger than *Badister.* Both labrum and mandibles clearly asymmetric. Terminal segment of palpi broad, triangular. Upper surface coarsely punctate and elytra opaque, not iridescent. Pronotum (Fig. 472) broad with rounded sides and hind-angles. Elytra without dorsal puncture. Male with 2 segments of fore tarsus dilated.

357. ***Licinus depressus*** (Paykull, 1790)
Pl. 8: 1.

Carabus depressus Paykull, 1790, Mon. Car. Suec.: 34.

9.5-11.8 mm. Uniformly black, or pronotum, antennae and legs rarely somewhat piceous; forebody a little more shiny than the dull elytra, especially in the female. Head with sparse and fine, pronotum with dense and coarse, in part confluent punctuation; elytra with fine but sharp, slightly punctate striae; intervals coarsely punctate. Wings reduced.

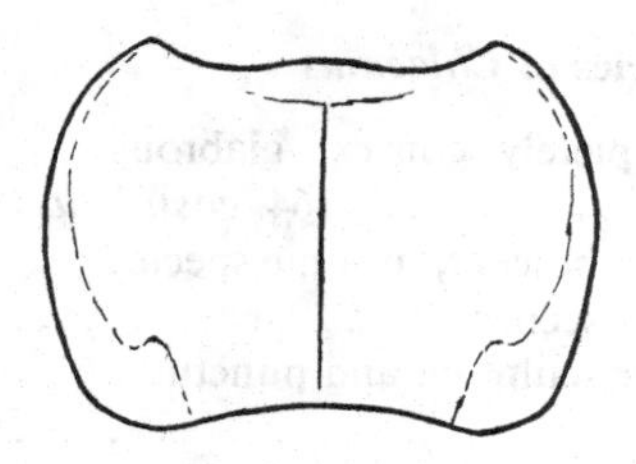

Fig. 472. Pronotum of *Licinus depressus* (Payk.).

Distribution. In Denmark very rare, only known from LFM, NEZ and B, but no records after 1955. - Sweden: usually rare and scattered in the south, north to southern Boh., Nrk., Upl. and Vstm., completely missing at higher altitudes. Somewhat more frequent in eastern Sk., Öl., Gtl. and in the Mälaren valley. - Norway: restricted to and rare in a few south-eastern districts. - Finland: very rare, some records from Al, found in Ab at Kustö in the 19th century, Turku in 1966, and in Pargas in 1979. - From W. Europe to the Caucasus and W. Siberia.

Biology. A xerophilous species, living on dry, sandy or gravelly soil, often mixed with clay or chalk. The species prefers somewhat shaded sites in open grassland. In C. Europe also in dry forest. The larvae, and probably also the adults, feed on snails. The beetles occur from early spring to autumn, but it most frequent in July-September. Larsson (1939) and Lindroth (1945) considered the species a spring breeder, having adult hibernation; Barndt (1976) showed that in Germany it is an autumn breeder.

Tribe Chlaeniini

Genus *Chlaenius* Bonelli, 1810

Chlaenius Bonelli, 1810, Obs. Ent. 1 (Tab. Syn.).
Type-species: *Carabus marginatus* Rossi, 1790 (= *Carabus velutinus* Duftschmid, 1812).
Agostenus Steven, 1829, Mus. Hist. Nat. Univ. Caes. Mosq. 2: 15.
Type-species: *Carabus sulcicollis* Paykull, 1798.
Pelasnus Steven, 1829, Mus. Hist. Nat. Univ. Caes. Mosq. 2: 15.
Type-species: *Carabus quadrisulcatus* Illiger, 1798.
Chlaeniellus Reitter, 1908, Fauna Germ. 1: 187.
Type-species: *Carabus vestitus* Paykull, 1790.

Rather large beetles with (in our species) elytra pubescent, at least on the even intervals. Wings full. Tibiae not pubescent. Fore tarsus of male with 3 dilated segments.

The species of *Chlaenius* are predacious, feeding on insect larvae, slugs, earthworms, etc. (Larochelle, 1974). They are excellent fliers and are sometimes found as stragglers in sea drift. The breeding habitats are close to the water. Hibernation of several species usually takes place on rather dry ground.

Key to species of *Chlaenius*

1 Uneven intervals of elytra completely convex, glabrous and impunctate .... 364. *costulatus* (Motschulsky)
- All elytral intervals punctate and pubescent, in some species with alternating pale and dark pubescence 2
2 (1) Pronotum (Fig. 477) with irregular sculpture and punctures uneven dispersed 3
- Pronotum evenly punctate and pubescent 4
3 (2) First elytral interval anteriorly (behind the scutellum) smooth and shiny. Pubescence of elytra almost constantly alternating in colour .... 363. *quadrisulcatus* (Paykull)
- All intervals evenly punctate and pubescent. Pubescence of elytra (in nordic specimens) uniform .... 362. *sulcicollis* (Paykull)
4 (2) Apex and side-margin of elytra yellow. Pronotum: fig. 476 361. *vestitus* (Paykull)
- Elytra unicolorous. Pronotum with rounded hind-angles 5
5 (4) Antennae entirely black. Only head with clear metallic lustre .... 358. *tristis* (Schaller)
- First antennal segment pale, at least underneath. Entire upper surface with bright metallic lustre 6
6 (5) Only first antennal segment pale; palpi infuscated. 359. *nigricornis* (Fabricius)
- Antennae with 2 or 3 basal segments and entire palpi pale .... 360. *nitidulus* (Schrank)

358. ***Chlaenius tristis*** (Schaller, 1783)
Fig. 473; pl. 7: 17.

Carabus tristis Schaller, 1783, Abh. Naturf. Ges. Halle 1: 318.
Carabus holosericeus Fabricius, 1787, Mant. Ins. 1: 199.

11-13 mm. Black, but head greenish or bluish, pronotum and elytra usually faintly bronzed. Also appendages pale. Habitus as in *nigricornis*, but pronotum (Fig. 473) narrower and with shallower basal foveae; the elytra more stretched. Pubescence of pronotum and elytra dense, greyish brown.

Distribution. Denmark: in Jutland only one old record, SJ: Tønder 1867. Several records, but with strongly decreasing frequency, from Lolland, Falster, Møn, Zealand, and Bornholm. Latest captures: LFM (Bøtø 1979), B (Rønne 1950). The species probably maintained Danish populations in certain periods of the 19th century. Repeatedly captured in sea drift which indicates migratory tendencies. - Sweden: very rare and local, scattered distributed in the south, north to Vg., southern Vrm. and Vstm., Upl. and Med. (Alnön). Strongly decreasing, only a few records after 1950. - Norway: very rare, recorded only from AK, Bø and TEy. - Finland: rare in Al, Ab and N (Hangö peninsula); single records from Ka, St, Ta and Sa. Also Vib and Kr (at river Svir) in the USSR. - From W. Europe to North Africa, Asia Minor, Iran and E. Siberia.

Biology. Very hygrophilous, living on lake shores near the water-edge, usually on clayey soil with luxuriant vegetation of reeds, sedges, etc., and on mossy ground. It is regularly found together with *Oodes helopioides* on soft soil, e.g. in *Phragmites*-vegetation; also frequently associated with *Blethisa.* It is sometimes encountered in sea drift along the Baltic Sea coast. *C. tristis* is a pronounced spring breeder, most numerous in May-June. The adults hibernate far from water, for instance in sandy pine forest.

359. ***Chlaenius nigricornis*** (Fabricius, 1787)
Fig. 474; pl. 7: 19.

Carabus nigricornis Fabricius, 1787, Mant. Ins. 1: 202.

10-12.5 mm. Forebody, at least pronotum, normally golden or coppery, elytra green; rarely entire body greenish or even as dark as to cause confusion with *tristis.* Only first antennal segment pale, palpi infuscated; Specimens with rufous femora have been called »*melanocornis* Dej.«.

Distribution. Denmark: rare and very local, occurring usually only in small numbers. The species shows migratory tendencies; is often found in sea drift. - Sweden: formerly widely distributed in the south up to 61°N, and also in Hls., Ång. and Nb. Af-

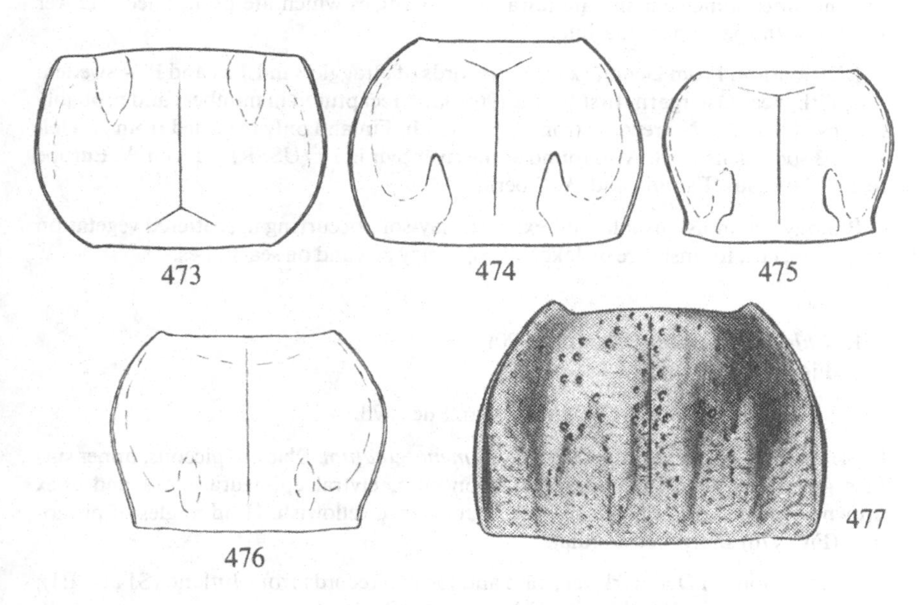

Figs 473-477. Pronotum of *Chlaenius.* - 473: *tristis* (Schall.); 474: *nigricornis* (F.); 475: *nitidulus* (Schrank); 476: *vestitus* (Payk.); 477: *quadrisulcatus* (Payk.).

ter 1950 rare and local, probably extinct in some places, but at least on Öl. still rather common. - Norway: uncommon, yet recorded from most of the south-eastern districts. - Finland: formerly widely distributed, occasionally abundant, north to Om: Kalajoki; after 1950 few records. Also records from Vib and Kr in the USSR. - From W. Europe to Asia Minor, the Caucasus and E. Siberia.

Biology. A shore-dweller, usually inhabiting the margin of eutrophic lakes, ponds and slow-running rivers, preferably on clayey or clay-mixed sandy soil with a rich, open, vegetation of *Phragmites, Carex,* etc. The beetles occur under heaps of reed, under stones, etc., but are also frequently encountered when running about in warm sunshine on bare spots. The species is occasionally found in sea drift. It is most numerous in May-June when propagation takes place.

360. ***Chlaenius nitidulus*** (Schrank, 1781)
Fig. 475.

Carabus nitidulus Schrank, 1781, Enum. Ins. Austr.: 213.
Carabus Schrankii Duftschmid, 1812, Fauna Austriae 2: 131.

10-20 mm. Very similar to the preceding species but usually with less colour contrast between forebody and elytra; antennae with 2 or 3 pale basal segments, palpi entirely pale; femora always infuscated. Pronotum (Fig. 475) with greatest width before middle and sides somewhat sinuate towards hind-angles which are pronounced. Lower surface with sparse punctuation.

Distribution. From Denmark only 3 records of stragglers in LFM and B. - Sweden: only Gtl., near Östergarn, first found 1965, later recaptured in numbers and probably now established. - No records from Norway. - In Finland only recorded from N: Helsinki, 3 specimens 1983. Also found at the river Svir in Kr (USSR). - From W. Europe to the Caucasus, Estonia and W. Siberia.

Biology. On moist, usually sun-exposed clay-soil, occurring in scattered vegetation of grasses, etc., for instance on lake shores, in clay pits and on sea-slopes.

361. ***Chlaenius vestitus*** (Paykull, 1790)
Fig. 476; pl. 7: 18.

Carabus vestitus Paykull, 1790, Mon. Car. Suec.: 73.

8.5-11 mm. Coloured very much as *Agonum marginatum.* Black to piceous, upper surface green. Extreme side-margin of pronotum, elytral epipleura, sides and apex brownish yellow. All appendages pale. Pubescence yellowish. Hind-angles of pronotum (Fig. 476) sharp, rectangular.

Distribution. In Denmark very rare and local; 3 records from Jutland (SJ and EJ); more records from the islands and Bornholm. Decreasing in number, except on B, where recorded from 10 localities after 1950. - Sweden: restricted to southern and western Skåne and very rare, most records being from before 1950. After 1960 only Sk.: Lomma, repeatedly found but uncommon. - Not in Norway or East Fennoscandia. - From W. Europe to Latvia, W. Siberia, the Caucasus, Syria, and (as separate subspecies) North Africa.

Biology. At the border of running or standing waters, for instance small ponds and clay pits; also on sea-slopes with out-welling freshwater. The species usually occurs on heavy clay-soil; also on clayey sand, sometimes in company with *Omophron limbatum*. During daytime the beetles hide in cracks in the clay, under stones, etc. It is a spring breeder, most numerous in June.

362. ***Chlaenius sulcicollis*** (Paykull, 1798)

Carabus sulcicollis Paykull, 1798, Fauna Suec. Ins. 1: 153.

14-15 mm. In the form and sculpture of pronotum similar to *quadrisulcatus* but the longitudinal furrows are lower and the punctures smaller. Black, upper surface usually with faint bronze tinge. Pronotum with 4-6 smooth and shiny, indistinctly delimited and anteriorly fused keels. The lateral bead broader. Elytra normally on all intervals evenly punctate and with dark golden pubescence (Siberian specimens with elytral pubescence as in *quadrisulcatus* are named »*gebleri* Ganglb.«). The raised pronotal bead sharply delimited to hind-angles.

Distribution. Denmark: very rare and with strongly decreasing frequency. Jutland: only NEJ: Gravlev enge, 1 specimen 1882. Several records from Falster, Møn and Zealand before 1900; after 1900 only LFM: Høje Møn, 2 specimens 1903. Bornholm: several records before 1900; after 1900 only Bagå, 2 specimens 1952. - Sweden: extremely rare, only a few and mostly old finds, Sk.: Äsperöd (in the 19th century), Sk.: Ystad (one specimen in each of the years 1910 and 1911), Sk.: Stenshuvud (one specimen in sea drift), Öl. (a few old specimens, without locality). - Not in Norway. - In Finland a few specimens were found in N: Hangö peninsula. - From France and N. Italy to Estonia and E. Siberia.

Biology. In wet habitats with rich vegetation, for instance at the border of rivers and lakes, and in wet meadows, occasionally on the seashore. After the termination of the reproductive period in spring, the adults seek out drier places, e.g. dry meadows and open pine forest, where hibernation takes place.

363. ***Chlaenius quadrisulcatus*** (Paykull, 1790)
Fig. 477.

Carabus quadrisulcatus Paykull, 1790, Mon. Car. Suec.: 109.
Tachypus caelatus Weber, 1801, Obs. Ent. 1: 42.

13-14 mm. Broader than *sulcicollis*, elytra shorter with sides more rounded. The convex keels of pronotum (Fig. 477) more sharply delimited and reaching further backwards; raised bead more diffusely delimited inwards. Elytra normally with alternating colour on the intervals, pubescence coloured as in *sulcicollis*, except in ab. *unicolor* H. Lindb.

Distribution. Denmark: only records of 4 accidental stragglers (LFM, NEZ and B). - Sweden: extremely rare. Sk.: Ringsjön (a few specimens, latest in 1882), Sm.: old records, the latest is Diö (1941); Jönköping area (not after 1919); Tranås near Svartån (1940), Öl. (one old record), Ög. (one old record), Vg. (one old record, and Svenljunga 1951), Upl.: Flororna N of Österbybruk (one specimen 1984). - Unrecorded from Norway and Finland, but a few records from Vib, and one from Kr: Pindushi, in the USSR. - From NE. France to E. Siberia, south to Czechoslovakia.

Biology. In similar habitats as *C. sulcicollis*. Reproduction takes place in spring, after which the adults seek out drier places, e.g. sandy meadows and open pine forest, where they hibernate. Not permanent in our area.

364. ***Chlaenius costulatus*** (Motschulsky, 1859)

Agostenus costulatus Motschulsky, 1859, Bull. Soc. Nat. Mosc. 32(2): 488.
Carabus quadrisulcatus Illiger, 1798, Käf. Preuss.: 176; *nec* Paykull, 1790.
Chlaenius Illigeri Ganglbauer, 1892, Käf. Mitteleur. 1: 391.

11-12 mm. A brightly coloured insect to some extent reminding of *Carabus nitens*. Elytra with alternating pubescent and glabrous intervals. Black, upper surface with strong metallic lustre in green, cupper or gold. Uneven elytral intervals convex, virtually impunctate; even intervals with dense punctuation and pubescence. Pronotum with rounded hind-angles, its punctuation coarse and sparse, pubescence almost obsolete; central furrow and basal foveae deep.

Distribution. Unknown in Denmark. - Sweden: recorded from 2 localities in Nb.: Bredmyrberget (one specimen 1962) and Pålängemyren (numerous specimens 1977, 1979, and later). - Not in Norway. - Finland: recorded from Ta: Lammi (19th century), Ok: Suomussalmi (1920-ies) and »Lapponia«. Also from Kr: Svir in the 19th century. - From NW. Germany to E. Siberia.

Biology. In wet habitats, usually at the border of lakes and ponds. A Scandinavian locality, Pålängemyren near Luleå, Nb., is described by Lundberg (1981). In this mire *C. costulatus* occurs in thin birch-growth on soft soil, in *Sphagnum*-hummocks surrounded by pools with *Carex, Menyanthes, Equisetum*, etc. The beetles have been collected in falltraps, together with *Elaphrus lapponicus*, mainly in June. They are supposed to hibernate in the hummocks.

Tribe Oodini

Genus *Oodes* Bonelli, 1810

Oodes Bonelli, 1810, Obs. Ent. 1 (Tab. Syn.).
Type-species: *Carabus helopioides* Fabricius, 1792.

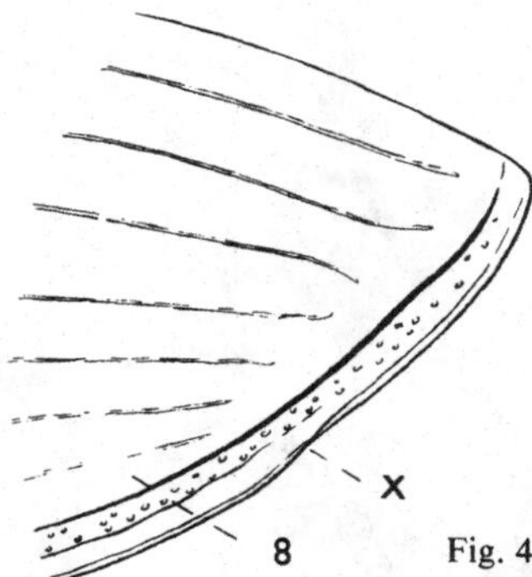

Fig. 478. Apex of left elytron of *Oodes helopioides* (F.), 8 = 8th interval; x = crossed epipleura.

A very characteristic genus (Fig. 479) with broad, laterally very rounded body. Colour dull black, not metallic. Our species are somewhat suggestive of a water-beetle. The reduction of setae is remarkable: none on pronotum or labial palpi; elytra with 2 dorsal punctures. The most striking feature is the keel running along apical margin of elytra (Fig. 478); epipleura crossed; striae very fine but complete. Wings full. Male with 3 dilated segments on fore tarsus.

The species of *Oodes* are very hygrophilous, occurring at the border of freshwater. The beetles live partly sub-aquatic; when disturbed, they usually enter the water and dive. They are carnivorous insects.

Key to species of *Oodes*

1 Broad, elytra less than 1.5 times as long as combined width of both .... 365. *helopioides* (Fabricius)

– Narrower, elytra about 1.5 times as long as combined width of both .... 366. *gracilis* Villa & Villa

365. ***Oodes helopioides*** (Fabricius, 1792)
Figs 478, 479; pl. 8: 2

Carabus helopioides Fabricius, 1792, Ent. Syst. 1: 155.

7.5-10 mm. Black with mouth-parts and first antennal segment, sometimes also tibiae and pronotum near hind-angles, piceous. Sides of meta-sterna evidently punctate. Female with broader pronotum.

Distribution. Denmark: rather common and distributed in eastern Jutland, and on the islands and Bornholm. In WJ only: Esbjerg, Skallingen, and Vedersø. No records from NWJ and NEJ. – Sweden: ranging from Sk. to Gstr. and Med. (Timrå), and common in the northern half of its range. In the southern highland only a few records. – Norway: very rare, only 2 records from Bø and VE. – Finland: found north to ObN,

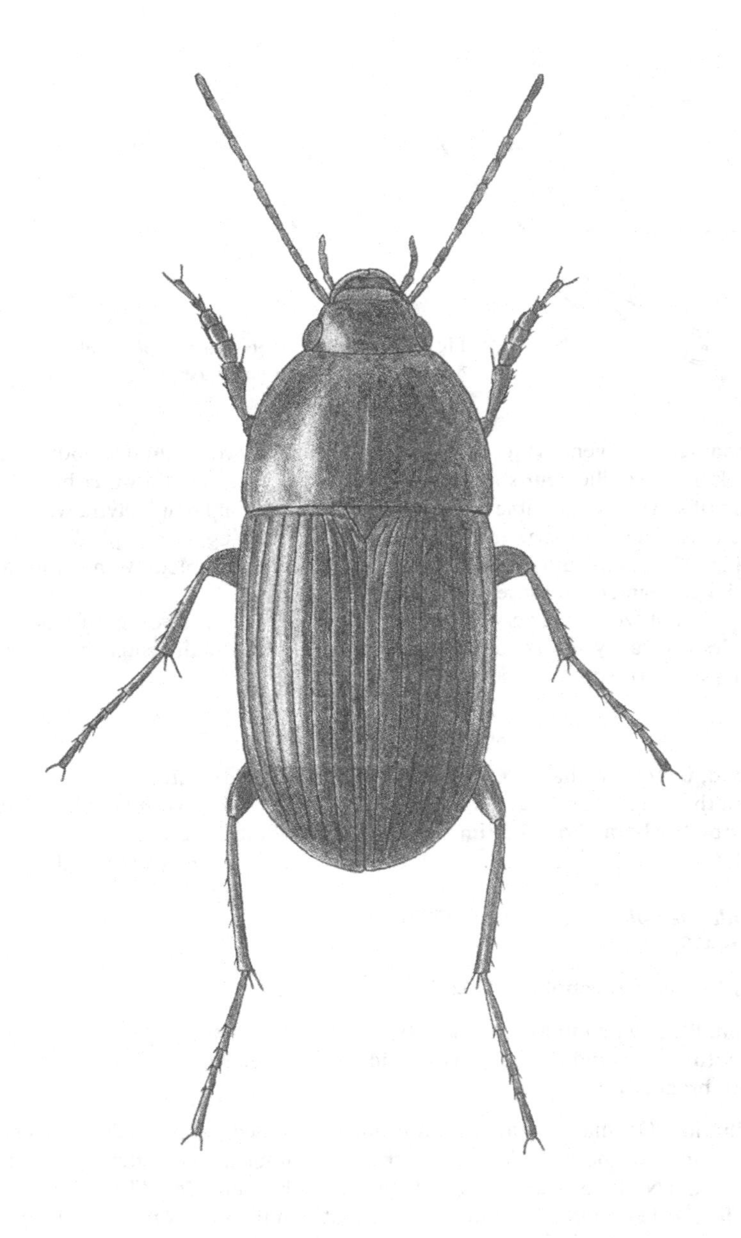

Fig. 479. *Oodes helopioides* (F.), length 7.5-10 mm. (After Victor Hansen).

but only locally common. Also Vib and Kr in the USSR. - From W. Europe to North Africa and C. Siberia.

Biology. At the border of lakes and slowly running rivers, on soft, muddy or silty soil. It is particularly typical of eutrophic, often polluted, lakes with rich and tall vegetation of e.g. *Phragmites, Typha* and *Glyceria.* The beetle often climbs along stems under the water surface. Most numerous in May-June, when propagation takes place.

366. ***Oodes gracilis*** Villa & Villa, 1833

Oodes gracilis Villa & Villa, 1833, Col. Eur. Dupl.: 33.

7.5-9 mm. General outline of body more parallel-sided, upper surface less dull, rather appearing greasy. Piceous black; pronotum around hind-angles, elytral epipleura, and legs brownish. Sides of pronotum more rounded with greatest width more anteriorly. Elytra more stretched. Sides of meta-sterna very finely punctate, or impunctate. Female very little deviating in width of pronotum. Fore tarsus of male less dilated.

Distribution. In our area only known from Sweden, where restricted to the Mälaren-area (Sdm., Upl. and Vstm.). In this area recorded from several localities, but is everywhere uncommon. All localities are in the close vicinity of the lake Mälaren; the only exception is Sdm.: Östermalma 1946. - From France and Spain to Asia Minor, the Caucasus and Turkestan, north to Estonia.

Biology. At the border of small eutrophic lakes and ponds, living in a habitat similar to that of the preceding species. Lindroth (1945) showed that the preferred temperature of *O. gracilis* is somewhat higher than that of *helopioides,* and that it is confined to such shore-habitats that are heated considerably during summer. It is a spring breeder.

Tribe Panagaeini

Genus *Panagaeus* Latreille, 1802

Panagaeus Latreille, 1802, Hist. Nat. Crust. Ins. 3: 91.
Type-species: *Carabus cruxmajor* Linnaeus, 1758.

Head squarish with excessively protruding eyes; neck constricted; terminal palpal segments inserted eccentrically (Fig. 480). Elytra patterned in red and black. Wings full. Fore tarsus of male with 2 dilated segments.

Key to species of *Panagaeus*

1 Pronotum (Fig. 481) broader, sides more or less sinuate posteriorly. Posterior red spot of elytra almost constantly reaching side-margin .... 367. *cruxmajor* (Linnaeus)
– Sides of pronotum (Fig. 482) very faintly, or not at all, sinuate posteriorly. Posterior elytral spot separated by black from side-margin .... 368. *bipustulatus* (Fabricius)

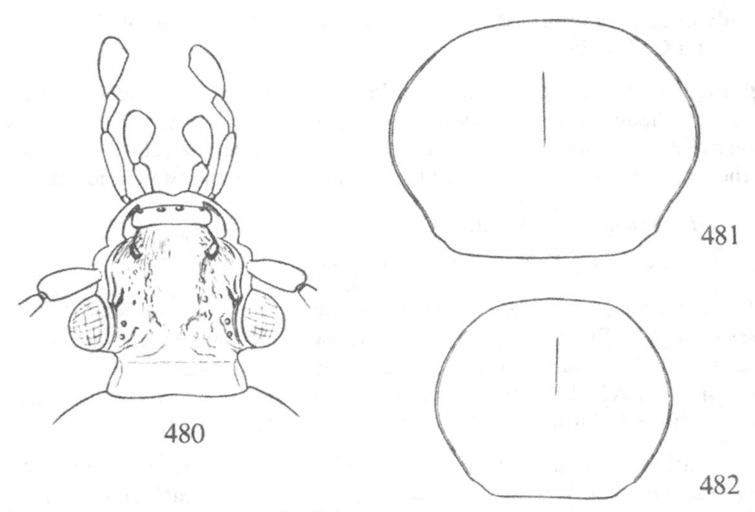

Fig. 480. Head of *Panagaeus cruxmajor* (L.).
Figs 481, 482. Pronotum of 481: *Panagaeus cruxmajor* (L.) and 482: *P. bipustulatus* (F.).

367. ***Panagaeus cruxmajor*** (Linnaeus, 1758)
Figs 480, 481, 483; pl. 8: 3.

Carabus crux major Linnaeus, 1758, Syst. Nat. ed. 10: 416.

7.5-9.1 mm. Black, each elytron with two orange-red spots, the black ground colour forming a cross, posterior red spot almost constantly reaching side-margin. Entire body with long erect pubescence. Pronotum (Fig. 481) with sides shallowly sinuate posteriorly. Upper surface with sparse and very coarse punctures. Elytral striae strongly, intervals more finely, punctate.

Distribution. Denmark: rare, and with decreasing frequency after 1950. Scattered occurrence in central and eastern parts of Jutland, on the islands, and on Bornholm. In WJ only recorded from Skallingen, in NWJ only from Hansted res. - Sweden: rather distributed in the southern half, north to Boh., S. Vrm., SE. Dlr., Gstr. and Hls. (Bergvik). Also at least 3 localities in the coastland of Nb. Has been strongly decreasing in the last decades. - Norway: very rare, only one record from HEs near the Swedish border. - Finland: rare and scattered, found north to Om. Also Vib and Kr in the USSR. - From E. Europe to North Africa, Asia Minor, N. Iran and E. Siberia.

Biology. A hygrophilous species, occurring at the margin of lakes and slowly running rivers, as well as in wet meadows; usually on soft, clayey soil with rich vegetation. It is most numerous in spring when breeding takes place, and in the autumn when the young beetles emerge. Overwintering adults are sometimes found under the bark of trees.

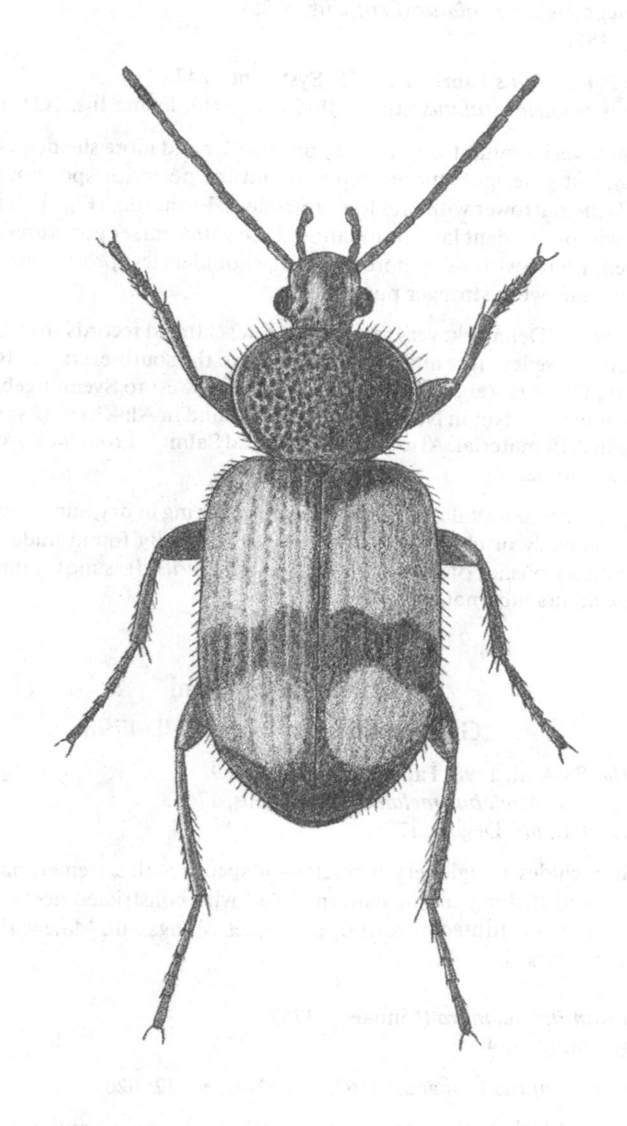

Fig. 483. *Panagaeus cruxmajor* (L.), length 7.5-9.1 mm. (After Victor Hansen).

368. ***Panagaeus bipustulatus*** (Fabricius, 1775)
Fig. 482.

Carabus 2 pustulatus Fabricius, 1775, Syst. Ent.: 247.
Panagaeus quadripustulatus Sturm, 1815, Deutschl. Fauna Ins. 5(3): 172.

6.5-7.5 mm. Very similar to *cruxmajor,* but smaller and more slender. Coloured in the same way, but pale spots more deep red, and the posterior spot not reaching side-margin. Head narrower with eyes less protrudent. Pronotum (Fig. 482) more perfectly circular, without evident lateral sinuation. Disc with coarser punctures and finer ones in between. Elytra with sides more rounded, shoulders less protruding, intervals with sparse but somewhat stronger punctures.

Distribution. Denmark: very rare; only a few scattered records from the islands and Bornholm. - Sweden: rare and local, restricted to the south-east: Sk., Bl., Sm. (on the coast), Öl., Gtl., Ög. (about 5 localities in the east, west to Svenningeby 190 m a.s.l.), Nrk. (Äsplunda). - Not in Norway. - Finland: found in Al: Kökar (2 specimens 1939), probably in drift material. Also Kr: Kuujärvi and Salmi. - From W. Europe to the Caucasus and Estonia.

Biology. More xerophilous than *P. cruxmajor,* living in dry, sun-exposed grassland, on sandy, gravelly or chalky soil. The species is usually found under stones, in dry moss, or under bushes of *Sarothamnus* and *Juniperus.* It is most numerous in May-June. The adults hibernate.

Tribe Odacanthini

Genus *Odacantha* Paykull, 1798

Odacantha Paykull, 1798, Faun. Suec. Ins. 1: 169.
Type-species: *Attelabus melanurus* Linnaeus, 1767.
Colliurus auct., *nec* Degeer, 1774.

The genus includes a single very characteristic species with extremely narrow forebody (Fig. 484), and striking colour pattern. Head with constricted neck. Apex of elytra truncate, striae substituted by rows of punctures. Wings full. Male with 3 dilated segments of fore tarsus.

369. ***Odacantha melanura*** (Linnaeus, 1767)
Fig. 484; pl. 8: 4

Attelabus melanurus Linnaeus, 1767, Syst. Nat. ed. 12: 620.

6.6-7.8 mm. Black with blue or green reflection, meso- and meta-sterna rufo-testaceous as the elytra, which have the entire apex black with a metallic lustre. Appendages very long, bright rufous but with antennae from fourth segment, most of palpi, apex of femora and tarsi dark.

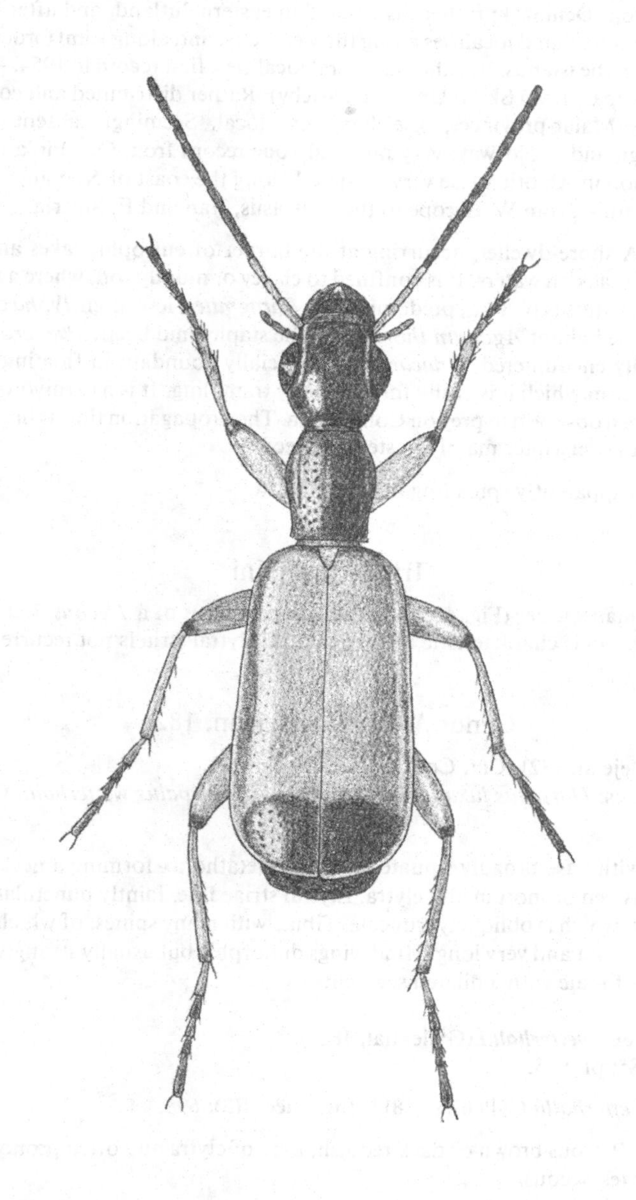

Fig. 484. *Odocantha melanura* (L.), length 6.6-7.8 mm. (After Victor Hansen).

Distribution. Denmark: rather distributed in eastern Jutland, and after 1950 also recorded from scattered localities along the west coast and along Limfjorden. Rather distributed on the islands. Bornholm: several localities, first record in 1954. - Sweden: in the eastern part from Sk. to Upl. (Älvkarleby). Rather distributed and common in Sk. and the Mälar-provinces, elsewhere very local. Seemingly absent from the southern highland. - Norway: very rare, only one record from Ø. - Finland: locally fairly common in Al; otherwise very scattered along the coast of N; many specimens from Sa: Lemi. - From W. Europe to the Caucasus, Iran and E. Siberia.

Biology. A shore-dweller, occurring at the border of eutrophic lakes and ponds, rarely along brackish waters. It is confined to clayey or muddy soil, where a rich vegetation of tall plants is present, predominantly *Phragmites*, less often *Typha* or *Glyceria*. In the same habitat *Agonum thoreyi* and the staphylinid beetle *Paederus riparius* L. are usually encountered. *Odacantha* is especially abundant in floating heaps of dead reeds, from which it is easily forced out by trampling. It is a carnivorous beetle, which has been observed to prey on Collembola. The propagation time is in spring; the young beetles overwinter mainly in stems of reed.

Note. It is apparently spreading in Scandinavia.

Tribe Masoreini

The single small species (Fig. 485) somewhat reminiscent of a *Trechus*, but the shape of the pronotum is characteristic and the sutural elytral stria is not recurrent.

Genus ***Masoreus*** Dejean, 1821

Masoreus Dejean, 1821, Cat. Coll. Col. B. Dejean: 15.
Type-species: *Harpalus luxatus* Serville, 1821 (= *Harpalus wetterhallii* Gyllenhal, 1813).

Pronotum with base broadly sinuate laterally. Metathorax forming a neck-like constriction between pronotum and elytra. Elytral striae fine, faintly punctulate but evident to apex, which is obliquely truncate. Tibiae with many spines, of which the most apical one is stout and very long. Hind wings dimorphic but usually strongly reduced. Fore tarsus of male with 3 dilated segments.

370. ***Masoreus wetterhallii*** (Gyllenhal, 1813)
Fig. 485; pl. 8: 5.

Harpalus Wetterhallii Gyllenhal, 1813, Ins. Suec. 1(3): 698.

4.7-5.6 mm. Piceous brown or dark reddish, base of elytra and often pronotum pale; appendages testaceous.

Distribution. Denmark: very scattered occurrence along the coasts of Jutland, more

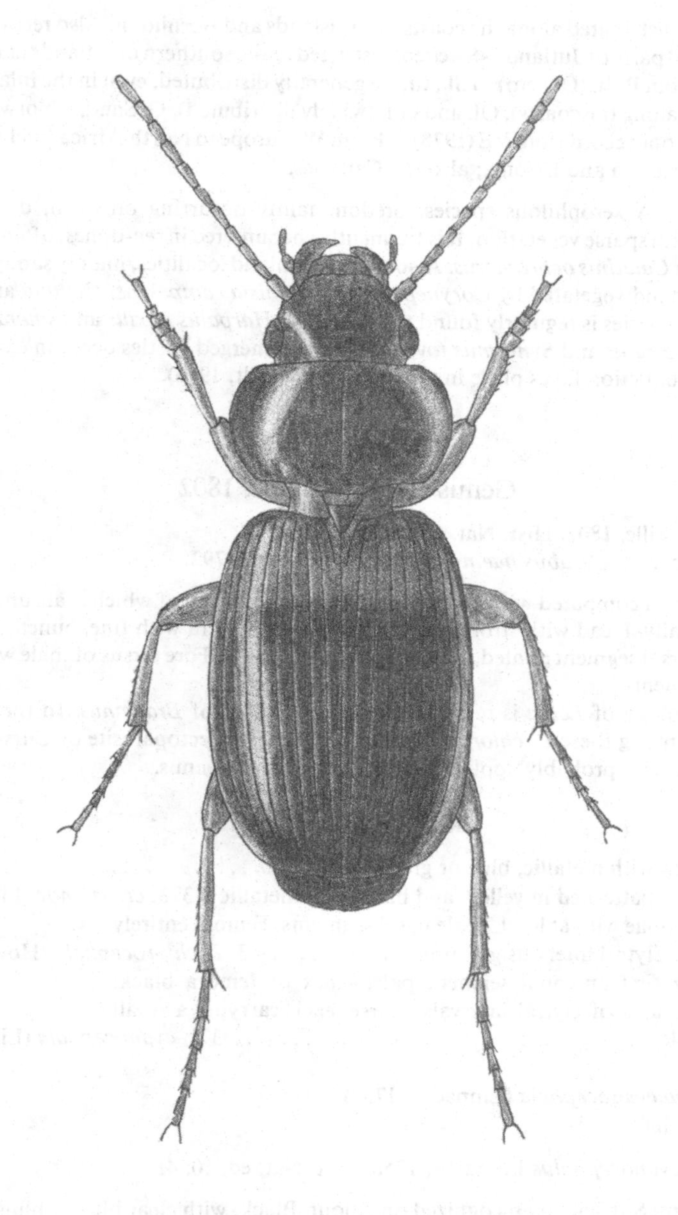

Fig. 485. *Masoreus wetterhallii* (Gyll.), length 4.7-5.6 mm. (After Victor Hansen).

generally distributed along the coasts of the islands and Bornholm. Also records from the central parts of Jutland. - Sweden: restricted to the southern coastlands and usually rare. From Boh. (Öckerö) to Sk. (more generally distributed, even in the inland), Bl. and Sm. (along the coasts), Öl. and Gtl. (widely distributed), G. Sand. - Norway: very rare, only one record from VE (1978). - From W. Europe to North Africa (incl. Egypt), Iran, W. Siberia and Estonia; also the Canaries.

Biology. A xerophilous species, predominantly occurring on open, dry, sandy ground with sparse vegetation. It is frequently encountered in sea-dunes, often in company with *Calathus ochropterus.* Also found on inland localities, mainly sandy heaths and grassland vegetated by *Corynephorus, Artemisia campestris, Calluna* and *Thymus.* The species is regularly found together with *Harpalus anxius* and *smaragdinus, Calathus erratus* and *Syntomus foveatus.* Newly emerged beetles occur in early summer; reproduction takes place in the autumn (Barndt, 1976).

Tribe Lebiini

Genus *Lebia* Latreille, 1802

Lebia Latreille, 1802, Hist. Nat. Crust. Ins. 3: 85.
Type-species: *Carabus haemorrhoidalis* Fabricius, 1792.

Elytra broad compared with pronotum (Fig. 486), the base of which is abruptly truncate laterally. Head with strongly constricted neck. Elytra with fine, punctate striae. Fourth tarsal segment dilated, claws dentate. Wings full. Fore tarsus of male with 3 dilated segments.

The biology of *Lebia* is remarkably similar to that of *Brachinus.* In the 3 cases known (among these *L. chlorocephala*), the larva is an ectoparasite of Chrysomelid pupae and this probably applies to all members of the genus.

Key to species of *Lebia*

1 Elytra with metallic, blue or green, lustre 2
- Elytra patterned in yellow and black, not metallic . 373. *cruxminor* (Linnaeus)
2(1) Antennae with at least 2 pale basal segments. Femora entirely pale. Elytral intervals glabrous .... 372. *chlorocephala* (Hoffmann)
- Only first antennal segment pale. Apex of femora black. Punctures of elytral intervals coarse, each carrying a small bristle .... 371. *cyanocephala* (Linnaeus)

371. ***Lebia cyanocephala*** (Linnaeus, 1758)
Fig. 486.

Carabus cyanocephalus Linnaeus, 1758, Syst. Nat. ed. 10: 415.

5.7-7.8 mm. Sufficiently recognized on colour. Black, with clear blue or bluish green reflection; pronotum , first antennal segment (at least underneath), and legs except apex of femora, rufous; scutellum, meso- and metathorax black. Elytra flat, widening posteriad, intervals with coarse punctures, each with a bristle.

Distribution. Denmark: very rare, only a few 19th century records from B. - Sweden: very rare and local in the south, north to Boh., Dls., Nrk., Upl. and Vstm.; no finds in the southern highland. Decreasing in frequency, almost all records are from before 1950. - Norway: rare, occurs only in the south-east. - Finland: very rare in the south, found north to Ta. Also Vib and southern Kr in the USSR. - From W. Europe to NW. China, south to Palestine and North Africa.

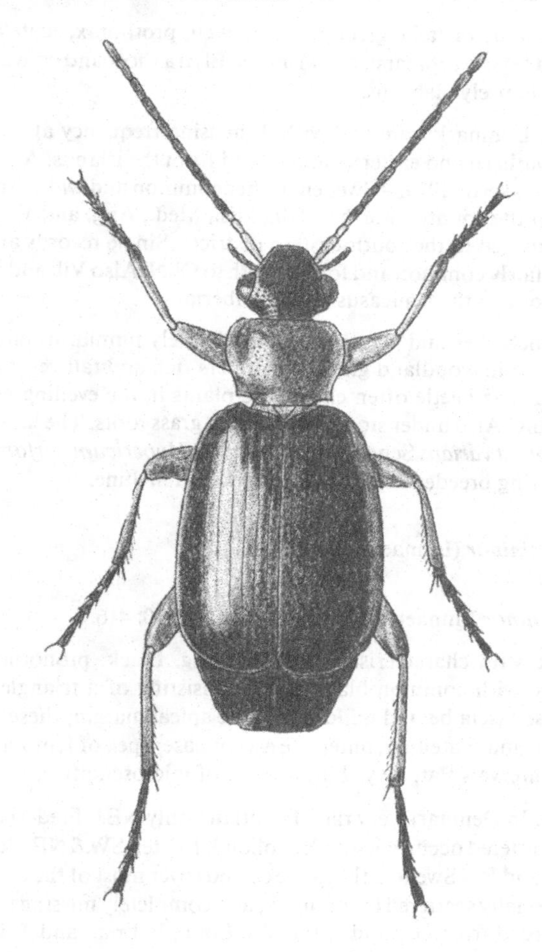

Fig. 486. *Lebia cyanocephala* (L.), length 5.7-7.8 mm. (After Victor Hansen).

Biology. The most xerophilous *Lebia*, living on open, dry hill-sides and fields, on sandy, gravelly or chalky soil. In Norway (1879) and in C. Europe, it has been captured on plants, mostly flowering Compositae. Its host is unknown. Most Scandinavian specimens are from July.

372. ***Lebia chlorocephala*** (Hoffmann, 1803)
Pl. 8: 7.

Carabus chlorocephalus Hoffmann, 1803, Ent. Hefte 2: 117.

5.8-8.1 mm. Strongly metallic green or blue-green, prothorax, scutellum, meso- and metanotum and legs, except tarsi, clear rufous. Elytra short and convex; intervals with fine punctures, entirely glabrous.

Distribution. Denmark: rare and with decreasing frequency after 1950. Scattered records from southern and eastern Jutland and from the islands. Also WJ: Lem 1965 and B: Paradisbakkerne 1979. - Sweden: rather common and widely distributed in the southern half of the country, north to Dlr., Hls., Med., Ång. and Vb. - Norway: rare and mainly restricted to the south-eastern districts. Single records also from Ry and SFi. - Finland: fairly common and found north to ObN. Also Vib and Kr of the USSR. - From W. Europe to the Caucasus and C. Siberia.

Biology. In meadows and grassland on moderately humid, usually clayey soil, in open country and in woodland glades. It prefers rich and tall vegetation of grasses, *Hypericum*, etc. The beetle often climbs the plants in the evening and may then be taken by sweeping. Also under stones and among grass roots. The larva parasitizes the pupa of *Chrysolina varians* Schall. which feeds on *Hypericum perforatum* (Lindroth, 1954). It is a spring breeder, most frequently found in June.

373. ***Lebia cruxminor*** (Linnaeus, 1758)
Pl. 8: 6.

Carabus crux minor Linnaeus, 1758, Syst. Nat. ed. 10: 416.

6-7 mm. Elytra with characteristic cross-marking. Black, pronotum rufous, elytra rufo-testaceous, with common black cross, consisting of a triangle at scutellum, a broad transverse fascia behind middle, and the apical margin; these are more-or-less confluent. Palpi and scutellum, antennae except base, apex of femora, and tarsi infuscated. Elytral intervals flat, very shiny, devoid of microsculpture.

Distribution. In Denmark very rare. In Jutland only NEJ: Frederikshavn 1978, under sea drift. Scattered occurrence on E. Lolland, Falster, SW.&NE. Zealand; a few old records from F and B. - Sweden: the range extends over most of the country, but usually rare, and especially scattered in the north, and completely missing in the high mountains. Not recorded from G. Sand., Hrj., Ås. Lpm., P. Lpm. and T. Lpm. - Norway: locally fairly common, but mainly restricted to the south-eastern districts. A single record from SFi. - Finland: rare but found north to LkW: Muonio. Also in Vib and Kr of the USSR. – From W. Europe to E. Siberia and Japan, south to Syria, Asia Minor and North Africa.

Biology. In meadows with rich vegetation on more or less dry gravelly soil, both in open country and in forest edges and glades. Sometimes captured on plants (e.g. *Achillea millefolium*) and bushes (e.g. flowering hawthorn). It has repeatedly been encountered together with the Chrysomelid *Galeruca tanaceti* (L.), which is probably the host. Reproduction takes place in spring; young beetles emerge in July-August.

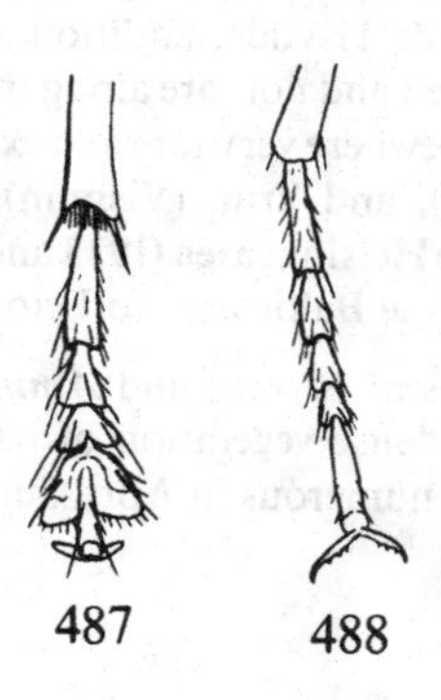

Figs 487, 488. Mid tarsus of 487: *Demetrias atricapillus* (L.) and 488: *Dromius longiceps* Dej.

Genus *Demetrias* Bonelli, 1810

Demetrias Bonelli, 1810, Obs. Ent. 1 (Tab. Syn.).
Type-species: *Carabus atricapillus* Linnaeus, 1758.

This genus is separated from *Dromius* of the *linearus* group by the deeply bilobed fourth segment of all tarsi (Fig. 487). All species are flat and narrow, with a testaceous ground colour. Head at least as wide as pronotum. Claws simple or with 1 to 3 small teeth. Wings full or dimorphic. Fore tarsus of male with 3 faintly dilated segments.

Key to species of *Demetrias*

1 Temples hairy. Elytral intervals with a single row of erect setae. Claws with 3 teeth .... 375. *atricapillus* (Linnaeus)
– Temples glabrous. Only third elytral interval with 4 setiferous punctures. Claws simple or with a single tooth 2
2(1) Claws with an internal tooth. Dark elytral markings restricted to apex and often to suture .... 374. *monostigma* Samouelle
– Claws smooth. Dark elytral pattern more expanded . 376. *imperialis* (Germar)

374. ***Demetrias monostigma*** (Samouelle, 1819)
Pl. 8: 8.

Demetrias monostigma Samouelle, 1819, Ent. Comp.: 156.

4.2-5.1 mm. Testaceous, head black, elytra apically across the suture with a rhomboid or elongate, rarely indistinct, dark spot, often narrowly prolonged forwards. Antennae usually somewhat infuscated apically. Claws with a single tooth. Wings normally highly reduced; however unique macropterous specimens are known from the European continent.

Distribution. Denmark: uncommon, but recorded with increasing frequency during the last decades. Jutland: along the western coast, and along the eastern coast south to Frederikshavn; also Læsø. On the islands on several localities along the southern Baltic coast, but also NWZ: Sjællands odde and NEZ: Tisvilde; additional inland records at lakes. B: Dueodde, 1979. - Sweden: distributed and not rare along the south and east coast of Skåne; also inland (e.g. Ringsjön). Elsewhere very rare and extremely local: Bl. (Mjällby), Öl. (Högsrum), Gtl. (Lina myr), and Vrm. (Visnum). - Not recorded from Norway. - Finland: two records from the Helsinki area (1938 and 1944). - From W. Europe to the Caucasus and Turkestan, in the Baltic area to Estonia.

Biology. In dunes at the seashore, occurring in tufts of *Elymus* and *Ammophila,* often associated with *Calathus ochropterus.* Also in dense vegetation of reeds and sedges on marshy lake-shores. A spring breeder, most numerous in April-June.

375. ***Demetrias atricapillus*** (Linnaeus, 1758)
Fig. 487.

Carabus atricapillus Linnaeus, 1758, Syst. Nat. ed. 10: 416.

4.5-5.6 mm. Easily recognized on the well developed pilosity (see the key). Testaceous as in *monostigma,* except that the elytra are entirely pale or only diffusely darker along the suture and, sometimes, at apex and/or around the scutellum. Head black. Pronotum with deeper basal foveae. Claws with 3 teeth. Wings fully developed.

Distribution. Denmark: uncommon, but with increasing frequency. In Jutland along the eastern coast from the German border north to Mols; also WJ: Fanø 1918. Rather distributed along the coasts of the islands and Bornholm; also some inland records. - Sweden: rather distributed but uncommon in SW.Skåne; first record dates from 1951, since spreading. In SE.Skåne wind-transported specimens are sometimes numerous on the shores of the Baltic Sea. - Not recorded from Norway or East Fennoscandia. - From W. Europe to North Africa, Asia Minor and Syria, north to Latvia.

Biology. In open country, usually occurring among grass and sedges or under heaps of decaying plants, on sandy as well as clayey soil. The species lives both on dry ground, for instance on the seashore among dune-grasses, and on marshy lake-shores. It is often captured by sweeping; sometimes found in sea drift. Propagation is in spring.

Note. It is a recent immigrant in Scandinavia and is still expanding.

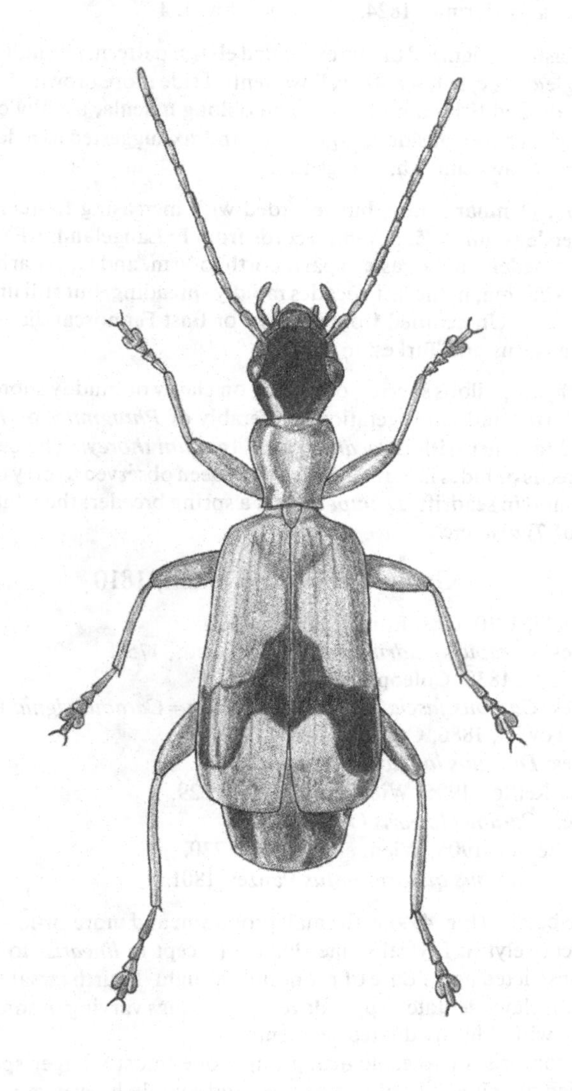

Fig. 489. *Demetrias imperialis* (Germ.), length 4.9-5.6 mm. (After Victor Hansen).

376. ***Demetrias imperialis*** (Germar, 1824)
Fig. 489; pl. 8: 9.

Dromias imperialis Germar, 1824, Ins. Spec. Nov. 1: 1.

4.9-5.6 mm. Easily recognized on the expanded elytral pattern, reminding of a heraldic »German Eagle«. Pale, rufescently yellow, ventral side more brown. Along the suture a dark stripe, dilated behind middle into an oblong macula, usually connected obliquely with a similar spot at side-margin. Mesothorax suggested as a dark triangle behind scutellum. Claws smooth. Wings full.

Distribution. Denmark: rare, but recorded with increasing frequency. In Jutland only SJ: Haderslev dam 1975. Several records from F: Langeland, LFM: Lolland, SZ, NEZ and B. - Sweden: in the eastern part, north to Vrm. and Upl. Earlier known only from the lake Mälaren, in the last decades rapidly spreading, but still uncommon. Not at high altitudes. - Unrecorded from Norway or East Fennoscandia. - From W. Europe to the Causasus and Turkestan.

Biology. A hygrophilous species, occurring on clayey or muddy shores of eutrophic lakes having a rich and tall vegetation, preferably of *Phragmites* or *Typha.* It is frequently found together with *Odacantha* and *Agonum thoreyi.* The beetle often runs about on the reeds or hides in leaf-sheaths; it has been observed to prey on Collembola. Sometimes found in seadrift. *D. imperialis* is a spring breeder; the adults hibernate in leaf-sheaths of *Typha,* etc.

Genus *Dromius* Bonelli, 1810

Dromius Bonelli, 1810, Obs. Ent. 1 (Tab. Syn.).
Type-species: *Carabus quadrimaculatus* Linnaeus, 1758.
Philorhizus Hope, 1838, Coleopt. Man. 2: 63.
Type-species: *Carabus fasciatus* Fabricius, 1801 (=*Carabus sigma* Rossi, 1790).
Paradromius Fowler, 1886, Col. Brit. Ins. 1: 141.
Type-species: *Dromius longiceps* Dejean, 1826.
Manodromius Reitter, 1905, Wien. ent. Ztg. 24: 229.
Type-species: *Carabus linearis* Olivier, 1795.
Calodromius Reitter, 1905, Wien. ent. Ztg. 24: 230.
Type-species: *Carabus quadrinotatus* Panzer, 1801.

Flat, elongate beetles (Fig. 499) with small pronotum and more-or-less parallel-sided, apically truncate elytra. Elytral striae shallow (except in *linearis)* to almost absent. Head with constricted neck. Base of pronotum straight. Fourth tarsal segment not dilated (Fig. 488), claws dentate. Appendages pale. Wings varying in some species. Fore tarsus of male with 3 feebly dilated segments.

The genus comprises two ecological groups: one (mostly larger species) contains members which are almost confined to trees and mainly hibernate under bark. The other group (Subg. *Philorhizus)* in terricolous, occurring in open and usually dry ground.

Key to species of *Dromius*

1 Above 5 mm (except small *linearis).* Elytral striae evident, seventh interval with 2 or more coarse punctures adjoining sixth stria. Elytral base without a pore puncture 2

– Less than 5 mm. Elytral striae obsolete, seventh interval impunctate. Elytral base, at level of apex of scutellum, with a small pore puncture 11

2 (1) Base of elytra margined laterally only (outside third entire stria). Forebody narrow (Figs 490, 491). Elytra with pale ground colour 3

– Elytra with complete raised basal bead. Elytra differently coloured 4

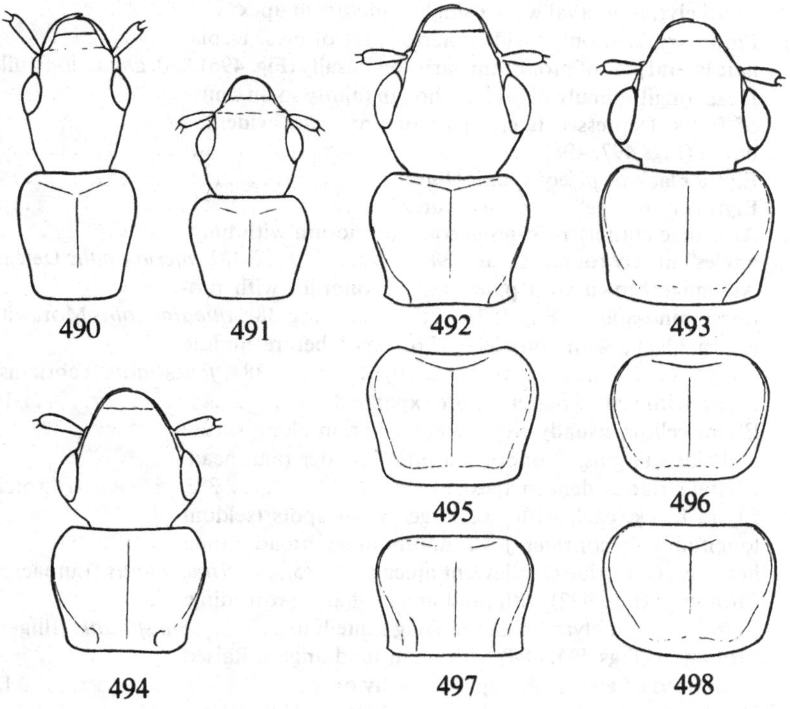

Figs 490-494. Head and pronotum of *Dromius.* – 490: *longiceps* Dej.; 491: *linearis* (Oliv.); 492: *spilotus* (Ill.); 493: *quadrisignatus* Dej.; 494: *sigma* (Rossi).

Figs 495-498. Pronotum of *Dromius.* – 495: *agilis* (F.); 496: *angustus* Brullé; 497: *quadraticollis* Mor.; 498: *meridionalis* Dej.

3 (2) Head (Fig. 490) very narrow with temples (in front of neck) much longer than diameter of eye. Elytra with shallow, faintly punctate striae .... 377. *longiceps* Dejean
– Head (Fig. 491) much shorter. Elytral striae sharp and clearly punctate .... 378. *linearis* (Olivier)
4 (2) Elytra black, each with two large yellow (sometimes longitudinally confluent) spots, the posterior of which occupies the entire apex (Fig. 499). Entire frons strongly longitudinally wrinkled .... 385. *quadrimaculatus* (Linnaeus)
– Elytra uniformly dark, or with pale spot on anterior half. Frons wrinkled only near eye, or faintly so also medially 5
5 (4) Third elytral interval with at least 5 coarse shallow punctures. Elytra quite dark or with diffuse pale macula in anterior half .... 379. *agilis* (Fabricius)
– Third elytral interval with a single puncture at apex 6
6 (5) Frons wrinkled only inside anterior part of eyes. Deplanate lateral part of pronotum narrower basally (Fig. 496) 380. *angustus* Brullé
– Head longitudinally wrinkled, though faintly so on centre of frons. Depressed lateral part of pronotum widening basad (Figs 497, 498) 7
7 (6) Elytra black or piceous, unicolorous 8
– Elytra entirely yellow or bicoloured 9
8 (7) Antennae entirely rufo-testaceous. Pronotum with hind-angles entirely rounded (Fig. 498) .... 382. *meridionalis* Dejean
– Antennae brown with paler base. Pronotum with produced hind-angles (Fig. 497) .... 381. *quadraticollis* Morawitz
9 (7) Elytra black, with rounded yellow spot before middle .... 384. *fenestratus* (Fabricius)
– Elytra with yellow pattern more expanded 10
10 (9) Elytra yellow, usually with piceous margin along suture and side margins. Pronotum a little broader than head. Elytral striae evident to apex .... 383. *schneideri* Crotch
– Elytra black, each with two large yellow spots (seldom longitudinally confluent). Pronotum much broader than head. Elytral striae obsolescent apically . 385. *quadrimaculatus* (Linnaeus)
11 (1) Pronotum (Fig. 492) with hind-angles sharp, protruding. Raised bead of elytral base reaching scutellum ...... 386. *spilotus* (Illiger)
– Pronotum (Figs 493, 494) with blunt hind-angles. Raised basal bead of elytra developed laterally only 12
12 (11) Head with strongly constricted neck (Fig. 493). Elytra with entire base dark .... 387. *quadrisignatus* Dejean
– Head with neck moderately obliquely constricted (Fig. 494). Base of elytra pale (except possibly suture) 13
13 (12) Elytra entirely pale or with narrowly infuscated suture.

	Macropterous	388. *melanocephalus* Dejean
–	Elytra with irregularly transverse band (rarely interrupted) just behind middle. Usually brachypterous	14
14 (13)	Abdomen as pale as the rest of the ventral side or slightly infuscated laterally. Pronotum entirely pale	389. *sigma* (Rossi)
–	Abdomen entirely piceous black. Pronotum often infuscated on disc	390. *notatus* Stephens

Subgenus *Paradromius* Fowler, 1886

377. ***Dromius longiceps*** Dejean, 1826
Fig. 490.

Dromius longiceps Dejean, 1826, Spec. Gén. Col. 2: 450.

5.3-6.5 mm. Like a *Demetrias* but with narrow tarsi. Very flat and stretched. Rufoferrugineous, head and abdomen darker, elytra with a posterior dark spot, widening apicad, across suture. Frons almost smooth. Head very narrow (Fig. 490). Elytra with shallow, faintly punctate striae. Wings full.

Distribution. Denmark: very rare. Only a few records before 1900; after that year recorded from EJ: Anholt 1974, NEJ: Læsø several times since 1968; F: Stigeø and Odense kanal 1950-54, F: Enebærodde 1971, LFM: Diget S of Keld skov 1957, LFM: Maltrup 1976, SZ: Køge 1971, NEZ: Hundige strand 1971, NEZ: Nivå strand 1967, and B: Bastemose 1977. - Sweden: very rare and local. In Sk., Bl. and Hall. recorded only from coastal localities. Further north in Ög. - Vrm. and Gstr. more of an inland species. Isolated record in Nb.: Haparanda Sandskär. - Not in Norway. - Finland: only found in two separate areas, Ab: Turku area and ObS: Hailuoto. - From W. Europe to the Caucasus, the Gorki area and Estonia.

Biology. On marshy lake-shores and river banks, usually occurring in dense vegetation of *Phragmites,* sometimes in company with *Odacantha* and *Demetrias imperialis.* It is often found in heaps of reed and among litter under bushes of willow or alder. In our area the species has also been encountered on seashores in tufts of *Elymus* and *Ammophila.* It is most numerous in spring, when propagation takes place, and in autumn, when the young beetles occur. The adults have been found hibernating under bark.

Note. The species is apparently expanding in Scandinavia (Lindroth, 1972; Bangsholt, 1979).

Subgenus *Manodromius* Reitter, 1905

378. ***Dromius linearis*** (Olivier, 1795)
Fig. 491; pl. 8: 14.

Carabus linearis Olivier, 1795, Ent. 3: 111.

4.4-6 mm. Narrow and parallel-sided as *longiceps,* but forebody much shorter (Fig. 491), and elytra strongly punctate. Reddish yellow, head and abdomen darker; elytra pale with suture and apex often infuscated. Frons with dense longitudinal wrinkles between the eyes. Wings dimorphic, normally quite reduced, but completely full-winged specimens are rare.

Distribution. Denmark: common and very distributed. - Sweden: common and generally distributed in the southern coastland from Boh. to Sk. to Upl.; a few inland records near the great lakes Vättern (in Ög.) and Vänern (in Vg.) - Not in Norway. - Finland: rather rare, only in the south-west. Also at river Svir in Kr. - From W. Europe to North Africa (incl. Egypt), Syria, the Caucasus and the Urals.

Biology. A xerophilous species, mainly occurring in dunes at the coast, in tufts of *Elymus* and *Ammophila,* often together with *Calathus ochropterus* and *Demetrias monostigma.* Also on dry, sandy grassland in inland localities. Often caught by sweeping. It is most numerous in May-September.

Subgenus *Dromius* s.str.

379. ***Dromius agilis*** (Fabricius, 1787)
Fig. 495.

Carabus agilis Fabricius, 1787, Mant. Ins. 1: 204.

6.0-6.2 mm. Best separated form all the following arboricolous species through the more numerous dorsal punctures. Rufo-piceous, elytra darker, sometimes paler at base or with diffuse pale macula (»*bimaculatus* Dejean«) in anterior half. Frons almost smooth at middle. Third elytral interval with at least 5 coarse, shallow punctures. Wings full.

Distribution. Denmark: uncommon but rather distributed in southern and eastern Jutland and on the islands and B. Apparently absent from large areas of WJ, NWJ and NEJ. - Sweden: rather common and widespread over the country. - Norway: occurs in most districts, but is very scattered. - Finland: common north to Li. Also in Vib, Kr and Lr of the USSR. - From W. Europe to W. Siberia.

Biology. An arboreal species, living on both coniferous and deciduous trees. In the mountains reaching the birch region. From spring to autumn the adult beetle occurs on stems and branches, in winter under bark, usually at the base of the trunks. The species is most easily found in its hibernation site, but is also frequently collected in summer by beating dead branches.

380. ***Dromius angustus*** Brullé, 1834
Fig. 496.

Dromius angustus Brullé, 1834, Revue Ent. (Silbermann) 3: 105.

6-6.8 mm. Narrower and with more slender antennae than in the related species. Testaceous brown, head and elytra usually darker, the latter often with pale base or diffuse spot, as in the form »*bimaculatus*« of *agilis*. Frons wrinkled only inside anterior part of eye, smooth at middle. Deplanate part of pronotum (Fig. 496) narrower basally. Elytra parallel-sided which gives a certain similarity to *schneideri*, but they are more dull due to strong microsculpture.

Distribution. Denmark: very rare, but recorded with increasing frequency. First Danish record is B: Blykobbe 1933. After 1960 several records from EJ, NEJ, SZ, NWZ, NEZ and B. - Sweden: distributed in the coastland from Boh. to Sk. to G. Sand. Has been strongly increasing since 1960, and now locally rather common. - Norway: rare, only in coastal areas in the extreme south. - No records from East Fennoscandia. - From W. Europe to Poland and Austria.

Biology. Arboreal, in Scandinavia almost confined to pine growings on sandy soil, in C. Europe also on spruce and broad-leaved trees. In summer the adults live on branches and twigs, from which they may be collected by beating; they hibernate under bark at the base of the trunks.

381. ***Dromius quadraticollis*** Morawitz, 1862
Fig. 497.

Dromius quadraticollis Morawitz, 1862, Melang. Biol. Bull. Acad. Sci. St. Petersb. 4: 199.
Dromius cordicollis Vorbringer, 1898, Ent. Nachr. 24: 286.

5.5-6.1 mm. At once recognized on the form of the pronotum. Rufo-piceous, centre of pronotum, elytra and antennae except the 3 basal segments more-or-less infuscated. Frons strongly wrinkled. Pronotum (Fig. 497) with sinuate sides and sharp hind-angles.

Distribution. Not in Denmark. - Sweden: Sk.: Sandhammaren, one specimen in sea drift 1969. - Not in Norway. - Finland: found in sea drift in N: Esbo, Långgrund (1982). A similar record from Vib: Tytärsaari. - From Estonia, Poland and Bulgaria to E. Siberia.

Biology. An arboreal species living on both coniferous and broad-leaved trees. In summer the beetle occurs on the branches, in winter under bark.

382. ***Dromius meridionalis*** Dejean, 1825
Fig. 498.

Dromius meridionalis Dejean, 1825, Spec. Gén. Col. 1: 242.

6-7 mm. Coloured as *agilis*; the maculated form called »*discus* Puel«. Wrinkles on frons developed along the entire inside of eye and at least suggested on centre. Pronotum (fig. 498) broader than in *agilis*, with the hind-angles somewhat more obtuse, and the deplanate lateral part wider posteriorly. Elytra less widened apicad.

Distribution. In our area only recorded from Denmark. Here very rare, the latest record being SJ: Vemmingbund 1944. Several still older records from the southern parts of the country. - From W. Europe to C. Siberia; also the Azores.

Biology. Arboreal, mostly on dead branches of deciduous trees; formerly on wattles on Lolland and Falster (LFM). Mainly in June.

383. ***Dromius schneideri*** Crotch, 1871

Carabus marginellus Fabricius, 1794, Ent. Syst. 4: 442, *nec* Herbst, 1784.
Dromius schneideri Crotch, 1871, List Descr. Col.: 6.

5.8-6.4 mm. The palest of the arboricolous species. Pale yellow, abdomen darker, head black, pronotum normally infuscated on centre; elytra piceous along sidemargins and apex. Immature specimens are best separated from *angustus* through the longitudinally wrinkled frons, from the *quadrimaculatus* form with confluent yellow spots by the apically evident elytral striae.

Distribution. Denmark: very rare; NEZ: Copenhagen area in numbers 1872-74; NEZ: Lyngby åmose one specimen 1980; NEZ: Tisvilde one specimen 1906; B: Hammeren one specimen 1910; B: Bykobbe pl. one specimen 1965. - Sweden: generally distributed and rather common in the south, north to about 61°N. In the northern half mainly in the coastland of Hls. - Nb. and rare; a few isolated records from Jmt. - Norway: local in the south-eastern districts; also one record from HO. - Finland: common in the southern and central parts, north to ObS. Also Vib and Kr in the USSR. - From NE.France to the Urals.

Biology. Arboreal, usually living on pine, but also on spruce and larch; rarely on deciduous trees (e.g. in Norway and Finland). In summer the species lives in the tree crowns, sometimes occurring in abundance; in winter under bark.

384. ***Dromius fenestratus*** (Fabricius, 1794)
Pl. 8: 10.

Carabus fenestratus Fabricius, 1794, Ent. Syst. 4: 443.

5.7-6.5 mm. With constant coloration. Piceous, under surface somewhat paler, pronotum with pale margins (exceptionally entirely rufo-piceous). Each elytron just before middle with a small rounded, strongly delimited, yellow spot; appendages yellow. Frons densely wrinkled inside the eyes and also, though somewhat more faintly, sculptured on the centre. Thereby separated from the similarly coloured individuals of *agilis* and *angustus.* Elytra more shiny than in *agilis,* sixth interval with at least 2 punctures, third interval with a single apical puncture.

Distribution. Denmark: very rare, but recorded with increasing frequency. First Danish record is WJ: Esbjerg 1924. Recent records from WJ, EJ, NWJ, NEJ, NWZ, and B. - Sweden: generally distributed and usually common in the south, north to 62°N. Rare and very local further north, northernmost records being from Nb. and

Ås. Lpm. – Norway: scattered throughout all districts, north to NT (64°N). – Finland: rather common in the southern and central parts, north to Ok. Also Vib in the USSR. – From Portugal to the Gorki area, USSR.

Biology. Arboreal, mostly on pine, rarely on spruce and broad-leaved trees. Like the

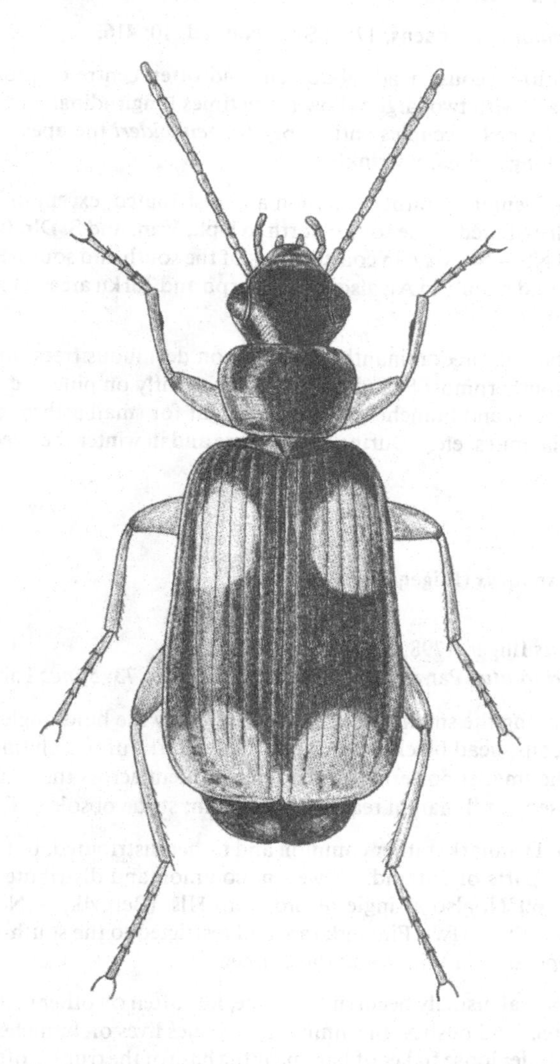

Fig. 499. *Dromius quadrimaculatus* (L.), length 5.2-6.4 mm. (After Victor Hansen).

preceding species it lives in the canopy in summer and hibernates under bark. It is most easily found in winter under loose flakes of bark at the base of the trunks.

385. ***Dromius quadrimaculatus*** (Linnaeus, 1758)
Fig. 499; pl. 8: 11.

Carabus 4-maculatus Linnaeus, 1758, Syst. Nat. ed. 10: 416.

5.2-6.4 mm. Rufo-piceous; head, abdomen, and often centre of pronotum darker. Elytra black, each with two large yellow (sometimes longitudinally confluent) spots, the posterior of which occupies entire apex (in *schneideri* the apex is dark). Entire frons strongly, longitudinally wrinkled.

Distribution. Denmark: rather common and distributed, except in WJ. - Sweden: common and distributed in the south, north to Upl., Vrm. and S. Dlr. Isolated records from Ång. and Nb. - Norway: in coastal areas of the south and south-east. - Finland: uncommon, found mainly in Al; also in Ab: Korpo and Turku area. - From W. Europe to the Caucasus.

Biology. Arboreal, predominantly occurring on deciduous trees, for instance oak and beech, in southernmost Scandinavia also frequently on pine and spruce. It is an inhabitant of stems and branches, hunting at night for small arthropods (other beetles, Collembola, mites, etc.). During the daytime and in winter the beetle hides under bark.

386. ***Dromius spilotus*** (Illiger, 1798)
Fig. 492.

Carabus spilotus Illiger, 1798, Verz. Käf. Preuss.: 234.
Carabus quadrinotatus Panzer, 1800, Fauna Ins. Germ. 73: 5; *nec* Fabricius, 1798.

3.0-4.6 mm. Among the small species distinguished by the hind-angles of pronotum (Fig. 492). Piceous, head black, pronotum usually dark rufous, elytra each with two yellow spots, the smaller posterior ones often confluent across the suture; apex dark. Elytra with raised basal margin reaching scutellum; striae obsolete. Fully winged.

Distribution. Denmark: rather common and rather distributed, but apparently absent from some parts of Jutland. - Sweden: common and distributed in the south, north to about 60°N; also a single record from Hls. (Bergvik). - Norway: local in coastal areas, north to STy. - Finland: rare and restricted to the south-western part. - From W. Europe to Asia Minor and the Crimea.

Biology. Arboreal, usually occurring on pine, less often on other conifers; rarely on broad-leaved trees and bushes. In summer the species lives on branches and twigs; in winter usually under loose flakes of bark near the base of the trunks, often in company with *Salpingus castaneus* Panz.

387. ***Dromius quadrisignatus*** Dejean, 1825
Fig. 493.

Dromius quadrisignatus Dejean, 1825, Spec. Gén. Col. 1: 236.

3.5-4 mm. Easily recognized on the strongly constricted neck (Fig. 493). Piceous, with darker head, pronotum rufous, elytra each with two large pale spots, the posterior of which occupies entire apex. Wings full. Hind-angles of pronotum obtuse. Basal bead of elytra incomplete, not reaching scutellum.

Distribution. Denmark: only a few 19th century records from Falster (LFM). - Sweden: Sk., Falsterbo since 1968, sometimes numerous and probably established. - Not in Norway of East Fennoscandia. - From W. Europe to Greece and Rumania.

Biology. Arboreal, living on different kinds of deciduous trees; also on dead branches and twigs on the ground. On Falster (Denmark: LFM) found on wattles.

Subgenus *Philorhizus* Hope, 1840

388. ***Dromius melanocephalus*** Dejean, 1825
Pl. 8: 13.

Dromius melanocephalus Dejean, 1825, Spec. Gén. Col. 1: 234.

3-3.4 mm. Brownish, head black, pronotum rufous, sometimes darker on disc; elytra pale testaceous with translucent dark triangle (the transparent mesothorax) at base, also suture often narrowly infuscated. Abdomen often as dark as in *notatus*. Elytra narrow, parallel-sided. Wings full.

Distribution. In Denmark rather common and distributed. - Sweden: rather common and distributed in SW.Skåne, eastwards to Skillinge. Elsewhere only recorded from Hall. (Hunnestad 1979) and Öl. (Strandtorp 1975). - Not in Norway or East Fennoscandia. - From W. Europe to Asia Minor and Syria.

Biology. In open country, usually in dry, sandy meadows and grassland, dwelling among moss, dead leaves, grasses, etc. Also among dune-grasses on the seashore, frequently associated with *D. linearis*. Mainly met with in spring, when breeding takes place, and in autumn, when the young beetles emerge.

389. ***Dromius sigma*** (Rossi, 1790)
Fig. 500; pl. 8: 12.

Carabus sigma Rossi, 1790, Fauna Etrusca 1: 226.

3.2-4 mm. Abdomen as pale as the rest of the ventral surface, or slightly infuscated laterally; head black, pronotum and elytra rufo-testaceous. Elytra with well-defined transverse dark fascia, not reaching side-margin, and only exceptionally prolonged inwards to apex. Microsculpture of pronotum weak, meshes in part about isodiametric. Full-winged specimens very rare. Penis: fig. 500.

Distribution. Denmark: rather common and distributed in the southern and eastern parts of Jutland, and on the islands. Absent from large areas of WJ, NWJ and NEJ. B: only Blykobbe. - Sweden: rather distributed and common in the south, north to 62°N; in the north mainly near the coasts and absent from large areas. - Norway: rare and scattered in the south-east. - Finland: very common in the south, north to LkW: Muonio. Also Vib and Kr in the USSR. - From W. Europe to Asia Minor, the Caucasus and W. Siberia.

Biology. Predominantly at the border of standing and running freshwater on somewhat shaded ground, usually in rich vegetation of *Phragmites* etc., dwelling among reeds and leaves; often under *Salix* or *Alnus*. Also in tussocks of *Elymus* and *Ammophila* at the sea. It is most numerous in April-June. The adults have been found in hollow reed stems during winter.

390. ***Dromius notatus*** Stephens, 1827
Fig. 501.

Dromius notatus Stephens, 1827, Ill. Brit. Ent. Mand. 1: 24.
Dromius nigriventris Thomson, 1857, Skand. Col. 1: 51.

3.2-3.7 mm. Ground colour more dirty testaceous dark. Disc of pronotum infuscated. Elytral fascia often ill-defined, prolonged to apex along side-margin, which has the bead more-or-less infuscated. Abdomen entirely piceous black. Elytra somewhat broader apically. Microsculpture of pronotum consisting of transverse meshes. Penis: fig. 501.

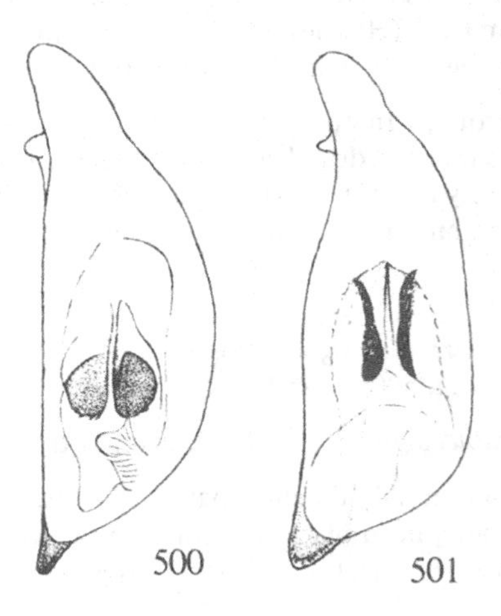

Figs 500, 501, Penis of 500: *Dromius sigma* (Rossi) and 501: *D. notatus* Stph.

Distribution. Denmark: rather common and distributed in the eastern and northern parts of Jutland, and in NEZ and B. Otherwise scattered, and apparently absent from central and western parts of Jutland. - Sweden: generally distributed and common in the south, north to S. Vrm., S. Dlr., Gstr. and S. Hls. - Norway: rather common in coastal areas south of 60°N. - Finland: rare in the south-western parts, north to Ta. - From W. Europe to Asia Minor, the Caucasus and Turkestan.

Biology. In open country on dry, sandy or gravelly soil, occurring in tussocks of grass, among plant residues, etc. It is most predominant in vegetation of *Ammophila* and *Elymus* at the sea, often occurring together with *D. linearis*. Almost throughout the year.

Genus *Syntomus* Hope, 1838

Syntomus Hope, 1838, Col. Man. 2: 64.
 Type-species: *Carabus truncatellus* Linnaeus, 1761.
Metabletus Schmidt-Goebel, 1846, Faun. Col. Birm. 3: 38.
 Type-species: *Lebia obscuroguttata* Duftschmid, 1812.

Separated from *Microlestes* by the slightly oblique elytral apex (Fig. 502) and by the pubescence of antennae starting from fourth segment. Claws evidently dentate. Wings normally reduced to a small scale. Fore tarsus of male with 3 faintly dilated segments.

The species of *Syntomus* are carnivorous beetles and have spring propagation.

Key to species of *Syntomus*

1 Black, or with faint bronze hue. Dorsal punctures of elytra small .... 391. *truncatellus* (Linnaeus)
- Upper surface with evident, bronze or brassy, lustre. Dorsal punctures foveate .... 392. *foveatus* (Fourcroy)

391. ***Syntomus truncatellus*** (Linnaeus, 1761)
Pl. 8: 15 (as *Metabletus t.)*

Carabus truncatellus Linnaeus, 1761, Fauna Suec. ed. 2: 224.

2.6-3.2 mm. Black, at moist with an extremely faint bronze hue. First antennal segment and legs (at least to some extent) rufo-piceous. The 2 or 3 dorsal punctures of third elytral interval faint, never foveate. Elytral microsculpture not granulate.

Distribution. Denmark: rather common and distributed in eastern Jutland north to Limfjorden, and on the islands and Bornholm. In the remaining part of Jutland only a few scattered coastal records. - Sweden: common and widespread over most of the country, but completely absent in the west from N. Dlr. to T. Lpm. - Norway: fairly common in the south-east and central areas, north to 62°N. - Finland: very common, found north to LkW. Also Vib and Kr in the USSR. - From W. Europe to the Caucasus and E. Siberia.

Biology. On open, sun-exposed, rather dry ground, on different types of soil, preferably with sparse vegetation of grasses. It is especially typical of dry meadow ground; also on arable land and in open woodland. Mainly in spring.

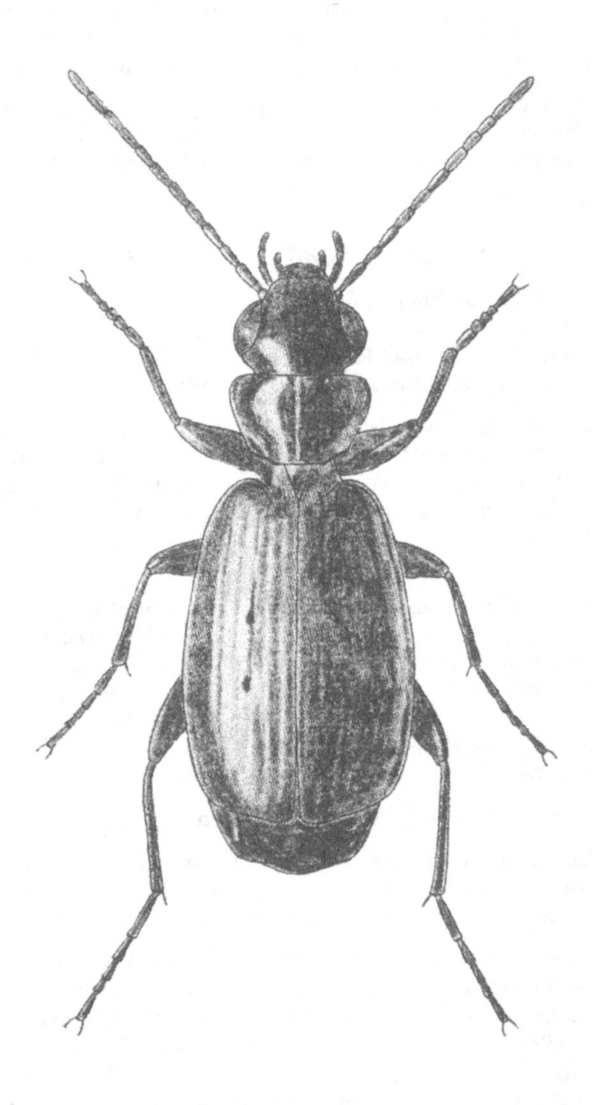

Fig. 502. *Syntomus foveatus* (Fourcr.), length 3.1-3.8 mm. (After Victor Hansen).

392. ***Syntomus foveatus*** (Fourcroy, 1785)
Fig. 502.

Buprestis foveatus Fourcroy, 1785, Ent. Paris. 1: 52.
Lebia foveola Gyllenhal, 1810, Ins. Suec. 1 (2): 183.

3.1-3.8 mm. Black, upper surface with evident, often strong, bronze or aeneous lustre, which because of the granulate microsculpture is somewhat dull. First antennal segment and legs hardly paler. Elytra longer than in preceding species, almost parallel-sided. Third interval with 2 (seldom 3) foveate punctures. Antennae more slender. Wings constantly reduced.

Distribution. In Denmark common and very distributed. - Sweden: common and distributed in the south, north to about 60°N. - Norway: very rare, only recorded from Ø and VE. - Finland: many records from Al; otherwise one record from Ab: Kaarina, and one from St: Yläne (in the 19th century). Also Vib and southern Kr in the USSR. - From W. Europe to the Caucasus and W. Siberia.

Biology. A stenotopic species, restricted to dry, sandy soil with sparse vegetation of grasses, *Calluna, Thymus,* etc. Both inland and in coastal areas, also in cultivated land. It is often active during the day; mainly met with in spring.

Genus *Lionychus* Wissmann, 1846

Lionychus Wissmann, 1846, Stettin. ent. Ztg. 7: 25.
Type-species: *Lebia quadrillum* Duftschmid, 1812.

Similar to *Syntomus* in general habitus, but at once recognized on the structure of the pronotum (Fig. 503). Elytra almost constantly with 2 or 4 yellow spots; apex obliquely truncate as in *Syntomus.* Claws smooth. Fore tarsus of male with 4 dilated segments. Wings full.

393. ***Lionychus quadrillum*** (Duftschmid, 1812)
Fig. 503; pl. 8: 16.

Lebia quadrillum Duftschmid, 1812, Fauna Austriae 2: 246.

3-4 mm. Raised lateral bead of pronotum continued to base inside the dentiform hind-angles (Fig. 503), so that the prosternal sides are partly visible from above. Black with faint bronze hue; appendages dark. Each elytron with a large spot behind shoulder and another smaller, sometimes obsolete, rarely totally missing, spot behind middle. These spots bright yellow. Totally immaculate specimens rare. Only 4 inner elytral striae well impressed, intervals with an irregular row of punctures.

Distribution. Not in Denmark. - In Sweden rare but spreading, and increasing in frequency; records from 4 districts. Vg.: Töreboda 1968; Nrk.: Örebro since 1945; Vstm.: widely distributed along the Bodelundsåsen, about 15 localities known from

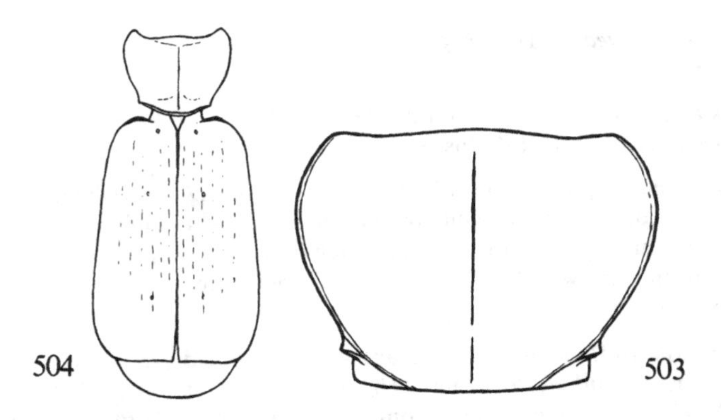

Fig. 503. Pronotum of *Lionychus quadrillum* (Dft.).
Fig. 504. Pronotum and elytra of *Microlestes minutulus* (Gz.).

Björnö in the south to near Fläckebo in the north; first record 1961 from near Västerås; also found near Strömsholm; Upl.: Ekerö 1968. - No records from Norway. - From Finland one record, Ab: Pargas 1983. - From W. Europe to Asia Minor and Rumania.

Biology. On dry, sandy or gravelly soil, for instance on railway banks and in gravel-pits. In C. Europe usually on sandy river banks and lake shores in mountains.

Note. The species was probably introduced to Sweden during the last war and is continuously spreading.

Genus *Microlestes* Schmidt-Goebel, 1846

Microlestes Schmidt-Goebel, 1846, Faun. Col. Birm. 3: 41.
Type-species: *Microlestes inconspicuus* Schmidt-Göbel, 1846.
Blechrus Motschulsky, 1847, Bull. Soc. Nat. Mosc. 20 (2): 219.
Type-species: *Lebia glabrata* Duftschmid, 1812.

Includes two small dark species separated from *Syntomus* by the transversely truncate elytral apex (Fig. 504). Pronotum more lobate medially. Elytra more parallel-sided, third interval with 2 extremely fine (or obsolete) punctures. Pubescence of antennae starting on third segment. Wings dimorphic. Elytra with faint striae. Claws faintly denticulate. Head with reticulate microsculpture. Male with dilated segments of fore tarsus. Penis much varying in shape.

The species are carnivorous, preying on Collembola, mites, etc. They reproduce in spring; the adults have been found hibernating under bark.

Key to species of *Microlestes*

1 Somewhat larger, 2.7-3.7 mm, elytra proportionately longer,

less widened behind the middle. Hind margin of pronotum less protruding .... 394. *minutulus* (Goeze)

- Somewhat smaller, 2.5-2.8 mm, elytra proportionately shorter, more widened behind the middle. Hind margin of pronotum more protruding .... 395. *maurus* (Sturm)

394. ***Microlestes minutulus*** (Goeze, 1777)
Figs 504, 505; pl. 8:17.

Carabus minutulus Goeze, 1777, Ent. Beytr. 1: 665.
Lebia glabratus Duftschmid, 1812, Fauna Austriae 2: 248.

2.7-3.7 mm. Black without metallic hue, but legs often a little paler brownish. Head with strong microsculpture on frons consisting of isodiametric (almost circular) meshes. Wings always full. Penis: fig. 505.

Distribution. Denmark: very rare and scattered. - Sweden: rare but spreading and locally more common in the northern part of the range. Recorded from Sk., Bl., Hall., Sm. (near the coast), Öl., Gtl., G. Sand., Ög., Vg., Nrk. (several localities after 1980), Sdm., Upl. (Värmdö 1968), and Vrm. (Karlstad 1969-70). - No records from Norway. - Finland: rare in the south, found north to Ta and Sa. Also Vib and southern Kr in the USSR. - From W. Europe to the Caucasus, E. Siberia and Japan.

Biology. On open, rather dry, sun-exposed ground, on sandy or gravelly, sometimes clay-mixed soil with sparse vegetation; often in gravel-pits, also on arable land. It has repeatedly been found in sea drift. Mainly in spring and early autumn.

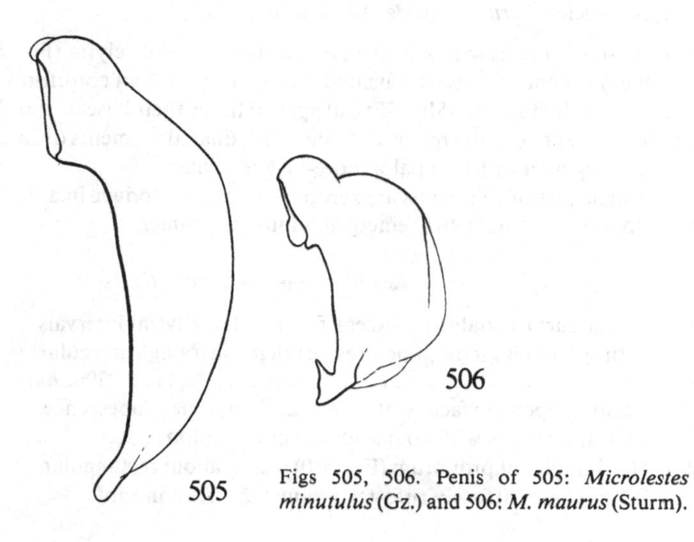

Figs 505, 506. Penis of 505: *Microlestes minutulus* (Gz.) and 506: *M. maurus* (Sturm).

395. ***Microlestes maurus*** (Sturm, 1827)
Fig. 506.

Dromias maurus Sturm, 1827, Deutschl. Fauna Ins. 5 (7): 55.

2.5-2.8 mm. Smaller than preceding species, forebody narrower, elytra shorter, slightly widened apically. Black with faint bronze hue; legs sometimes piceous. Microsculpture of frons weaker, consisting of longitudinal, in part fusing, lines. Dimorphic, wings always with deflexed apex, but often too small to be functional. Penis: fig. 506.

Distribution. Denmark: rare. In Jutland scattered in the eastern part; also Mors in NWJ 1941-43. Rather distributed on the islands and Bornholm. - Sweden: rare and very local in the south, north to Ög. and Vg. - No records from Norway. - Finland: only four records, Ab: Turku, Ka: Imatra (twice), and Sa: Punkaharju. Also Kr: Salmi. - From W. Europe to Asia Minor, the Caucasus and C. Siberia, north to Estonia.

Biology. In open country on dry, sandy or gravelly, often clay-mixed soil with sparse vegetation; also in slightly shaded sites, e.g. under bushes. Mainly in spring.

Genus *Cymindis* Latreille, 1806

Cymindis Latreille, 1806, Gen. Crust. Ins. 1: 190.
Type-species: *Buprestis humeralis* Fourcroy, 1785.
Tarsostinus Motschulsky, 1864, Bull. Soc. Nat. Mosc. 37 (2): 240.
Type-species: *Cymindis lateralis* Fischer v. Waldheim, 1821.
Tarulus Bedel, 1906, Cat. rais. Col. N.Afr. 1: 253.
Type-species: *Tarus zargoides* Wollaston, 1863.

Middle-sized species with stretched and distinctly striate elytra (Fig. 511). Upper surface hairy (in one species only behind the eyes). Pronotum cordiform with sides sinuate posteriorly (Figs 507-510). Elytral apex oblique, their base or shoulder pale. Claws pectinate. Wings usually reduced. Male with 3 dilated segments of fore tarsus and also terminal segment of labial palpi more-or-less dilated.

The members of this genus are xerophilous and reproduce in autumn. Larvae have been found in spring, newly emerged adults in summer.

Key to species of *Cymindis*

1 Upper surface glabrous, except for temples. Elytral intervals with sparse and faint punctures, arranged in a single irregular row .... 396. *humeralis* (Fourcroy)
- Entire upper surface with erect and oblique pubescence. Elytral intervals with coarse, dense punctuation 2
2 (1) Hind-angles of pronotum (Fig. 510) sharp, about rectangular. Elytral base diffusely rufo-testaceous; the bead inwards

reaching at most to third stria . . . . 399. *vaporariorum* (Linnaeus)
- Hind-angles of pronotum less sharp (Figs 507, 508). Elytra with more-or-less well defined shoulder-spot; basal bead inwards reaching first stria 3
3 (2) Pronotum much wider than head, with broadly depressed sides (Fig. 509) . . . . 398. *macularis* Fischer v. Waldheim
- Pronotum (Fig. 508) with much less rounded and widened sides . . . . 397. *angularis* Gyllenhal

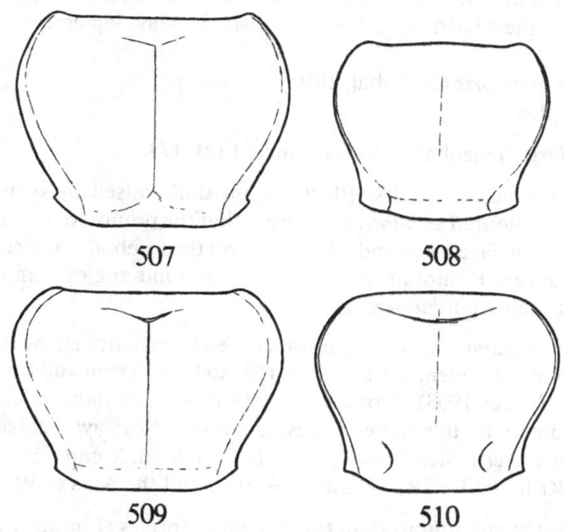

Figs 507-510. Pronotum of *Cymindis*. - 507: *humeralis* (Fourcr.); 508: *angularis* Gyll.; 509: *macularis* Fisch. Waldh.; 510: *vaporariorum* (L.).

396. ***Cymindis humeralis*** (Fourcroy, 1785)
Fig. 507.

Buprestis humeralis Fourcroy, 1785, Ent. Paris. 1: 57.

8.5-11.0 mm. The largest of our *Cymindis*-species, with surface shiny due to a weak microsculpture. Black, lower surface more-or-less rufo-testaceous; sides of pronotum usually brown; elytra with oblong, well delimited yellow shoulder-spot, continuing as a narrow margin along sides. Appendages ferrugineous. Forebody with fine and sparse punctuation, except strong on temples behind eyes. Hind-angles of pronotum (Fig. 507) dentiform. Elytra with deep, finely punctate striae; intervals strongly convex with faint punctures. Wings reduced. Male with terminal segment of labial palpi strongly axe-shaped.

Distribution. Denmark: only 2 records, SZ: Højstrup on Stevns 1917, and NWZ: Rørvig ca 1870. - Sweden: Sk., Bl., Sm., Öl. and Gtl. On central Öland collected repeatedly; elsewhere very rare and local. Most finds on the mainland are old, the latest are Sk.: Stenshuvud 1967 and Sm.: Emmaboda 1968. - Not recorded from Norway or East Fennoscandia. - From France and Italy to the Crimea and Lithuania.

Biology. A heat-preferent species, living on open, dry ground with short vegetation, in Scandinavia confined to coastal regions. On Öl. it is regularly found on the alvar, often together with *C. angularis,* in short but dense vegetation of grasses, species of *Thymus* and *Sedum, Hieracium pilosella,* etc. In C. Europe also on heather-covered ground, often at the edge of pine-forests. Mostly in May-September.

397. ***Cymindis angularis*** Gyllenhal, 1810
Fig. 508; pl. 8: 18.

Cymidis angularis Gyllenhal, 1810, Ins. Suec. 1 (2): 173.

6.5-9 mm. Our smallest species. Black, elytra dull caused by dense macro- and microsculpture. Coloured as *humeralis,* except that the pronotum is entirely brownish red. Entire upper surface hairy and with dense, on the forebody very coarse, punctuation. Wings reduced. Pronotum (Fig. 508) narrow, hind-angles at most rectangular. Labial palpi of male as in *humeralis.*

Distribution. Denmark: very rare, only recorded from scattered localities in Jutland and on Bornholm. - Sweden: scattered from Sk. to Boh., Vstm. and Upl.; also isolated in N. Vrm. (Sysslebäck 1963). Along the west coast and the Baltic coast, including on Öl. and Gtl., sometimes in numbers; elsewhere rare. - Norway: local along the south coast and in the eastern inland, north to 62°N. - Finland: found north to Oa and Kb. Also Vib and Kr in the USSR. - From C. Europe and the Alps to W. Siberia.

Biology. Less thermophilous than the preceding species (Lindroth 1949), living in dry, sun-exposed country on sandy or gravelly, often chalky soil, preferably in coastal regions. It is a common species on the grassy parts of the alvar of Öl. and Gtl., occurring together with *C. humeralis, Harpalus azureus, Panagaeus bipustulatus,* etc., but is also found on *Calluna*-heaths. The habitat is usually dominated by short vegetation of grasses, *Thymus, Calluna,* etc. The beetles are frequently found aggregated under stones. Mainly in May-September.

398. ***Cymindis macularis*** Fischer von Waldheim, 1824
Fig. 509.

Cymindis macularis Fischer von Waldheim, 1824, Ent. Imp. Russ. 2: 25.

7.7-10 mm. Piceous to reddish brown, head and abdomen darkest, elytra with rufotestaceous shoulder macula, which may be diffuse. In addition, also often suture and a small spot near the apex pale; appendages pale yellow. Upper surface with coarser and denser punctuation and pubescence than in *angularis.* First antennal segment

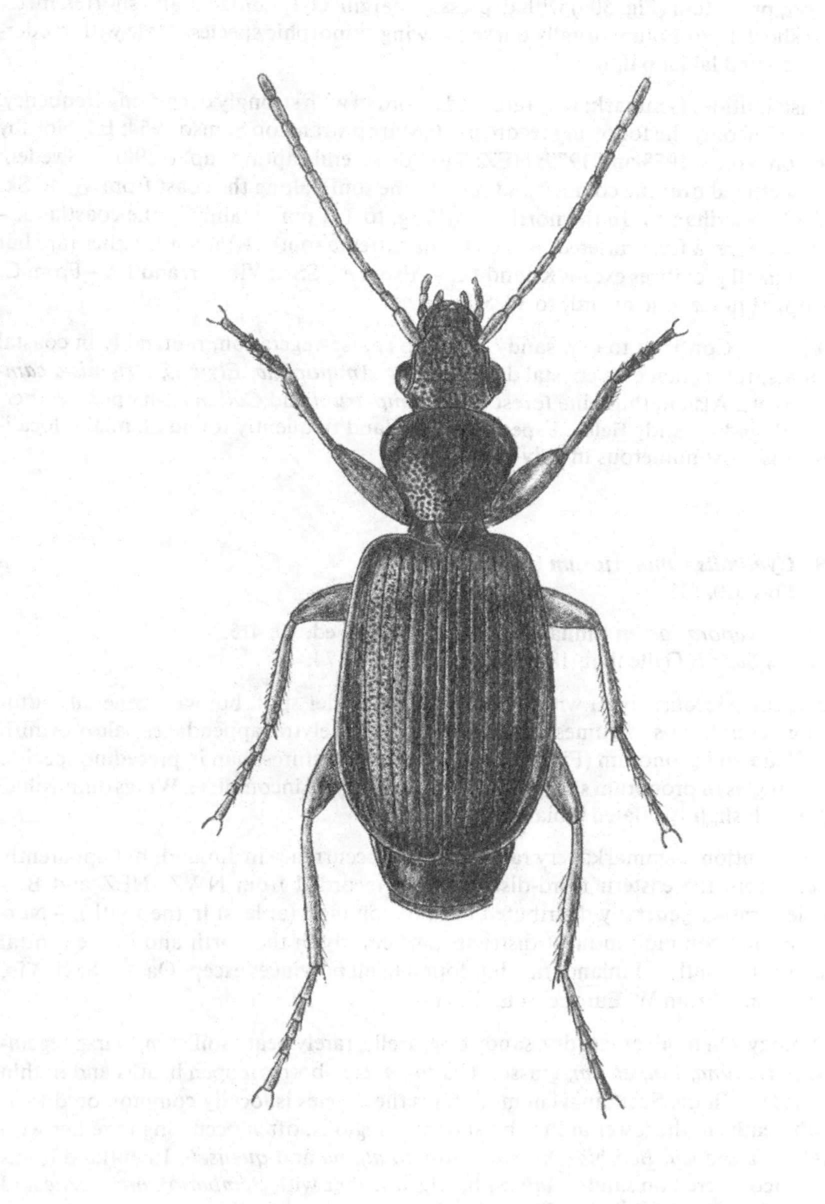

Fig. 511. *Cymindis vaporariorum* (L.), length 8-9.6 mm. (After Victor Hansen).

longer, pronotum (Fig. 509) with depressed margin, elytra broader and shorter, metatarsi shorter. Pronotum usually darker. A wing-dimorphic species. Male with moderately dilated labial palpi.

Distribution. Denmark: very rare, and recorded with strongly decreasing frequency. After 1950 only the following records: EJ: Mårup strand on Samsø 1954; EJ: Nordby hede on Samsø 1958 and 1979; NEZ: Tisvilde several captures up to 1981. - Sweden: very scattered over the country and rare. In the south along the coast from Vg. to Sk. to Upl. (Sandhamn). In the north from Ång. to T. Lpm., mainly in the coastland. - Norway: rare, a few scattered records in the extreme south. - Finland: rather rare but found in all provinces except Ks and Le. - Also the USSR: Vib, Kr, and Lr. - From C. Europe (France: one record) to W. Siberia.

Biology. Confined to dry, sandy soil with sparse vegetation, preferably in coastal habitats, for instance on coastal dunes among *Ammophila, Elymus, Artemisia campestris*, etc. Also in thin pine forest under *Empetrum* and *Calluna*, on open heather-ground, and in sandy fields. Especially in Finland frequently found on inland localities. It is most numerous in July-August.

399. ***Cymindis vaporariorum*** (Linnaeus, 1758)
Figs 510, 511.

Carabus vaporariorum Linnaeus, 1758, Syst. Nat. ed. 10: 415.
Cymidis basalis Gyllenhal, 1810, Ins. Suec. 1 (2): 174.

8-9.6 mm. Piceous, elytra without delimited shoulder spot, but with generally rufo-ferrugineous base, sometimes almost entirely brown elytra; appendages pale brownish red. Head and pronotum (Fig. 511) with coarser punctures than in preceding species. Hind-angles of pronotum sharp. Basal margin of elytra incomplete. Wings dimorphic. Male with slightly dilated labial palpi.

Distribution. Denmark: very rare. Scattered occurrence in Jutland, but apparently absent from the eastern fjord-district. Also recorded from NWZ, NEZ and B. - Sweden: rather generally distributed but not common (at least in the south). - Norway: locally common in most districts, particularly in the north and in the central areas of the south. - Finland: rare but found in all provinces except Oa. - USSR: Vib, Kr and Lr. - From W. Europe to E. Siberia.

Biology. On more or less dry, sandy or gravelly, rarely peaty soil with sparse vegetation of *Calluna, Empetrum*, grasses, *Cladonia*, etc., both on open heaths and in thin pine forest. In the Scandinavian mountains the species is locally common on dwarf-shrub heaths in the lower arctic and subarctic regions, often occurring together with *Miscodera arctica, Bembidion grapii, Amara alpina* and *quenseli*. In Jutland it has been encountered on sandy *Calluna*-heaths together with *Bembidion nigricorne* and *Amara famelica*. Mainly in May-September.

Subfamily Brachininae

The Bombardier beetles are usually referred to a separate subfamily because they differ from all other Carabidae in the high number of visible abdominal segments: 7 in the female, 8 in the male (Fig. 512). This structure is probably correlated with the famous crepitating power of these beetles, enabling them to direct the defense spray accurately.

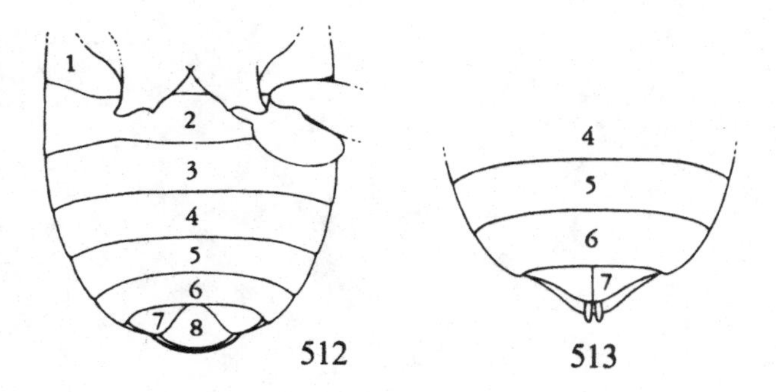

Figs 512, 513. Underside of abdomen in Brachininae. – 512: ♂; 513: ♀.

Genus *Brachinus* Weber, 1801

Brachinus Weber, 1801, Obs. Ent. 1: 22.
Type-species: *Carabus crepitans* Linnaeus, 1758.
Brachynus auct., incorrect spelling.

The general appearance is uniform and very characteristic in all members of this world-wide genus. Forebody very narrow (Fig. 514), with pronotum not or hardly wider than head. Elytra very broad, almost squarish, with obsolete striae and apical membraneous fringe. Forebody rufous, elytra (in our species) uniformly dark, usually with metallic reflection. Entire upper surface, including antennae, pubescent. Wings full. Fore tarsus of male with 3 dilated segments.

When attacked or disturbed, a *Brachinus* will eject a defensive spray from its pygidial glands that open at the tip of the abdomen. The spray is discharged as a hot poisonous cloud, and is accompanied by an audible sound. Eisner (1958) showed that the spray mechanism protects *Brachinus* against arthropod predators, e.g. other carabids and ants. Details of the explosive chemistry has been worked out by Schildknecht *et al.* (1961, 1968, 1970): one part of the anal gland system produces hydroquinone and hydrogen peroxide, another part secretes an enzyme mixture of catalases and peroxidases. When the two secretions are mixed in a specialized »firing chamber« hydroquinone is oxidised into the repellent substance quinone, and hydrogen peroxide is split into water and oxygen, which causes the reaction mixture to be expelled.

The development of 4 North American species is known. In all cases, the larva is an ectoparasite of beetle pupae (as in *Lebia)*.

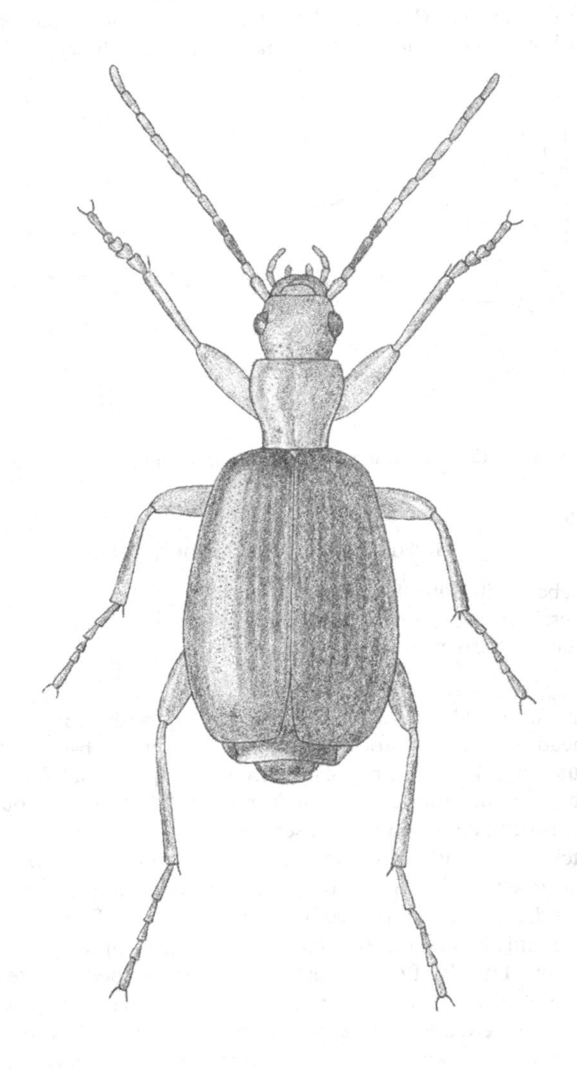

Fig. 514. *Brachinus crepitans* (L.), length 6.1-9.8 mm. (After Victor Hansen).

400. ***Brachinus crepitans*** (Linnaeus, 1758)
Fig. 514; pl. 8: 19.

Carabus crepitans Linnaeus, 1758, Syst. Nat. ed. 10: 414.

6.1-9.8 mm. Black or piceous, normally with bluish or greenish reflection. Forebody rufous, underside of hindbody piceous. Antennal segments 3 and 4 more-or-less infuscated; legs pale. Elytral striae shallow but evident; apical membrane pubescent.

Distribution. Denmark: very rare, only known from a limited area around Sose on southern Bornholm. - Sweden: in the south-eastern part of the country, north to Upl. In Sk., Hall. and Sm. only a few old records; on Öl. and Gtl. distributed. - Not recorded from Norway. - In Finland a specimen was recorded in sea drift in N: Hangö in 1945. - From W. Europe to Asia Minor, Syria, the Caucasus, and Turkestan, north to Latvia and S. Estonia (one old record).

Biology. A heat-preferent species, living on open, sun-exposed, rather dry ground, usually on clayey or chalky soil with dense but short vegetation of grasses etc., e.g. at the margin of stony ditches. It is locally common on the alvar of Öl. and Gtl. Often gregarious and associated with *Agonum dorsale,* sometimes also with *Harpalus azureus, rupicola* and *melleti.* It is most numerous in spring and autumn; oviposition has been observed in June, first instar larvae in July. Its host is unknown.

				DENMARK												
		Germany	G. Britain	SJ	EJ	WJ	NWJ	NEJ	F	LFM	SZ	NWZ	NEZ	B	Sk.	Bl.
Stomis pumicatus (Pz.)	171	●	●	●	●	●	●	●	●	●	●	●	●	●	●	●
Pterostichus punctulatus (Schall.)	172	●		●	●	●	●		●	●	●	●	●	●	●	
P. lepidus (Leske)	173	●	●	●	●	●	●	●	●	●	●	●	●	●	●	●
P. kugelanni (Pz.)	174	●	●										●			
P. cupreus (L.)	175	●	●	●	●	●	●	●	●	●	●	●	●	●	●	●
P. versicolor (Sturm)	176	●	●	●	●	●	●	●	●	●	●	●	●	●	●	●
P. longicollis (Dft.)	177	●	●		●							●				
P. vernalis (Pz.)	178	●	●	●	●	●		●	●	●	●	●	●	●	●	●
P. macer (Marsh.)	179	●	●	●		●			●	●		●				
P. aterrimus (Hbst.)	180	●	●		●			●		●	●	●	●	●	●	●
P. aethiops (Pz.)	181	●	●													
P. madidus (F.)	182	●	●		●											
P. oblongopunctatus (F.)	183	●	●	●	●	●	●	●	●	●	●	●	●	●	●	●
P. adstrictus Eschtz.	184		●													
P. quadrifoveolatus Letz.	185	●		●	●	●			●	●	●		●	●	●	●
P. niger (Schall.)	186	●	●	●	●	●	●	●	●	●	●	●	●	●	●	●
P. melanarius (Ill.)	187	●	●	●	●	●	●	●	●	●	●	●	●	●	●	●
P. nigrita (Payk.)	188	●	●	●	●	●	●	●	●	●	●	●	●	●	●	●
P. anthracinus (Ill.)	189	●	●	●	●	●	●		●	●	●	●	●	●	●	●
P. gracilis (Dej.)	190	●	●	●	●	●			●	●	●	●	●	●	●	●
P. minor (Gyll.)	191	●	●	●	●	●	●	●	●	●	●	●	●	●	●	●
P. strenuus (Pz.)	192	●	●	●	●	●	●	●	●	●	●	●	●	●	●	●
P. diligens (Sturm)	193	●	●	●	●	●	●	●	●	●	●	●	●	●	●	●
P. brevicornis Kirby	194															
P. middendorffi (J. Sahlbg.)	195															
P. melas (Creutz.)	196				●											
P. burmeisteri Heer	197												●			i
Abax parallelepipedus (P.&M.)	198	●	●	●	●	●			●	●	●	●	●		●	
Calathus fuscipes (Gz.)	199	●	●	●	●	●	●	●	●	●	●	●	●	●	●	●
C. erratus (Sahlbg.)	200	●	●	●	●	●	●	●	●	●	●	●	●	●	●	●
C. ambiguus (Payk.)	201	●	●	●	●	●	●	●	●	●	●	●	●	●	●	●
C. melanocephalus (L.)	202	●	●	●	●	●	●	●	●	●	●	●	●	●	●	●
C. ochropterus (Dft.)	203	●		●	●	●	●	●	●	●	●	●	●	●	●	●
C. micropterus (Dft.)	204	●	●	●	●	●	●	●				●	●	●	●	●
C. rotundicollis Dej.	205	●	●	●	●	●	●	●	●	●	●	●	●	●	●	
Sphodrus leucophthalmus (L.)	206	●	●	●	●	●		●		●	●	●	●		●	
Laemostenus terricola (Hbst.)	207	●	●	●	●	●	●	●	●	●	●	●	●	●	●	●
Dolichus halensis (Schall.)	208	●		●					●	●	●	●	●	●	●	

SWEDEN

	Hall.	Sm.	Öl.	Gtl.	G. Sand.	Ög.	Vg.	Boh.	Dlsl.	Nrk.	Sdm.	Upl.	Vstm.	Vrm.	Dlr.	Gstr.	Hls.	Med.	Hrj.	Jmt.	Ång.	Vb.	Nb.	Ås. Lpm.	Ly. Lpm.	P. Lpm.	Lu. Lpm.	T. Lpm.
171	●	●	●	●		●	●	●		●	●	●	●		●													
172	●		●	●																								
173	●	●	●	●		●	●	●	●	●	●	●	●	●	●		●	●	●	●	●	●	●	●	●	●	●	
174																												
175	●	●	●	●		●	●	●	●	●	●	●	●	●	●	●												
176	●	●	●	●	●	●	●	●	●	●	●	●	●	●	●	●	●	●	●	●	●	●	●	●	●		●	●
177																												
178	●	●	●	●		●	●	●	●	●	●	●	●	●	●	●	●	●			●	●	●					
179																												
180	●	●	●	●		●	●	●		●		●	●															
181																												
182																												
183	●	●	●	●	●	●	●	●	●	●	●	●	●	●	●	●	●	●		●	●	●	●	●	●	●	●	
184		●					●						●	●	●		●	●	●	●	●	●	●	●	●	●	●	●
185	●	●				●	●				●	●																
186	●	●	●	●	●	●	●	●	●	●	●	●	●	●	●	●	●	●		●	●	●	●					
187	●	●	●	●		●	●	●	●	●	●	●	●	●	●	●	●	●		●	●	●	●					
188	●	●	●	●	●	●	●	●	●	●	●	●	●	●	●	●	●	●	●	●	●	●	●	●	●	●	●	●
189	●	●	●	●	●	●	●		●	●	●	●	●	●	●	●												
190		●	●	●	●	●	●		●	●	●	●	●	●	●	●												
191	●	●	●	●		●	●	●	●	●	●	●	●	●	●	●	●	●	●	●	●	●	●	●	●	●	●	
192	●	●	●	●		●	●	●	●	●	●	●	●	●	●	●	●	●		●	●	●	●	●		●	●	
193	●	●	●	●		●	●	●	●	●	●	●	●	●	●	●	●	●	●	●	●	●	●	●	●	●	●	●
194																												
195																												
196																												
197																												
198																												
199	●	●	●	●	●	●	●	●	●	●	●	●	●	●	●	●												
200	●	●	●	●	●	●	●	●	●	●	●	●	●	●	●	●	●	●		●	●	●	●		●			
201	●	●	●	●	●	●	●	●		●	●	●																
202	●	●	●	●	●	●	●	●	●	●	●	●	●	●	●	●	●	●	●	●	●	●	●	●	●	●	●	●
203	●	●	●	●	●																							
204	●	●	●	●	●	●	●	●	●	●	●	●	●	●	●	●	●	●	●	●	●	●	●	●	●	●	●	●
205																												
206	●	●					●					●	●															
207	●	●	●	●		●	●	●		●	●	●	●		●													
208																												

		NORWAY														
		Ø + AK	HE (s + n)	O (s + n)	B (ø + v)	VE	TE (y + i)	AA (y + i)	VA (y + i)	R (y + i)	HO (y + i)	SF (y + i)	MR (y + i)	ST (y + i)	NT (y + i)	Ns (y + i)
Stomis pumicatus (Pz.)	171	●														
Pterostichus punctulatus (Schall.)	172															
P. lepidus (Leske)	173	●	●	●	●	●	●	●	●	●	●	●	●	●	●	
P. kugelanni (Pz.)	174															
P. cupreus (L.)	175	●	●		●	●	●	●	●		●					
P. versicolor (Sturm)	176	●	●	●	●	●	●	●	●	●	●	●	●	●	●	
P. longicollis (Dft.)	177															
P. vernalis (Pz.)	178	●	●	●	●	●	●	●	●	●	●	●	●			
P. macer (Marsh.)	179															
P. aterrimus (Hbst.)	180								●	●						
P. aethiops (Pz.)	181															
P. madidus (F.)	182															
P. oblongopunctatus (F.)	183	●	●	●	●	●	●	●	●	●	●	●	●	●	●	●
P. adstrictus Eschtz.	184		●	●			●		●	●		●	●	●	●	●
P. quadrifoveolatus Letz.	185							●								
P. niger (Schall.)	186	●	●	●	●	●	●	●	●	●	●	●	●	●	●	●
P. melanarius (Ill.)	187	●	●	●	●	●	●	●	●	●	●	●	●	●	●	
P. nigrita (Payk.)	188	●	●	●	●	●	●	●	●	●	●	●	●	●	●	●
P. anthracinus (Ill.)	189															
P. gracilis (Dej.)	190															
P. minor (Gyll.)	191	●	●	●	●	●	●	●	●	●	●	●		●	●	
P. strenuus (Pz.)	192	●	●	●	●	●	●	●	●	●	●	●	●	●	●	●
P. diligens (Sturm)	193	●	●	●	●	●	●	●	●	●	●	●	●	●	●	●
P. brevicornis Kirby	194															
P. middendorffi (J. Sahlbg.)	195															
P. melas (Creutz.)	196															
P. burmeisteri Heer	197															
Abax parallelepipedus (P.&M.)	198	●					●									
Calathus fuscipes (Gz.)	199	●	●	●	●	●	●	●	●	●	●	●	●	●	●	●
C. erratus (Sahlbg.)	200	●	●	●	●	●	●	●	●	●	●	●	●	●	●	●
C. ambiguus (Payk.)	201	●				●			●							
C. melanocephalus (L.)	202	●	●	●	●	●	●	●	●	●	●	●	●	●	●	●
C. ochropterus (Dft.)	203								●	●						
C. micropterus (Dft.)	204	●	●	●	●	●	●	●	●	●	●	●	●	●	●	●
C. rotundicollis Dej.	205															
Sphodrus leucophthalmus (L.)	206															
Laemostenus terricola (Hbst.)	207	●				●			●	●	●			●		
Dolichus halensis (Schall.)	208															

	Nn (ø+v)	TR (y+i)	F (v+i)	F (n+ø)	Al	Ab	N	Ka	St	Ta	Sa	Oa	Tb	Sb	Kb	Om	Ok	Ob S	Ob N	Ks	LkW	LkE	Le	Li	Vib	Kr	Lr
					FINLAND																				USSR		
171						●	●	●																	●		
172																											
173					●	●	●	●	●	●	●	●	●	●	●	●	●	●	●	●	●				●	●	●
174																											
175					●	●	●	●	●	●	●	●	●	●	●	●	●	●							●	●	
176					●	●	●	●	●	●	●	●	●	●	●	●	●	●	●	●	●				●	●	
177																											
178					●	●	●	●	●	●	●	●	●	●	●	●	●	●							●	●	
179																											
180						●	●				●	●	●		●										●	●	
181						●	●																		●	●	
182							●																				
183	●	●	●		●	●	●	●	●	●	●	●	●	●	●	●	●	●	●	●	●	●	●	●	●	●	●
184	●	●	●			●	●		●	●	●	●	●	●	●	●	●	●	●	●	●	●	●	●		●	●
185					●	●	●				●				●										●	●	
186	●	●			●	●	●	●	●	●	●	●	●	●	●	●	●	●	●						●	●	
187					●	●	●	●	●	●	●	●	●	●	●	●	●					●			●	●	●
188	●	●			●	●	●	●	●	●	●	●	●	●	●	●	●	●	●	●	●	●		●	●	●	●
189																										●	
190					●	●	●			●					●										●	●	
191					●	●	●	●	●	●	●	●	●	●	●	●	●		●	●					●	●	
192	●	●			●	●	●	●	●	●	●	●	●	●	●	●	●	●	●	●					●	●	●
193	●	●	●	●	●	●	●	●	●	●	●	●	●	●	●	●	●	●	●	●	●	●	●	●	●	●	●
194																											●
195																											●
196																											
197																											
198																											
199					●	●	●	●			●														●	●	
200	●	●			●	●	●	●	●	●	●	●	●	●	●	●	●	●	●	●		●			●	●	
201					●	●	●				●														●	●	
202	●	●	●	●	●	●	●	●	●	●	●	●	●	●	●	●	●	●	●	●	●	●	●	●	●	●	●
203																											
204	●	●	●	●	●	●	●	●	●	●	●	●	●	●	●	●	●	●	●	●	●	●	●	●	●	●	●
205																											
206						●	●		●	●				●			●			●							
207						●	●					●													●	●	
208																											

				DENMARK												
		Germany	G. Britain	SJ	EJ	WJ	NWJ	NEJ	F	LFM	SZ	NWZ	NEZ	B	Sk.	Bl.
Synuchus vivalis (Ill.)	209	●	●	●	●	●	●	●	●	●	●	●	●	●	●	●
Olisthopus rotundatus (Payk.)	210	●	●	●	●	●	●	●	●	●	●	●	●	●	●	●
Agonum dorsale (Pont.)	211	●	●	●	●	●	●	●	●	●	●	●	●	●	●	●
A. obscurum (Hbst.)	212	●	●	●	●	●	●	●	●	●	●	●	●	●	●	●
A. albipes (F.)	213	●	●	●	●		●		●	●	●	●	●	●	●	●
A. livens (Gyll.)	214	●	●	●	●			●	●	●	●	●	●	●	●	●
A. mannerheimii (Dej.)	215															
A. assimile (Payk.)	216	●	●	●	●	●	●	●	●	●	●	●	●	●	●	●
A. krynickii (Sperk)	217	●								●	●		●		●	
A. longiventre (Mann.)	218	●														
A. marginatum (L.)	219	●	●	●	●	●	●	●	●	●	●	●	●	●	●	●
A. sexpunctatum (L.)	220	●	●	●	●	●	●	●	●	●	●	●	●	●	●	●
A. impressum (Pz.)	221	●														
A. ericeti (Pz.)	222	●	●	●	●	●		●							●	
A. gracilipes (Dft.)	223	●	●		●			●	●	●	●	●	●	●	●	●
A. muelleri Hbst.)	224	●	●	●	●	●	●	●	●	●	●	●	●	●	●	●
A. sahlbergii (Chaud.)	225		●													
A. dolens (Sahlbg.)	226	●				●			●		●			●	●	
A. versutum Sturm	227	●	●	●	●	●	●	●		●	●	●	●	●	●	●
A. viduum (Pz.)	228	●	●	●	●	●	●	●	●	●	●	●	●	●	●	●
A. moestum (Dft.)	229	●	●	●	●	●		●	●	●	●	●	●	●	●	●
A. lugens (Dft.)	230	●			●				●	●	●		●	●	●	
A. bogemannii (Gyll.)	231															
A. quadripunctatum (Deg.)	232	●	●		●	●		●		●			●	●	●	●
A. micans (Nic.)	233	●	●	●	●				●	●	●	●			●	●
A. piceum (L.)	234	●	●	●	●	●	●	●	●	●	●	●	●	●	●	●
A. gracile Sturm	235	●	●	●	●	●	●	●	●	●	●		●	●	●	●
A. fuliginosum (Pz.)	236	●	●	●	●	●	●	●	●	●	●	●	●	●	●	●
A. thoreyi Dej.	237	●	●	●	●	●	●	●	●	●	●	●	●	●	●	●
A. consimile (Gyll.)	238															
A. munsteri (Hell.)	239	●		●												
A. exaratum (Mann.)	240															
Amara strenua Zimm.	241	●	●	●	●	●								●		
A. tricuspidata Dej.	242	●												●	●	
A. plebeja (Gyll.)	243	●	●	●	●	●	●	●	●	●	●	●	●	●	●	●
A. similata (Gyll.)	244	●	●	●	●	●	●	●	●	●	●	●	●	●	●	●
A. ovata (F.)	245	●	●	●	●	●	●	●	●	●	●	●	●	●	●	●
A. montivaga Sturm	246	●	●													

	Hall.	Sm.	Öl.	Gtl.	G. Sand.	Ög.	Vg.	Boh.	Dlsl.	Nrk.	Sdm.	Upl.	Vstm.	Vrm.	Dlr.	Gstr.	Hls.	Med.	Hrj.	Jmt.	Ång.	Vb.	Nb.	Ås. Lpm.	Ly. Lpm.	P. Lpm.	Lu. Lpm.	T. Lpm.
209	●	●	●	●	●	●	●	●	●	●	●	●	●	●	●	●	●	●		●	●	●	●				●	
210	●	●	●	●	●	●	●	●	●	●	●	●	●	●	●							●						
211	●	●	●	●		●	●	●	●	●	●	●	●															
212	●	●	●	●	●	●	●	●	●	●	●	●	●	●	●	●	●		●									
213	●	●	●	●		●	●	●	●	●	●	●	●	●	●													
214	●	●	●	●	●	●	●			●	●	●	●	●		●	●	●										
215														●									●	●				
216	●	●	●			●	●	●	●	●	●	●	●	●	●	●	●	●		●	●	●						●
217		●	●	●																								
218												●			●	●												
219	●	●	●	●	●	●	●	●	●		●	●	●	●														
220	●	●	●	●		●	●	●	●	●	●	●	●	●	●	●	●	●		●	●	●	●	●	●	●	●	●
221																												
222	●	●				●	●	●	●	●	●	●	●	●	●	●	●		●	●	●	●	●		●		●	●
223	●		●	●	●	●	●					●					●											
224	●	●	●	●		●	●	●	●	●	●	●	●	●	●	●	●	●		●	●	●	●	●				●
225																												
226	●	●				●	●	●	●		●	●	●	●	●		●		●	●	●	●	●	●	●	●	●	●
227	●	●	●	●		●	●	●	●	●	●	●	●	●	●	●	●	●		●	●	●	●				●	
228	●	●	●	●		●	●	●	●	●	●	●	●	●	●	●	●	●	●	●	●	●	●		●	●	●	
229	●	●	●	●		●	●	●	●	●	●	●	●	●	●	●	●											
230			●	●							●	●	●															
231		●					●	●									●				L	A	P	P	O	N	I	A
232	●	●	●	●	●	●	●			●	●	●	●	●	●	●	●	●		●	●	●	●	●	●	●	●	
233		●				●	●	●			●	●	●	●	●	●	●			●	●	●	●			●		
234	●	●	●	●	●	●	●	●	●	●	●	●	●	●	●	●	●	●	●	●	●	●	●	●	●		●	●
235	●	●	●	●	●	●	●	●	●	●	●	●	●	●	●	●	●	●	●	●	●	●	●	●	●	●	●	●
236	●	●	●	●		●	●	●	●	●	●	●	●	●	●	●	●	●	●	●	●	●	●	●	●	●	●	●
237	●	●	●	●	●	●	●	●	●	●	●	●	●	●	●	●	●				●	●	●					
238																			●	●			●	●	●	●	●	●
239						●								●	●		●		●	●		●	●			●	●	
240																												
241																												
242																												
243	●	●	●	●		●	●	●	●	●	●	●	●	●	●	●	●	●		●	●	●	●	●	●		●	
244	●	●	●	●		●	●	●	●	●	●	●	●	●	●	●	●		●	●	●	●	●		●		●	
245	●	●	●	●	●	●	●	●		●	●	●	●	●	●	●	●			●	●	●	●	●	●		●	
246			●				●	●	●	●				●	●								●					

		NORWAY														
		Ø + AK	HE (s + n)	O (s + n)	B (ø + v)	VE	TE (y + i)	AA (y + i)	VA (y + i)	R (y + i)	HO (y + i)	SF (y + i)	MR (y + i)	ST (y + i)	NT (y + i)	Ns (y + i)
Synuchus vivalis (Ill.)	209	●	●	●	●	●	●	●	●	●	●	●	●	●	●	
Olisthopus rotundatus (Payk.)	210	●						●	●	●	●	●	●	●		
Agonum dorsale (Pont.)	211	●			●	●	●	●	●							
A. obscurum (Hbst.)	212	●			●	●	●	●	●							
A. albipes (F.)	213	●				●	●	●	●	●	●			●		
A. livens (Gyll.)	214															
A. mannerheimii (Dej.)	215			●												
A. assimile (Payk.)	216	●	●	●	●	●	●	●	●	●		●		●	●	●
A. krynickii (Sperk)	217															
A. longiventre (Mann.)	218															
A. marginatum (L.)	219	●			●	●				●						
A. sexpunctatum (L.)	220	●	●	●	●	●	●	●	●	●		●			●	
A. impressum (Pz.)	221															
A. ericeti (Pz.)	222	●	●	●	●				●	●	●	●		●		
A. gracilipes (Dft.)	223			●			●									
A. muelleri Hbst.)	224	●	●	●	●	●	●	●	●	●	●	●	●	●	●	
A. sahlbergii (Chaud.)	225															
A. dolens (Sahlbg.)	226	●	●	●	●	●	●	●						●	●	
A. versutum Sturm	227	●	●	●	●	●	●	●	●							
A. viduum (Pz.)	228	●	●	●	●	●	●	●	●	●			●	●	●	
A. moestum (Dft.)	229	●														
A. lugens (Dft.)	230															
A. bogemannii (Gyll.)	231															
A. quadripunctatum (Deg.)	232	●	●	●			●	●						●		●
A. micans (Nic.)	233	●	●	●	●	●	●						●	●	●	
A. piceum (L.)	234	●		●	●	●		●	●	●	●			●	●	
A. gracile Sturm	235	●	●	●	●	●	●	●	●	●	●		●	●		●
A. fuliginosum (Pz.)	236	●	●	●	●	●	●	●		●	●	●	●	●	●	●
A. thoreyi Dej.	237	●				●				●						●
A. consimile (Gyll.)	238			●	●		●	●			●	●				●
A. munsteri (Hell.)	239		●												●	
A. exaratum (Mann.)	240															
Amara strenua Zimm.	241															
A. tricuspidata Dej.	242															
A. plebeja (Gyll.)	243	●	●	●	●	●	●	●	●	●	●	●	●	●	●	●
A. similata (Gyll.)	244	●	●		●	●	●	●	●	●	●					
A. ovata (F.)	245	●	●				●		●	●	●					●
A. montivaga Sturm	246	●	●		●	●				●						

					FINLAND																				USSR		
	Nn (ø + v)	TR (y + i)	F (v + i)	F (n + ø)	Al	Ab	N	Ka	St	Ta	Sa	Oa	Tb	Sb	Kb	Om	Ok	Ob S	Ob N	Ks	LkW	LkE	Le	Li	Vib	Kr	Lr
209					●	●	●	●	●	●	●	●	●	●	●	●	●	●	●	●					●	●	
210					●	●	●			●	●														●	●	
211						●	●																				
212					●	●	●	●	●	●	●		●	●	●										●	●	
213						●	●	●																	●		
214					●	●	●	●	●	●	●		●		●										●	●	
215					●	●	●		●	●	●		●		●	●	●	●	●		●				●	●	●
216					●	●	●	●	●	●	●	●	●												●	●	
217																											
218																											
219					●	●	●	●																	●	●	
220				●	●	●	●	●	●	●	●	●	●	●	●	●	●	●	●	●	●	●		●	●	●	
221																										●	
222				●		●	●	●	●	●	●	●	●	●	●	●	●	●	●	●	●			●	●	●	●
223					●		●																				
224					●	●	●	●	●	●	●	●	●	●	●	●				●					●	●	
225																										●	●
226					●	●	●		●	●	●	●	●	●	●	●	●	●	●	●	●	●	●	●	●	●	●
227					●	●	●	●	●	●	●		●	●	●	●	●	●	●						●	●	
228					●	●	●	●	●	●	●	●	●	●	●	●	●	●	●	●	●				●	●	
229					●	●	●		●	●															●	●	
230					●																						
231						●				●											●	●				●	
232					●	●	●	●	●	●	●	●	●	●	●	●	●	●	●	●	●	●		●	●	●	●
233						●	●	●	●		●	●													●		
234					●	●	●	●	●	●	●	●	●	●	●	●	●	●	●	●	●	●	●		●	●	●
235	●	●		●	●	●	●	●	●	●	●	●	●	●	●	●	●	●	●	●	●	●		●	●	●	●
236	●	●	●	●	●	●	●	●	●	●	●	●	●	●	●	●	●	●	●	●	●	●	●	●	●	●	●
237				●	●	●	●	●	●	●	●	●	●	●	●	●		●			●				●	●	
238	●	●	●	●																●	●		●	●		●	●
239						●	●			●	●			●	●		●					●		●	●	●	●
240																											●
241																											
242																											
243					●	●	●	●	●	●	●	●	●	●	●	●	●	●				●			●	●	
244					●	●	●	●	●	●	●	●	●	●	●	●		●							●	●	
245					●	●	●	●	●	●	●	●	●	●	●	●	●		●	●		●		●	●	●	●
246						●	●	●		●	●	●	●	●	●										●	●	

				DENMARK											
		Germany	G. Britain	SJ	EJ	WJ	NWJ	NEJ	F	LFM	SZ	NWZ	NEZ	B	Sk.	Bl.
Amara nitida Sturm	247	●	●	●	●			●		●			●		●	●
A. communis (Pz.)	248	●	●	●	●	●	●	●	●	●	●	●	●	●	●	●
A. convexior Stph.	249	●	●	●	●	●	●	●	●	●	●	●	●		●	●
A. nigricornis Thoms.	250															
A. lunicollis Schiødte	251	●	●	●	●	●	●	●	●	●	●	●	●	●	●	●
A. curta Dej.	252	●	●							●			●		●	●
A. littorea Thoms.	253	●													●	
A. aenea (Deg.)	254	●	●	●	●	●	●	●	●	●	●	●	●	●	●	●
A. spreta Dej.	255	●	●	●	●	●	●	●	●	●	●	●	●	●	●	●
A. famelica Zimm.	256	●	●	●	●	●	●	●					●		●	
A. eurynota (Pz.)	257	●	●	●	●	●	●	●	●	●	●	●	●		●	●
A. familiaris (Dft.)	258	●	●	●	●	●	●	●	●	●	●	●	●	●	●	●
A. anthobia Villa & Villa	259	●	●		●	●					●		●	●	●	
A. lucida (Dft.)	260	●	●	●	●	●	●	●	●	●	●	●	●	●	●	●
A. tibialis (Payk.)	261	●	●	●	●	●	●	●	●	●	●	●	●	●	●	●
A. erratica (Dft.)	262															
A. interstitialis Dej.	263															
A. ingenua (Dft.)	264	●		●	●	●	●	●	●	●	●		●	●	●	●
A. fusca Dej.	265	●	●		●		●	●	●	●	●	●	●	●	●	
A. cursitans (Zimm.)	266	●	●	●	●	●	●	●	●	●	●	●	●	●	●	●
A. municipalis (Dft.)	267	●		●	●	●	●	●		●	●	●	●	●	●	
A. quenseli (Schönh.)	268	●	●	●	●	●		●	●	●	●	●	●	●	●	●
A. bifrons (Gyll.)	269	●	●	●	●	●	●	●	●	●	●	●	●	●	●	●
A. infima (Dft.)	270	●	●	●	●	●		●	●			●	●	●	●	●
A. praetermissa (Sahlbg.)	271	●	●	●	●	●	●	●		●			●	●	●	●
A. brunnea (Gyll.)	272	●		●	●	●	●	●	●	●	●	●	●	●	●	●
A. crenata Dej.	273	●								?				●	●	
A. apricaria (Payk.)	274	●	●	●	●	●	●	●	●	●	●	●	●	●	●	●
A. fulva (Müll.)	275	●	●	●	●	●	●	●	●	●	●	●	●	●	●	●
A. majuscula (Chaud.)	276	●			●	●		●	●	●	●		●	●	●	●
A. consularis (Dft.)	277	●	●	●	●	●	●	●	●	●	●	●	●	●	●	●
A. aulica (Pz.)	278	●	●	●	●	●	●	●	●	●	●	●	●	●	●	●
A. convexiuscula (Marsh.)	279	●	●	●	●	●	●	●	●	●	●	●	●	●	●	●
A. torrida (Pz.)	280															
A. alpina (Payk.)	281		●													
A. hyperborea Dej.	282															
A. equestris (Dft.)	283		●	●	●	●	●	●	●	●	●	●	●	●	●	●
Zabrus tenebrioides (Gz.)	284		●	●	●	●			●	●					●	

SWEDEN

	Hall.	Sm.	Öl.	Gtl.	G. Sand.	Ög.	Vg.	Boh.	Dlsl.	Nrk.	Sdm.	Upl.	Vstm.	Vrm.	Dlr.	Gstr.	Hls.	Med.	Hrj.	Jmt.	Ång.	Vb.	Nb.	Ås. Lpm.	Ly. Lpm.	P. Lpm.	Lu. Lpm.	T. Lpm.
247	●	●	●	●		●	●	●	●	●	●	●	●		●	●	●	●		●			●					
248	●	●	●	●	●	●	●	●	●	●	●	●	●	●	●	●	●	●	●	●	●	●	●	●	●		●	●
249	●	●	●	●		●	●	●			●	●	●		●													
250															●		●		●	●		●	●	●	●	●	●	●
251	●	●	●	●	●	●	●	●	●	●	●	●	●	●	●	●	●		●	●	●	●	●	●	●		●	●
252	●	●	●	●		●	●	●	●	●	●	●	●	●	●	●	●											
253	●	●	●	●		●	●			●	●	●					●			●			●					●
254	●	●	●	●	●	●	●	●	●	●	●	●	●	●	●		●											
255	●						●																					
256	●	●	●	●		●	●	●	●	●	●	●	●	●	●	●	●				●	●	●					
257	●	●	●	●		●	●	●	●	●	●	●	●	●	●	●	●	●		●	●	●	●		●		●	●
258	●	●	●	●	●	●	●	●	●	●	●	●	●	●	●	●	●	●		●	●	●	●	●	●	●	●	●
259			●																									
260	●	●	●	●	●	●	●	●				●																
261	●	●	●	●	●	●	●	●		●	●	●	●	●	●		●	●		●	●	●	●		●		●	
262																						●	●		●	●	●	●
263															●		●			●	●	●	●	●	●	●	●	●
264	●	●	●	●		●	●	●	●	●	●	●	●	●	●	●	●	●		●	●	●	●	●	●		●	
265																												
266	●	●	●	●	●		●	●	●	●	●	●	●	●	●													
267	●	●	●	●		●	●			●	●	●	●	●	●		●	●		●	●	●	●	●	●	●	●	●
268	●		●	●	●		●			●				●	●	●	●	●	●	●	●	●	●	●	●	●	●	●
269	●	●	●	●	●	●	●	●	●	●	●	●	●	●	●	●	●	●		●	●	●	●	●	●		●	●
270	●	●	●				●			●		●		●	●		●											
271	●	●	●	●	●	●	●	●	●	●	●	●	●	●	●		●	●	●	●	●	●	●	●	●	●	●	●
272	●	●	●	●	●	●	●	●	●	●	●	●	●	●	●	●	●	●	●	●	●	●	●	●	●	●	●	●
273				●																								
274	●	●	●	●	●	●	●	●	●	●	●	●	●	●	●	●	●	●	●	●	●	●	●	●	●	●	●	●
275	●	●	●	●	●	●	●	●	●	●	●	●	●	●	●	●	●	●	●	●	●	●	●	●	●		●	●
276	●	●	●	●	●	●	●	●		●	●	●		●			●			●							●	
277	●	●	●	●	●	●	●	●	●	●	●	●	●	●	●	●	●			●	●	●	●					
278	●	●	●	●	●	●	●	●	●	●	●	●	●	●	●	●	●	●	●	●	●	●	●	●	●	●	●	
279	●	●	●	●	●	●	●	●		●		●																
280																							●		●	●	●	●
281															●				●	●				●	●	●	●	●
282																										●		
283	●	●	●	●		●	●	●	●	●	●	●	●	●	●		●					●						
284																												

		NORWAY														
		Ø + AK	HE (s + n)	O (s + n)	B (ø + v)	VE	TE (y + i)	AA (y + i)	VA (y + i)	R (y + i)	HO (y + i)	SF (y + i)	MR (y + i)	ST (y + i)	NT (y + i)	Ns (y + i)
Amara nitida Sturm	247	●	●	●	●		●	●	●	●	●	●	●	●		
A. communis (Pz.)	248	●	●	●	●	●	●	●	●	●	●	●	●	●	●	●
A. convexior Stph.	249	●	●	●	●	●	●	●	●		●			●	●	●
A. nigricornis Thoms.	250		●	●					●					●		●
A. lunicollis Schiødte	251	●	●	●	●	●	●	●	●	●	●	●	●	●	●	●
A. curta Dej.	252	●		●	●	●	●	●				●	●		●	
A. littorea Thoms.	253	●	●	●	●											
A. aenea (Deg.)	254	●	●	●	●	●	●	●	●	●						
A. spreta Dej.	255	●							●	●						
A. famelica Zimm.	256	●	●		●		●									
A. eurynota (Pz.)	257	●	●	●	●	●	●	●	●	●		●	●	●	●	●
A. familiaris (Dft.)	258	●	●	●	●	●	●	●	●	●	●	●	●	●	●	●
A. anthobia Villa & Villa	259															
A. lucida (Dft.)	260								●							
A. tibialis (Payk.)	261	●	●	●	●	●	●	●	●					●	●	
A. erratica (Dft.)	262															●
A. interstitialis Dej.	263		●	●			●							●	●	●
A. ingenua (Dft.)	264	●	●	●	●	●	●	●	●	●		●	●	●	●	●
A. fusca Dej.	265															
A. cursitans (Zimm.)	266	●					●		●							
A. municipalis (Dft.)	267	●	●	●		●	●		●	●				●	●	
A. quenseli (Schönh.)	268	●	●	●	●		●	●	●	●	●	●		●	●	●
A. bifrons (Gyll.)	269	●	●	●	●	●	●	●	●	●	●	●	●	●	●	●
A. infima (Dft.)	270		●	●	●			●	●	●						
A. praetermissa (Sahlbg.)	271	●	●	●	●		●	●	●	●	●	●		●		●
A. brunnea (Gyll.)	272	●	●	●	●		●	●	●	●	●	●	●	●		●
A. crenata Dej.	273															
A. apricaria (Payk.)	274	●	●	●	●	●	●	●	●	●	●	●	●	●	●	●
A. fulva (Müll.)	275	●	●	●	●	●	●	●	●	●				●		
A. majuscula (Chaud.)	276		●													
A. consularis (Dft.)	277	●	●	●	●	●	●	●	●	●	●	●	●	●	●	
A. aulica (Pz.)	278	●	●	●	●	●	●	●	●	●	●	●	●	●	●	●
A. convexiuscula (Marsh.)	279	●			●	●	●									
A. torrida (Pz.)	280													●		●
A. alpina (Payk.)	281		●	●	●		●				●	●		●		●
A. hyperborea Dej.	282															
A. equestris (Dft.)	283	●		●	●		●									
Zabrus tenebrioides (Gz.)	284															

					FINLAND																				USSR		
	Nn (ø + v)	TR (y + i)	F (v + i)	F (n + ø)	Al	Ab	N	Ka	St	Ta	Sa	Oa	Tb	Sb	Kb	Om	Ok	Ob S	Ob N	Ks	LkW	LkE	Le	Li	Vib	Kr	Lr
247					●	●	●	●	●	●	●	●	●	●	●	●	●	●	●						●	●	
248	●	●			●	●	●	●	●	●	●	●	●	●	●	●	●	●	●	●	●	●		●	●	●	●
249						●	●																		●	●	
250		●		●		●											●		●	●	●	●		●		●	●
251	●	●	●	●	●	●	●	●	●	●	●	●	●	●	●	●	●	●	●	●	●	●		●	●	●	●
252					●	●	●	●	●	●	●														●	●	
253						●	●		●	●		●	●		●						●				●	●	
254					●	●	●	●	●	●	●	●	●	●	●		●								●	●	
255						●	●																		●	●	
256					●	●	●	●	●	●	●	●	●	●	●	●									●	●	
257					●	●	●	●	●	●	●	●	●	●	●	●	●	●	●	●	●	●			●	●	●
258	●	●			●	●	●	●	●	●	●	●	●	●	●	●	●	●	●	●	●				●	●	●
259																											
260						●																			●		
261					●	●	●	●	●	●	●	●	●	●	●	●	●	●							●	●	
262	●	●	●	●													●		●	●	●	●	●	●		●	●
263	●	●	●	●		●	●				●	●	●	●	●	●	●	●	●	●	●	●		●	●	●	●
264					●	●	●	●	●	●	●	●	●	●	●	●	●	●		●					●	●	
265																											
266						●	●			●	●															●	
267					●	●	●	●		●	●	●				●		●							●	●	
268	●	●	●	●		●	●	●	●	●	●	●			●	●	●	●	●	●	●	●	●	●	●	●	●
269	●	●			●	●	●	●	●	●	●	●	●	●	●	●	●	●	●						●	●	
270							●			●		●													●	●	
271	●	●	●	●	●	●	●				●			●	●		●	●	●	●	●	●	●	●	●	●	●
272	●	●	●	●	●	●	●	●	●	●	●	●	●	●	●	●	●	●	●	●	●		●	●	●	●	●
273					●																						
274	●	●	●	●	●	●	●	●	●	●	●	●	●	●	●	●	●	●	●	●	●	●	●	●	●	●	●
275					●	●	●	●	●	●	●	●	●	●	●	●	●	●	●	●	●	●		●	●	●	●
276					●	●	●	●		●	●			●											●	●	
277					●	●	●	●	●	●	●	●	●	●	●										●	●	
278	●	●			●	●	●	●	●	●	●	●	●	●	●	●	●	●	●	●	●	●	●	●	●	●	●
279						●	●																		●		
280	●	●	●	●														●	●	●	●	●	●	●		●	●
281	●	●	●	●																●		●	●	●			●
282																								●			●
283					●	●	●	●	●	●	●		●	●	●										●	●	
284																											

				DENMARK												
		Germany	G. Britain	SJ	EJ	WJ	NWJ	NEJ	F	LFM	SZ	NWZ	NEZ	B	Sk.	Bl.
Harpalus rupicola Sturm	285	●	●							●	●	●		●		
H. nitidulus (Stph.)	286	●		●	●				●	●	●	●	●	●	●	
H. puncticollis (Payk.)	287	●	●	●				●							●	
H. melletii Heer	288	●	●	●	●				●	●	●		●	●	●	
H. rufibarbis (F.)	289	●	●	●	●	●	●	●	●	●	●	●	●	●	●	●
H. schaubergerianus (Puel)	290	●	●													
H. puncticeps (Stph.)	291	●	●	●	●		●	●	●	●	●	●	●	●	●	●
H. azureus (F.)	292	●	●	●					●	●				●		
H. signaticornis (Dft.)	293	●								●				●	●	
H. griseus (Pz.)	294	●			●				●	●	●	●	●	●	●	●
H. rufipes (Deg.)	295	●		●	●	●	●	●	●	●	●	●	●	●	●	●
H. calceatus (Dft.)	296	●	●		●				●	●	●		●	●	●	●
H. affinis (Schrk.)	297	●	●	●	●	●	●	●	●	●	●	●	●	●	●	●
H. distinguendus (Dft.)	298	●								●				●	●	●
H. smaragdinus (Dft.)	299	●	●	●	●	●	●	●	●	●	●	●	●	●	●	●
H. serripes (Quensel)	300	●	●	●	●	●		●	●	●	●	●	●	●	●	●
H. melancholicus Dej.	301	●	●		●					●			●	●	●	
H. autumnalis (Dft.)	302	●													●	
H. solitaris Dej.	303	●		●	●	●	●	●		●	●	●	●	●	●	●
H. latus (L.)	304	●	●	●	●	●	●	●	●	●	●	●	●	●	●	●
H. nigritarsis Sahlbg.	305															
H. luteicornis (Dft.)	306	●													●	●
H. xanthopus Gemm. & Har.	307	●			●				●	●	●		●	●	●	●
H. quadripunctatus Dej.	308	●	●	●	●	●	●	●	●	●	●	●	●	●	●	●
H. rubripes (Dft.)	309	●	●	●	●	●	●	●	●	●	●	●	●	●	●	●
H. rufipalpis Sturm	310	●		●	●	●			●	●	●		●	●	●	
H. neglectus Aud.-Serv.	311	●	●	●	●	●	●	●	●	●	●	●	●	●	●	●
H. servus (Dft.)	312	●	●			●			●	●	●	●	●	●	●	
H. tardus (Pz.)	313	●	●	●	●	●	●	●	●	●	●	●	●	●	●	●
H. anxius (Dft.)	314	●	●	●	●	●	●	●	●	●	●	●	●	●	●	●
H. picipennis (Dft.)	315	●									●				●	
H. pumilus (Sturm)	316	●		●	●				●	●	●	●	●	●	●	
H. froelichii Sturm	317	●	●	●	●			●	●	●	●	●	●	●	●	
H. hirtipes (Pz.)	318	●		●	●			●	●	●	●	●	●	●	●	●
H. flavescens (Pill. & Mitt.)	319	●												●	●	
Diachromus germanus (L.)	320	●	●	●						●	●					
Anisodactylus poeciloides (Stph.)	321	●	●							●	●		●		●	
A. binotatus (F.)	322	●	●	●	●	●	●	●	●	●	●	●	●	●	●	●

SWEDEN

	Hall.	Sm.	Öl.	Gtl.	G. Sand.	Ög.	Vg.	Boh.	Dlsl.	Nrk.	Sdm.	Upl.	Vstm.	Vrm.	Dlr.	Gstr.	Hls.	Med.	Hrj.	Jmt.	Ång.	Vb.	Nb.	Ås. Lpm.	Ly. Lpm.	P. Lpm.	Lu. Lpm.	T. Lpm.
285			●	●							●	●																
286			●	●	●																							
287			●			●	●			●	●	●	●															
288		●	●	●																								
289	●	●	●	●	●	●	●	●	●	●	●	●	●	●	●	●		●										
290																												
291	●		●	●																								
292		●	●	●																								
293																												
294	●	●	●	●	●		●																					
295	●	●	●	●	●	●	●	●	●	●	●	●	●	●	●	●	●	●		●	●	●						
296	●	●	●	●	●		●	●				●																
297	●	●	●	●	●	●	●	●	●	●	●	●	●	●	●	●	●	●	●	●	●	●	●	●	●	●	●	
298		●				●	●				●	●	●	●	●		●											
299	●	●	●	●		●	●	●	●	●	●	●	●	●														
300		●	●	●								●																
301			●	●																								
302																												
303	●	●	●			●	●		●	●	●	●		●	●		●	●		●		●	●	●	●	●	●	●
304	●	●	●	●		●	●	●	●	●	●	●	●	●	●	●	●	●	●	●	●	●	●	●	●	●	●	●
305																							●			●	●	
306	●			●			●	●	●				●															
307		●	●			●	●	●		●	●	●	●	●	●		●	●	●	●	●	●	●		●	●	●	●
308	●	●	●	●	●	●	●	●	●	●	●	●	●	●	●	●	●	●	●	●	●	●	●	●	●	●	●	●
309	●	●	●	●	●	●	●	●	●	●	●	●	●	●							●							
310	●		●	●		●																						
311	●		●	●	●		●																					
312	●		●	●																								
313	●	●	●	●	●	●	●	●	●	●	●	●	●	●	●	●	●	●										
314	●		●	●		●				●		●																
315			●																									
316			●	●	●	●																						
317	●		●																									
318	●		●	●																								
319			●																									
320																												
321																												
322	●	●	●	●	●	●	●	●	●	●	●	●	●	●	●	●	●				●							

		NORWAY														
		Ø + AK	HE (s + n)	O (s + n)	B (ø + v)	VE	TE (y + i)	AA (y + i)	VA (y + i)	R (y + i)	HO (y + i)	SF (y + i)	MR (y + i)	ST (y + i)	NT (y + i)	Ns (y + i)
Harpalus rupicola Sturm	285															
H. nitidulus (Stph.)	286															
H. puncticollis (Payk.)	287	●	●	●	●	●	●									
H. melletii Heer	288															
H. rufibarbis (F.)	289	●	●	●	●	●	●	●		●	●					
H. schaubergerianus (Puel)	290															
H. puncticeps (Stph.)	291															
H. azureus (F.)	292															
H. signaticornis (Dft.)	293															
H. griseus (Pz.)	294	●			●											
H. rufipes (Deg.)	295	●	●	●	●	●	●	●	●	●	●	●	●	●	●	
H. calceatus (Dft.)	296	●														
H. affinis (Schrk.)	297	●	●	●	●	●	●	●	●	●	●	●	●	●	●	●
H. distinguendus (Dft.)	298	●	●		●											
H. smaragdinus (Dft.)	299	●		●	●	●	●		●							
H. serripes (Quensel)	300															
H. melancholicus Dej.	301															
H. autumnalis (Dft.)	302															
H. solitaris Dej.	303	●	●	●	●			●	●	●		●	●			
H. latus (L.)	304	●	●	●	●	●	●	●	●	●	●	●	●	●	●	●
H. nigritarsis Sahlbg.	305															
H. luteicornis (Dft.)	306	●			●											
H. xanthopus Gemm. & Har.	307	●		●	●						●	●		●		●
H. quadripunctatus Dej.	308	●	●	●	●	●	●	●	●	●	●	●	●	●	●	●
H. rubripes (Dft.)	309	●	●	●			●	●	●			●				●
H. rufipalpis Sturm	310															
H. neglectus Aud.-Serv.	311															
H. servus (Dft.)	312															
H. tardus (Pz.)	313	●	●	●	●	●	●	●	●	●		●				
H. anxius (Dft.)	314															
H. picipennis (Dft.)	315															
H. pumilus (Sturm)	316															
H. froelichii Sturm	317															
H. hirtipes (Pz.)	318															
H. flavescens (Pill. & Mitt.)	319															
Diachromus germanus (L.)	320															
Anisodactylus poeciloides (Stph.)	321															
A. binotatus (F.)	322	●	●		●	●	●	●	●	●				●		

					FINLAND																				USSR		
	Nn (ø + v)	TR (y + i)	F (v + i)	F (n + ø)	Al	Ab	N	Ka	St	Ta	Sa	Oa	Tb	Sb	Kb	Om	Ok	Ob S	Ob N	Ks	LkW	LkE	Le	Li	Vib	Kr	Lr
285																											
286						●	●	●																	●		
287					●	●	●	●		●	●														●	●	
288																											
289					●	●	●	●		●	●														●	●	
290							●																				
291																									●		
292																											
293																											
294						●	●																		●	●	
295					●	●	●	●	●	●	●	●	●	●	●	●	●			●					●	●	
296						●	●																		●	●	
297					●	●	●	●	●	●	●	●	●	●	●	●	●	●	●	●	●	●			●	●	●
298						●					●														●		
299					●	●	●	●		●	●			●	●										●	●	
300																											
301																											
302																											
303			●	●		●	●		●	●			●				●		●	●	●	●		●	●	●	●
304		●			●	●	●	●	●	●	●	●	●	●	●	●	●	●	●	●	●	●	●	●	●	●	●
305																					●						
306					●	●	●	●	●	●	●														●	●	
307		●			●	●	●	●			●		●	●	●										●	●	●
308	●	●	●	●	●	●	●	●	●	●	●	●	●	●	●	●	●	●	●	●	●	●	●	●	●	●	●
309					●	●					●															●	
310																											
311																											
312																											
313					●	●	●	●	●	●	●	●	●	●	●	●	●			●					●	●	
314											●				●										●		
315																											
316																											
317																											
318																											
319						●																				●	
320																											
321																											
322					●	●	●	●	●	●	●	●	●	●	●	●									●	●	

				DENMARK												
		Germany	G. Britain	SJ	EJ	WJ	NWJ	NEJ	F	LFM	SZ	NWZ	NEZ	B	Sk.	Bl.
Anisodactylus nemorivagus (Dft.)	323	●	●										●			
A. signatus (Pz.)	324	●						●		●						
Dicheirotrichus gustavii Crotch	325	●	●	●	●	●	●	●	●	●	●	●	●		●	
D. rufithorax (Sahlbg.)	326	●													●	
Trichocellus mannerh. (F.Sahlbg.)	327															
T. cognatus (Gyll.)	328	●	●	●	●	●	●	●		●		●	●	●	●	●
T. placidus (Gyll.)	329	●	●	●	●	●	●	●	●	●	●	●	●	●	●	●
Bradycellus ruficollis (Stph.)	330	●	●	●	●	●	●	●	●	●	●	●	●	●	●	●
B. ponderosus Lindr.	331															
B. verbasci (Dft.)	332	●	●	●	●	●	●	●	●	●	●	●	●	●	●	●
B. harpalinus (Serv.)	333	●	●	●	●	●	●	●	●	●	●	●	●	●	●	●
B. csikii Laczó	334	●	●		●								●		●	●
B. caucasicus (Chaud.)	335	●		●	●	●	●	●	●	●	●	●	●	●	●	●
Stenolophus teutonus (Schrank)	336	●	●	●						●	●			●	●	
S. skrimshiranus Stph.	337	●	●	●					●	●	●	●	●	●	●	
S. mixtus (Hbst.)	338	●	●	●	●	●		●	●	●	●	●	●	●	●	●
Acupalpus flavicollis (Sturm)	339	●	●	●	●	●			●	●		●	●	●	●	●
A. brunnipes (Sturm)	340	●	●												●	
A. meridianus (L.)	341	●	●	●	●	●	●	●	●	●	●	●	●	●	●	●
A. parvulus (Sturm)	342	●	●	●	●	●	●	●	●	●	●	●	●	●	●	●
A. dubius Schilsky	343	●	●		●	●	●	●	●				●		●	●
A. exiguus Dej.	344	●	●	●	●		●	●	●	●	●		●	●	●	●
A. consputus (Dft.)	345	●	●	●	●	●		●	●	●	●	●	●	●	●	●
A. elegans (Dej.)	346	●	●										●			
Perigona nigriceps (Dej.)	347	●	●	●	●	●							●		●	●
Badister unipustulatus Bon.	348	●	●	●					●	●	●		●	●	●	
B. bullatus (Schrank)	349	●	●	●	●	●	●	●	●	●	●	●	●	●	●	●
B. meridionalis Puel	350	●	●											●	●	
B. lacertosus Sturm	351	●		●	●			●	●	●	●	●	●		●	●
B. sodalis (Dft.)	352	●	●	●	●			●	●	●	●	●	●		●	●
B. dorsiger (Dft.)	353	●		●					●	●	●					
B. dilatatus Chaud.	354	●	●	●	●	●	●	●	●	●	●		●	●	●	●
B. peltatus (Pz.)	355	●	●		●				●	●	●	●	●	●	●	●
B. anomalus (Perr.)	356	●	●						●	●	●		●	●	●	
Licinus depressus (Payk.)	357	●	●							●			●	●	●	●
Chlaenius tristis (Schall.)	358	●	●	●						●	●	●	●	●	●	●
C. nigricornis (F.)	359	●	●	●	●	●	●	●	●	●	●	●	●	●	●	●
C. nitidulus (Schrank)	360	●	●							●				●		

SWEDEN

	Hall.	Sm.	Öl.	Gtl.	G. Sand.	Ög.	Vg.	Boh.	Dlsl.	Nrk.	Sdm.	Upl.	Vstm.	Vrm.	Dlr.	Gstr.	Hls.	Med.	Hrj.	Jmt.	Ång.	Vb.	Nb.	Ås. Lpm.	Ly. Lpm.	P. Lpm.	Lu. Lpm.	T. Lpm.
323																												
324																												
325	●	●	●	●		●	●	●																				
326		●		●		●	●			●	●	●	●	●	●													
327																												
328	●	●	●	●	●	●	●		●	●		●	●	●	●		●		●	●	●	●	●	●	●	●	●	●
329	●	●	●	●	●	●	●	●	●	●	●	●	●	●	●	●	●	●		●	●	●	●	●		●	●	
330	●	●	●		●	●	●	●	●	●	●	●	●	●	●	●	●				●	●						
331																												
332	●	●	●					●																				
333	●	●	●	●			●	●																				
334			●	●			●																					
335	●	●	●	●	●	●	●	●	●	●	●	●	●	●	●	●	●	●	●	●	●	●	●		●	●	●	●
336																												
337																												
338		●	●	●	●		●			●	●	●																
339	●	●	●	●		●	●	●	●	●	●	●	●	●	●	●	●											
340																												
341	●	●	●	●		●	●	●		●	●	●	●	●														
342	●	●	●	●	●	●	●	●	●	●	●	●	●	●	●	●	●				●	●	●					
343			●																									
344	●	●	●	●		●	●		●		●			●														
345	●	●	●	●	●	●	●				●	●																
346																												
347	●	●	●	●			●				●	●		●		●						●	●		●			
348		●	●	●							●	●	●															
349	●	●	●	●	●	●	●	●	●	●	●	●	●	●	●													
350			●	●																								
351		●	●	●	●	●	●			●	●	●	●	●	●	●	●											
352			●	●		●	●			●		●																
353																												
354	●	●	●	●	●	●	●	●			●	●	●							●								
355	●	●	●	●	●	●	●			●	●	●	●	●	●	●												
356			●	●																								
357	●	●	●	●	●	●	●	●			●	●	●															
358		●	●	●	●	●	●				●	●	●	●				●										
359	●	●	●	●		●	●	●	●	●	●	●	●	●	●	●	●				●		●					
360				●																								

		NORWAY														
		Ø + AK	HE (s + n)	O (s + n)	B (ø + v)	VE	TE (y + i)	AA (y + i)	VA (y + i)	R (y + i)	HO (y + i)	SF (y + i)	MR (y + i)	ST (y + i)	NT (y + i)	Ns (y + i)
Anisodactylus nemorivagus (Dft.)	323															
A. signatus (Pz.)	324															
Dicheirotrichus gustavii Crotch	325	●				●	●	●	●	●			●	●	●	●
D. rufithorax (Sahlbg.)	326															
Trichocellus mannerh. (F.Sahlbg.)	327															
T. cognatus (Gyll.)	328	●	●	●	●	●	●		●	●	●			●	●	●
T. placidus (Gyll.)	329	●	●	●	●	●	●	●	●	●	●	●	●	●	●	●
Bradycellus ruficollis (Stph.)	330	●			●	●	●	●	●	●	●	●				
B. ponderosus Lindr.	331															
B. verbasci (Dft.)	332															
B. harpalinus (Serv.)	333								●	●						
B. csikii Laczó	334															
B. caucasicus (Chaud.)	335	●	●	●	●	●	●	●	●	●	●	●	●	●	●	●
Stenolophus teutonus (Schrank)	336															
S. skrimshiranus Stph.	337															
S. mixtus (Hbst.)	338	●				●		●								
Acupalpus flavicollis (Sturm)	339	●	●		●	●	●									
A. brunnipes (Sturm)	340															
A. meridianus (L.)	341	●	●		●	●										
A. parvulus (Sturm)	342	●	●		●	●	●	●	●							
A. dubius Schilsky	343															
A. exiguus Dej.	344															
A. consputus (Dft.)	345															
A. elegans (Dej.)	346															
Perigona nigriceps (Dej.)	347	●														
Badister unipustulatus Bon.	348															
B. bullatus (Schrank)	349	●	●	●	●	●	●	●	●	●				●		
B. meridionalis Puel	350															
B. lacertosus Sturm	351											●	●			
B. sodalis (Dft.)	352															
B. dorsiger (Dft.)	353															
B. dilatatus Chaud.	354	●														
B. peltatus (Pz.)	355															
B. anomalus (Perr.)	356															
Licinus depressus (Payk.)	357	●			●	●	●	●								
Chlaenius tristis (Schall.)	358	●			●		●									
C. nigricornis (F.)	359	●	●	●	●	●	●									
C. nitidulus (Schrank)	360															

	Nn (ø+v)	TR (y+i)	F (v+i)	F (n+ø)	Al	Ab	N	Ka	St	Ta	Sa	Oa	Tb	Sb	Kb	Om	Ok	Ob S	Ob N	Ks	LkW	LkE	Le	Li	Vib	Kr	Lr
					FINLAND																				USSR		
323								●			●			●											●	●	
324																											
325	●	●	●	●																						●	●
326						●	●	●	●	●	●		●	●	●										●	●	
327																											●
328	●	●	●	●		●	●		●	●	●	●	●			●	●	●	●	●	●	●	●	●		●	●
329	●	●	●		●	●	●	●	●	●	●	●	●	●	●	●	●	●	●	●	●				●	●	
330					●	●	●		●	●	●					●									●	●	
331																										●	
332																											
333																									●		
334																											
335	●	●	●		●	●	●	●	●	●	●	●	●	●	●	●	●	●	●	●	●	●	●	●	●	●	●
336																											
337																											
338						●	●			●	●														●	●	
339					●		●	●	●	●	●				●										●	●	
340																											
341						●	●	●																	●		
342					●	●	●	●	●	●	●	●	●	●	●	●			●		●				●	●	
343																											
344					●	●	●	●	●	●	●	●													●	●	
345						●	●																				
346																											
347							●			●	●			●					●	●							
348																									●		
349					●	●	●	●																	●	●	
350																											
351						●	●	●	●	●	●			●	●										●	●	
352																										●	
353																											
354					●	●	●			●	●														●	●	
355					●	●	●	●			●			●											●	●	
356																											
357					●	●																					
358					●	●	●	●	●	●	●														●	●	
359					●	●	●	●	●	●	●		●	●	●	●									●	●	
360							●																			●	

				DENMARK											
		Germany	G. Britain	SJ	EJ	WJ	NWJ	NEJ	F	LFM	SZ	NWZ	NEZ	B	Sk.	Bl.
Chlaenius vestitus (Payk.)	361	●	●	●	●				●	●	●		●	●	●	
C. sulcicollis (Payk.)	362	●						●		●	●		●	●	●	
C. quadrisulcatus (Payk.)	363	●								●			●	●	●	
C. costulatus (Motsch.)	364	●														
Oodes helopioides (F.)	365	●	●	●	●	●			●	●	●	●	●	●	●	●
O. gracilis Villa & Villa	366															
Panagaeus cruxmajor (L.)	367	●	●	●	●	●	●		●	●	●	●	●	●	●	●
P. bipustulatus (F.)	368	●	●						●	●	●		●	●	●	●
Odacantha melanura (L.)	369	●	●	●	●	●	●		●	●	●	●	●	●	●	●
Masoreus wetterhallii (Gyll.)	370	●	●	●	●	●	●	●	●	●	●	●	●	●	●	●
Lebia cyanocephala (L.)	371	●	●											●	●	●
L. chlorocephala (Hoffm.)	372	●	●	●	●	●			●	●	●	●	●	●	●	●
L. cruxminor (L.)	373	●	●					●	●	●	●	●	●	●	●	●
Demetrias monostigma Sam.	374	●	●	●		●	●	●	●	●	●	●	●	●	●	●
D. atricapillus (L.)	375	●	●	●	●	●			●	●	●	●	●	●	●	
D. imperialis (Germ.)	376	●	●	●					●	●	●		●	●	●	●
Dromius longiceps Dej.	377	●	●		●			●	●	●	●		●	●	●	●
D. linearis (Oliv.)	378	●	●	●	●	●	●	●	●	●	●	●	●	●	●	●
D. agilis (F.)	379	●	●	●	●	●	●		●	●	●	●	●	●	●	●
D. angustus Brullé	380	●	●		●			●			●	●	●	●	●	●
D. quadraticollis Mor.	381	●													●	
D. meridionalis Dej.	382	●	●	●	●				●	●						
D. schneideri Crotch	383	●											●	●	●	●
D. fenestratus (F.)	384	●			●	●	●	●				●	●	●	●	●
D. quadrimaculatus (L.)	385	●	●	●	●	●	●	●	●	●	●	●	●	●	●	●
D. spilotus (Ill.)	386	●		●	●	●	●	●	●	●	●	●	●	●	●	●
D. quadrisignatus Dej.	387	●	●							●					●	
D. melanocephalus Dej.	388	●	●	●	●	●	●	●	●	●	●	●	●	●	●	
D. sigma (Rossi)	389	●	●	●	●	●		●	●	●	●	●	●	●	●	●
D. notatus Stph.	390	●	●	●	●	●	●	●	●	●	●	●	●	●	●	●
Syntomus truncatellus (L.)	391	●	●	●	●	●		●	●	●	●	●	●	●	●	●
S. foveatus (Fourcr.)	392	●	●	●	●	●	●	●	●	●	●	●	●	●	●	●
Lionychus quadrillum (Dft.)	393	●	●													
Microlestes minutulus (Gz.)	394	●		●	●	●			●	●	●		●	●	●	●
M. maurus (Sturm)	395	●	●	●	●		●		●	●	●	●	●	●	●	●
Cymindis humeralis (Fourcr.)	396	●									●	●			●	●
C. angularis Gyll.	397				●	●	●	●						●	●	●
C. macularis Fisch.-W.	398	●			●	●	●	●	●			●	●	●	●	●

SWEDEN

	Hall.	Sm.	Öl.	Gtl.	G. Sand.	Ög.	Vg.	Boh.	Dlsl.	Nrk.	Sdm.	Upl.	Vstm.	Vrm.	Dlr.	Gstr.	Hls.	Med.	Hrj.	Jmt.	Ång.	Vb.	Nb.	Ås. Lpm.	Ly. Lpm.	P. Lpm.	Lu. Lpm.	T. Lpm.
361																												
362			●																									
363		●	●			●	●					●																
364																							●					
365	●	●	●	●		●	●	●	●	●	●	●	●	●	●	●		●										
366											●	●	●															
367	●	●	●	●		●	●	●		●	●	●	●	●	●	●	●						●					
368		●	●	●		●				●																		
369		●	●	●		●	●			●	●	●	●															
370	●	●	●	●	●		●	●																				
371		●	●	●		●	●	●	●	●	●	●	●	●														
372	●	●	●	●		●	●	●	●	●	●	●	●	●	●	●	●	●			●	●						
373	●	●	●	●		●	●	●	●	●	●	●	●	●	●	●	●	●		●	●	●	●		●		●	
374			●	●										●														
375																												
376			●	●		●				●	●	●	●	●														
377	●	●				●	●			●	●	●	●	●		●							●					
378	●	●	●	●	●	●	●	●			●	●																
379	●	●	●	●		●	●	●	●	●	●	●	●	●	●	●	●	●		●	●	●	●	●	●	●	●	●
380	●	●	●	●	●		●	●																				
381																												
382																												
383	●	●	●	●		●	●	●	●	●	●	●	●	●	●	●	●			●	●	●	●					
384	●	●	●	●	●	●	●	●	●	●	●	●	●	●	●	●	●	●		●	●	●	i	●				
385	●	●	●	●		●	●	●	●	●	●	●	●	●	●						●		i					
386	●	●	●	●	●	●	●	●	●	●	●	●	●	●	●	●	●											
387																												
388	●		●																									
389	●	●		●		●	●	●	●	●	●	●	●	●	●	●	●	●		●	●	●	●		●			●
390	●	●	●	●	●	●	●	●	●	●	●	●	●	●	●	●	●											
391	●	●	●	●	●	●	●	●	●	●	●	●	●	●	●	●	●	●		●	●	●	●		●		●	
392	●	●	●			●	●	●		●	●	●		●														
393							●			●		●	●															
394	●	●	●	●	●	●	●			●	●	●		●														
395	●	●	●	●		●	●																					
396		●	●	●																								
397	●	●	●	●	●	●	●	●		●	●	●	●	●														
398	●		●	●	●	●	●					●									●	●	●		●			●

		NORWAY														
		Ø + AK	HE (s + n)	O (s + n)	B (ø + v)	VE	TE (y + i)	AA (y + i)	VA (y + i)	R (y + i)	HO (y + i)	SF (y + i)	MR (y + i)	ST (y + i)	NT (y + i)	Ns (y + i)
Chlaenius vestitus (Payk.)	361															
C. sulcicollis (Payk.)	362															
C. quadrisulcatus (Payk.)	363															
C. costulatus (Motsch.)	364															
Oodes helopioides (F.)	365				●	●										
O. gracilis Villa & Villa	366															
Panagaeus cruxmajor (L.)	367		●													
P. bipustulatus (F.)	368															
Odacantha melanura (L.)	369	●														
Masoreus wetterhallii (Gyll.)	370					●										
Lebia cyanocephala (L.)	371	●	●	●	●	●	●	●								
L. chlorocephala (Hoffm.)	372	●	●	●	●	●	●	●		●		●				
L. cruxminor (L.)	373	●	●	●	●	●	●	●				●				
Demetrias monostigma Sam.	374															
D. atricapillus (L.)	375															
D. imperialis (Germ.)	376															
Dromius longiceps Dej.	377															
D. linearis (Oliv.)	378															
D. agilis (F.)	379	●	●	●	●	●	●	●	●		●			●	●	●
D. angustus Brullé	380					●	●	●	●	●	●					
D. quadraticollis Mor.	381															
D. meridionalis Dej.	382															
D. schneideri Crotch	383	●	●		●	●	●	●	●							
D. fenestratus (F.)	384	●	●	●	●	●	●	●	●	●	●	●	●	●	●	
D. quadrimaculatus (L.)	385	●				●	●	●	●	●						
D. spilotus (Ill.)	386	●				●	●	●	●	●	●	●	●	●		
D. quadrisignatus Dej.	387															
D. melanocephalus Dej.	388															
D. sigma (Rossi)	389	●	●	●	●	●										
D. notatus Stph.	390	●			●	●	●	●	●	●						
Syntomus truncatellus (L.)	391	●	●	●	●	●	●	●	●	●		●				
S. foveatus (Fourcr.)	392	●														
Lionychus quadrillum (Dft.)	393															
Microlestes minutulus (Gz.)	394															
M. maurus (Sturm)	395															
Cymindis humeralis (Fourcr.)	396															
C. angularis Gyll.	397	●	●	●	●	●	●	●	●	●	●					
C. macularis Fisch.-W.	398	●			●				●	●						

					FINLAND																			USSR		
	Nn (ø + v)	TR (y + i)	F (v + i)	F (n + ø)	Al	Ab	N	Ka	St	Ta	Sa	Oa	Tb	Sb	Kb	Om	Ok	Ob S	Ob N	Ks	LkW	LkE	Le	Li	Vib	Kr	Lr
361																											
362							●																				
363																									●	●	
364										●							●									●	
365					●	●	●	●	●	●	●	●	●	●	●	●			●						●	●	
366																											
367					●	●	●		●	●	●		●		●	●									●	●	
368					●																					●	
369					●		●				●														●	●	
370																											
371					●	●	●		●	●															●	●	
372					●	●	●	●	●	●	●	●	●	●	●	●		●	●						●	●	
373					●	●	●	●	●	●	●	●	●	●	●	●	●	●	●		●				●	●	
374							●																				
375																											
376																											
377						●												●									
378					●	●	●																			●	
379		●		●	●	●	●	●	●	●	●	●	●	●	●	●	●	●	●	●	●		●	●	●	●	●
380																											
381							●																		●		
382																											
383					●	●	●	●	●	●	●	●	●	●	●	●	●	●							●	●	
384					●	●	●	●	●	●	●	●	●	●	●	●	●								●		
385					●	●																					
386					●	●	●																				
387																											
388																											
389					●	●	●	●	●	●	●	●	●	●	●	●		●	●	●	●				●	●	
390					●	●	●		●																		
391					●	●	●	●	●	●	●	●	●	●	●	●	●	●	●	●	●				●	●	
392					●	●			●																●	●	
393						●																					
394					●	●	●	●		●	●														●	●	
395						●		●			●															●	
396																											
397					●	●	●	●	●	●	●	●			●										●	●	
398					●	●	●	●	●	●	●	●	●	●	●	●	●	●	●		●	●		●	●	●	●

		Germany	G. Britain	DENMARK SJ	EJ	WJ	NWJ	NEJ	F	LFM	SZ	NWZ	NEZ	B	Sk.	Bl.
Cymindis vaporariorum (L.)	399	●	●	●	●	●	●	●				●	●	●	●	●
Brachinus crepitans (L.)	400	●	●											●	●	
Family Rhysodidae																
Rhysodes sulcatus (F.)																●
Correction to part I																
Miscodera arctica (Payk.)	74				●			●						●		

		NORWAY Ø + AK	HE (s + n)	O (s + n)	B (ø + v)	VE	TE (y + i)	AA (y + i)	VA (y + i)	R (y + i)	HO (y + i)	SF (y + i)	MR (y + i)	ST (y + i)	NT (y + i)	Ns (y + i)
Cymindis vaporariorum (L.)	399	●	●	●	●		●	●		●	●	●	●	●		●
Brachinus crepitans (L.)	400															
Family Rhysodidae																
Rhysodes sulcatus (F.)																
Correction to part I																
Miscodera arctica (Payk.)	74	●	●	●	●		●	●	●	●	●	●		●		●

SWEDEN

	Hall.	Sm.	Öl.	Gtl.	G. Sand.	Ög.	Vg.	Boh.	Dlsl.	Nrk.	Sdm.	Upl.	Vstm.	Vrm.	Dlr.	Gstr.	Hls.	Med.	Hrj.	Jmt.	Ång.	Vb.	Nb.	Ås. Lpm.	Ly. Lpm.	P. Lpm.	Lu. Lpm.	T. Lpm.
399	●	●	●	●		●	●	●	●	●	●	●	●	●	●	●	●	●	●	●	●	●	●	●	●	●	●	●
400	●	●	●	●		●					●	●	●															
		●																										
74																												

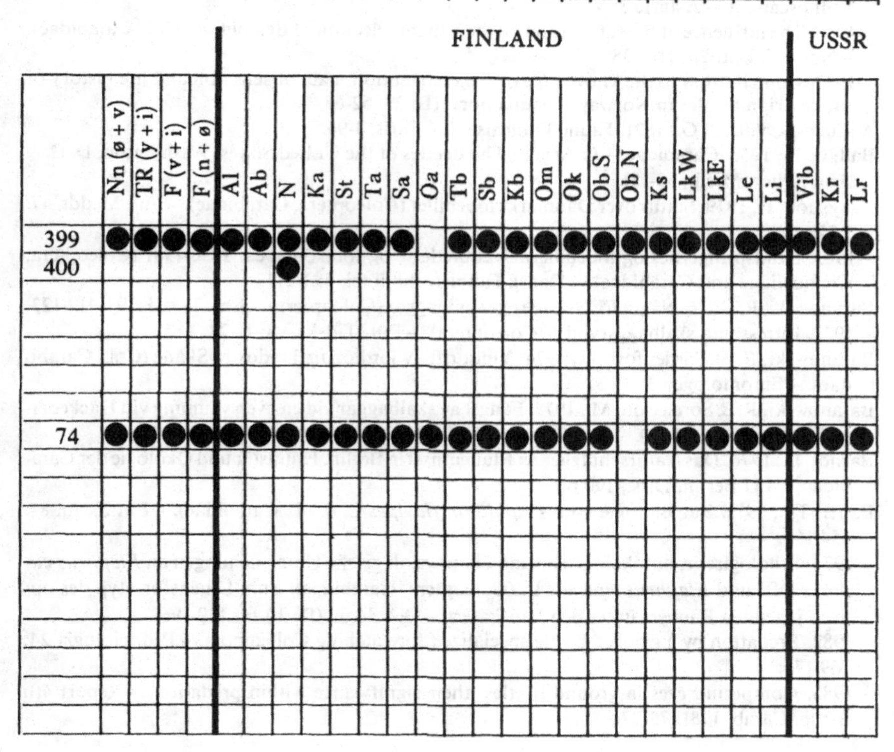

FINLAND **USSR**

	Nn (ø+v)	TR (y+i)	F (v+i)	F (n+ø)	Al	Ab	N	Ka	St	Ta	Sa	Oa	Tb	Sb	Kb	Om	Ok	Ob S	Ob N	Ks	LkW	LkE	Le	Li	Vib	Kr	Lr
399	●	●	●	●	●	●	●	●	●	●	●		●	●	●	●	●	●	●	●	●	●	●	●	●	●	●
400							●																				
74	●	●	●	●	●	●	●	●	●	●	●	●	●	●	●	●	●	●		●	●	●	●	●	●	●	●

Literature

Ahrens, A., 1812. Beiträge zur Kenntniss deutscher Käfer. - Neue Schr. Naturf. Ges. Halle 2(2): 1-40.

- 1830. Beschreibung einiger deutschen Arten der Gattung *Clivina*. - Ent. Arch. (Thon) 2: 57-61.

Andersen J., 1960. Bidrag til kunnskapen om norske billers utbredelse og levevis. - Norsk ent. Tidsskr. 11: 135-140.

- 1962. Billefunn fra forskellige deler av Norge, spesielt fra indre Sør-Trøndelag (Coleoptera). - Norsk ent. Tidsskr. 12: 49-54.

- 1966. Økologiske undersøkelser av noen stenotope elvebreddbiller av slekten *Bembidion* Latr. (Carabidae). - Thesis, Univ. Oslo, 137 pp.

- 1968. The effect of inundation and choice of hibernation sites of Coleoptera living on river banks. - Norsk ent. Tidsskr. 15: 115-133.

- 1970a. Habitat choice and life history of the Bembidiini (Col., Carabidae) on river banks in central and northern Norway. - Norsk ent. Tidsskr. 17: 17-65.

- 1970b. New records of *Bembidion mckinleyi scandinavicum* Lth (Coleoptera, Carabidae) in Fennoscandia. - Astarte 3: 37-39.

- 1978. The influence of the substratum on the habitat selection of Bembidiini (Col., Carabidae). - Norw. J. Ent. 25: 119-138.

- 1982. Contribution to the knowledge of the distribution, habitat selection and life history of the riparian beetles in Norway. - Fauna norv. (B) 29: 62-68.

Audinet-Serville, J. G., 1821. Faune Française. I. - Paris, 1-96.

Ball, G. E., 1960. Carabidae. - *In* Arnett, The Beetles of the United States. Washington, D. C. & Ann Arbor, Mich., 55-181.

Bangsholt, F., 1979. Status over Danmarks løbebiller (Coleoptera, Carabidae). - Ent. Meddr. 47: 1-21.

- 1983. Sandspringernes og løbebillernes udbredelse og forekomst ca. 1830-1981 (Coleoptera: Cicindelidae and Carabidae). - Dansk Faunist. Bibliotek 4: 1-271.

Baranowski, R., 1976. Några för Sverige nya skalbaggar (Coleoptera). - Ent. Tidsskr. 97: 117-123.

- 1977. Intressanta skalbaggsfynd 1 (Coleoptera). - Ent. Tidskr. 98: 11-28.

Baranowski, R. & Gärdenfors, U., 1974. Vinddrift av jordlöpare i sydöstra Skåne (Col., Carabidae). - Entomologen 3: 35-52.

Baranowski, R. & Sörensson, M., 1978. Fångst av skalbaggar vid en översvämning vid Häckeberga i Skåne (Coleoptera). - Ent. Tidskr. 99: 19-23.

Barndt, D., 1976. Das Naturschutzgebiet Pfaueninsel in Berlin. Faunistik und Ökologie der Carabiden. - FU Berlin, Diss., 190 pp.

Bauer, Th., 1971. Zur Biologie von *Asaphidion flavipes* L. (Col., Carabidae). - Ent.Zschr. 81: 154-164.

- 1974. Ethologische, autökologische und ökophysiologische Untersuchungen an *Elaphrus cupreus* Dft. und *Elaphrus riparius* L. (Coleoptera, Carabidae). Zum Lebensformtyp des optisch jagenden Räubers unter den Laufkäfern - Oecologia (Berl.) 14: 139-196.

- 1982. Predation by a carabid beetle specialized for catching Collembola. - Pedobiologia 24: 169-179.

- 1983. Compound eyes in ground beetles: their significance within predation. - Report 4th Symp. Carab. 1981: 75-77.

Bedel, L., 1879. Faune des Coléoptères du bassin de la Seine. I. Carnivora - Palpicornia. - Annls Soc. Ent. France, ser. 5, 9: 1-128.
- 1892-96. Catalogue raisonné des coléoptères du Nord de l'Afrique. - Paris, 1-320.
Bonelli, F. A., 1810. Ad Maximilianorum Spinola Tabula Synoptica Exhibens Genera Carabicorum in Sectiones et Stirpes disposita. - Torino, 1 p.
- 1813. Observations entomologiques. Deuxieme partie. - Mem. Acad. Imp. Sci. Turin 20: 433-484.
Born, P., 1898. *Carabus catenulatus* nov. var. *Wockei*. - Soc. Ent. 13: 74-75.
- 1926. Die Carabenfauna Norwegens. - Norsk ent. Tidsskr. 2: 57-76.
Brandmayr, P. & al., 1980. Larval versus imaginal taxonomy and the systematic status of the ground beetle taxa *Harpalus* and *Ophonus*. - Ent. Gener. 6: 335-353.
Brandmayr-Zetto, T. & Brandmayr, P., 1975. Biologia di *Ophonus puncticeps* Steph. Cenni sulla fitofagia delle larve e loro etologia (Coleoptera, Carabidae). - Ann. Fac. Sci. Agr. Univ. Torino 9: 421-430.
Breuning, S. v., 1932-1937. Monographie der Gattung *Carabus* L. - Best.-Tab. eur. Col. 104-110: 1-1610.
Briggs, J. B., 1965. Biology of some ground beetles (Col., Carabidae) injurious to strawberries. - Bull. Ent. Res. 56: 79-93.
Brown, W. J., 1932. New species of Coleoptera. - Canad. Ent. 64: 3-12.
Brullé, A., 1834. Observations critiques sur la synonymie des Carabiques. - Rev. Ent. (Silbermann) 2: 89-114.
Brüggemann, F., 1873. Systematisches Verzeichniss der bisher in der Gegend von Bremen gefundenen Käferarten. - Abh. Naturw. Ver. Bremen 3: 441-524.
Burakowski, B., 1957. *Amara (Amara) pseudocommunis* sp.n. from Central Europe (Coleoptera, Carabidae). - Ann. Zool. Warszawa 16: 343-348.
Casey, T. L., 1918. A Review of the North American Bembidiinae. - Mem. Coleopt. 8: 1-223.
Chaudoir, M., 1837. Description de quelques genres nouveaux et de quelques espèces nouvelles ou inédites de Carabiques. - Bull. Soc. Nat. Moscou 10 (3): 1-20.
- 1838. Tableu d'une nouvelle subdivision du genre *Feronia* Dejean suivi d'une characteristique de trois nouveaux genres de carabiques. - Bull. Soc. Nat. Moscou 11: 3-32.
- 1843: Genres nouveaux de la famille des Carabiques. (Continuation). - Bull. Soc. Nat. Moscou 16: 383-427.
- 1844. Trois mémoires sur la famille des Carabiques. II. Supplément a la faune entomologique de la Russie et des pays limitrophes. Carabiques nouveaux. - Bull. Soc. Nat. Moscou 17: 433-453.
- 1846. Enumeration des Carabiques et Hydrocanthares, recueillis pendant un voyage au Caucase et dans les provinces transcaucasiennes. - Kiew, 49-209.
- 1850. Supplément a la faune des Carabiques de la Russie. - Bull. Soc. Nat. Moscou 23 (2): 62-206.
- 1859. Beitrag zur Kenntniss der europäischen Feroniden. - Ent. Zeit., Stettin, 20: 113-131.
Clairville, J. Ph., 1806. Entomologie Helvétique. Vol. II. - Zuric, 1-251.
Claridge, M. F., 1974. Stridulation and defensive behaviour in the ground beetle *Cychrus caraboides* (L.). - J. Ent. (A) 49: 7-15.
Conradi-Larsen, E. M. & Sømme, L., 1973. The overwintering of *Pelophila borealis* Payk. II. Aerobic and anaerobic metabolism. - Norsk ent. Tidsskr. 20: 325-332.
Creutzer, C., 1799. Entomologische Versuche. - Wien, 1-142.
Crotch, G. R., 1869. Berichtigungen und Zusätze zum Catalogus Coleopterorum synonymicus et systematicus. - Col. Hefte 5: 105-112.

- 1871. List of all the Coleoptera described A. D. 1758-1821 referred to their modern genera. - Cambridge, 1-24.
Crowson, R. A., 1955. The natural classification of the families of Coleoptera. - Oxford, 1-187.
Csiki, E., 1929. Carabidae II. Harpalinae III. - Coleopt. Catal. 104: 347-528.
Curtis, J., 1829. British Entomology. Vol. VI, tt. 242-289. - London.
Cuvier, G. L., 1817. Le Règne Animal. Tome III. - Paris, 1-653.
Dalla Torre, C. W., 1877. Synopsis der Insecten Oberösterreichs. - Jahres-Ber. Ver. Naturk. Oesterr. Ens Linz 8: 17-74.
Daniel, K. & Daniel, J., 1890. *Tachycellus oreophilus,* eine der montanen Region angehörende, neue, deutsche Art. - Dtsch. ent. Zschr. 1890: 209-211.
Dawson, J. F., 1849. Descriptions of five new species of Coleoptera. - Ann. Mag. Nat. Hist. (2) 3: 213-216.
- 1854. Geodephaga Britannica. - London, 1-224.
Dawson, N., 1965. A comparative study of the ecology of eight species of fenland Carabidae (Coleoptera). - J. anim. Ecol. 34: 299-314.
De Zordo, I., 1979. Ökologische Untersuchungen an Wirbellosen des Zentralalpinen Hochgebirges (Obergurgl, Tirol), III. Lebenszyklen und Zönotik von Coleopteren. - Veröff. Univ. Innsbruck, 118. Alpin-Biol. Stud. 11: 1-132.
Degeer, C., 1774. Mémoires pour servir a l'histoire des Insectes. Tome 4me. - Stockholm, 1-457.
Dejean, P., 1821. Catalogue des Coléoptères de la collection de M. le baron Dejean. - Paris, 1-136.
- 1825. Species général des Coléoptères.- Tome ler. - Paris, 1-463.
- 1826. Species général des Coléoptères. Tome 2nd. - Paris, 1-501.
- 1828. Species général des Coléoptères. Tome 3me. - Paris, 1-556.
- 1829. Species général des Coléoptères. Tome 4me. - Paris, 1-520.
- 1831. Species général des Coléoptères. Tome 5me. - Paris, 1-883.
Den Boer, P. J., 1980. Wing polymorphism and dimorphism in ground beetles as stages in an evolutionary process (Coleoptera: Carabidae). - Ent. Gener. 6: 107-134.
Den Boer, P. J. (ed.), 1971. Dispersal and dispersal power of Carabid beetles. - Misc. Pap. Landb. hogesch. Wageningen 8: 1-151.
Dicker, G. H. L., 1951. *Agonum dorsale* Pont. (Col. Carabidae): An unusual egg-laying habit and some biological notes. - Ent. mon. Mag. 87: 33-34.
Donovan, E., 1806. The Natural History of British Insects. Vol. IX. - London, 1-100.
Drapiez, A., 1820. Description de cinq espèces d'insectes nouveaux. - Ann. Gén. Sci. Phys., Bruxelles 7: 275-280.
Dreisig, H., 1980. Daily activity, thermoregulation and water loss in the tiger beetle *Cicindela hybrida.* - Oecologia (Berl.) 44: 376-389.
Drift, J. van der, 1951. Analysis of the animal community in a beech forest floor. - Tijdschr. Entomol. 94: 1-168.
- 1959. Field studies on the surface fauna of forests. - Bijdr. Dierkde. 29: 79-103.
Duftschmid, K., 1812. Fauna Austriae. II. Teil. - Linz und Leipzig, 1-311.
Duméril, A. M. C., 1806. Zoologie analytique. - Paris, 1-345.
Eisner, T., 1958. The protective role of the spray mechanism of the bombardier beetle, *Brachynus ballistarius* Lec. - J. Insect Physiol. 2: 215-220.
Erichson, W. F., 1837. Die Käfer der Mark Brandenburg. Band I., Abt. I. - Berlin, 1-384.
Erwin, T. L., 1974a. Studies of the Subtribe Tachyina (Coleoptera: Carabidae: Bembidiini), Part II: A revision of the New World - Australian genus *Pericompsus* LeConte. - Smithson. Contr. Zool. 162: 1-96.
- 1974b. Studies of the subtribe Tachyina (Coleoptera: Carabidae: Bembiidini) Supplement A: lectotype designations for New World species, two new genera and notes on generic concepts.

- Proc. Ent. Soc. Washington 76: 123-155.
- 1979. Thoughts on the evolutionary history of ground beetles: hypotheses generated from comparative faunal analyses of lowland forest sites in temperate and tropical regions. - *In* Erwin, T. L. & al. (ed.), Carabid Beetles, their evolution, natural history, and classification, W. Junk Publ., 539-592.
- 1983. Taxon pulses, vicariance, and dispersal: an evolutionary synthesis illustrated by carabid beetles. - *In* Nelson, G. & Rosen, D. E. (ed.), Vicariance biogeography. A critique. Colombia Univ. Press, New York, 159-183.

Erwin, T. L. & Sims, L. L., 1984. Carabid Beetles of the West Indies (Insecta: Coleoptera): A Synopsis of the Genera, and Checklists of the Tribes of Caraboidea, and of the West Indian Species. - Quaest. Ent. 20: 351-466.

Eschscholtz, F., 1823. Species Insectorum novae descriptae (Carabicinen). Mem. Soc. Nat. Moscou 6: 95-108.
- 1830. Nova genera Coleopterorum Faunae Europae. - Bull. Soc. Nat. Moscou 2: 63-66.

Evans, P. D. & al., 1971. Observations on the biology and submergence behaviour of some littoral beetles. - J. mar. biol. Ass. U.K. 51: 375-386.

Faasch, H., 1968. Beobachtungen zur Biologie und zum Verhalten von *Cicindela hybrida* L. und *Cicindela campestris* L. und experimentelle Analyse ihres Beutefangverhaltens. - Zool. Jb. Syst. 95: 477-522.

Fabricius, J. C., 1777. Genera Insectorum. - Chilonii, 1-310.
- 1775. Systema Entomologiae. - Flensburgi et Lipsiae, 1-832.
- 1779. Reise nach Norwegen mit Bemerkungen aus der Naturhistorie und Ökonomie. - Hamburg, 1-388.
- 1787. Mantissa Insectorum. - Hafniae, 1-348.
- 1792. Entomologia systematica emendata et aucta. Tom. I: 1. - Hafniae, 1-330.
- 1794. Entomologica Systematica emendata et aucta. Tom. IV. - Hafniae, 1-478.
- 1796. Index alphabeticus in J. C. Fabricii Entomologiam Systematicam. - Hafniae, 1-175.
- 1801. Systema Eleutheratorum, Tom. I. - Kiliae, 1-506.

Fairmaire, L. & Laboulbène, A., 1854. Faune entomologique française ou description des Insectes qui se trouvent en France. Coléoptères. Tome Ier. - Paris, 1-665.

Faldermann, F., 1835. Coleopterorum ab ill. Bungio in China boreali, Mongolia et montibus Altaicis collectorum, nec non ab ill. Turczaninoffio et Stschukino e provincia Irkutzk missorum illustrationes. - Mem. Acad. St. Petersb. 2: 337-464.

Fall, H. C., 1926. A list of the Coleoptera taken in Alaska and adjacent parts of Yukon Territory in the summer of 1924. - Pan-Pacif. Ent. 2: 127-154, 191-208.

Ferenz, H. J., 1975. Anpassungen von *Pterostichus nigrita* F. (Coleoptera, Carabidae) an subarktische Bedingungen. - Oecologia (Berl.) 19: 49-57.

Fischer von Waldheim, G., 1824. Entomographia Imperii Rossici. Vol. II. - Mosquae, 1-264.

Focarile, A., 1964. Gli *Asaphidion* del gruppo *flavipes* (L.), con particolare riguardo alla fauna Italiana. - Mem. Soc. Ent. Ital. 43: 97-120.

Forsskåhl, B., 1972. The invertebrate fauna of the Kilpisjärvi area, Finnish Lapland. 9. Carabidae. With special notes on ecology and breeding biology. - Acta Soc. Fauna Flora fenn. 80: 99-119.

Fourcroy, A. F., 1785. Entomologia Parisiensis, sive Catalogus Insectorum quae in Agro Parisiensi reperiuntur. Pars I. - Parisiis, 1-544.

Fowler, W. W., 1886. The Coleoptera of the British Islands. Vol. I. Adephaga - Hydrophilidae. - London, 1-269.

Freude, H., 1976. Adephaga, 1. Carabidae. - *In* Freude, H., Harde, K. W. & Lohse, G. A., Die Käfer Mitteleuropas 2: 1-302.

Fröhlich, J. A., 1799. Einige neue Gattungen und Arten von Käfern. - Naturforscher, Halle, 28: 1-65.
Géhin, J. B., 1885. Catalogue synonymique et systematique des Coléoptères de la Tribu des Carabides. - Remiremont et Prague, 1-104.
Ganglbauer, L., 1892. Die Käfer von Mitteleuropa. Band I. - Wien, 1-557.
Gebler, F., 1833. Notae et addidamenta ad catalogum Coleopterorum Sibiriae occidentalis et confinis Tartariae operis, C. F. von Ledebours Reise in das Altaigebirge und die songarische Kirgisensteppe. - Bull. Soc. Nat. Moscou 6: 262-309.
- 1847. Verzeichniss der im Kolywano-Woskresenkischem Hüttenbezirke Südwest-Sibiriens beobachteten Käfer. - Bull. Soc. Nat. Moscou 20 (1): 263-361.
Gebler, F. A., 1829. Notae et additamenta ad catalogum Sibiriae occidentalis et confinis Tartariae. - Ledebours Reise 2.
Gemminger, M. & Harold, E., 1868. Catalogus Coleopterorum hucusque descriptorum synonymicus et systematicus. Tom. I. - Monachii, 1-432.
Germar, E. F., 1822. Fauna Insectorum Europae. Fasc. VII. - Halae, 1-25.
- 1824. Insectorum species novae aut minus cognitae. - Halae, 1-624.
Goeze, J. A. E., 1777. Entomologische Beyträge zu des Ritter Linné zwölften Ausgabe des Natursystems. Teil I. - Leipzig, 1-736.
Gozis, M. des, 1882. Memoire sur les pores sétigères prothoraciques dans la tribu des Carnivores. - Mitt. Schweiz. Ent. Ges. 6: 285-300.
- 1886. Recherche de l'espèce typique de quelques anciens genres. Rectifications synonymiques et notes diverses. - Montlucon, 1-36.
Grenier, A.,1863. Catalogue des Coléoptères de France. - Paris, 1-135.
Grossecappenberg, W. & al., 1978. Beiträge zur Kenntnis der terrestrischen Fauna des Gildenhauser Venns bei Bentheim. I. Die Carabidenfauna der Heiden, Ufer und Moore. - Abh. Landesmus. Naturk. Münster 40: 12-34.
Grüm, L., 1979. Mortality rates of the mobile and immobile stages in the life-cycle of Carabids. - *In* Carabid beetles: their evolution, natural history and classification: Proc. 1st Internat. Symp. Carabidol., Washington, 1976. - W. Junk, The Hague.
Guérin-Mèneville, F. E., 1823. Note topographique sur quelques insectes coléoptères et description de deux espèces des genres *Badister* et *Bembidion*. - Bull. Soc. Philom. Paris 1823: 121-124.
Gyllenhal, L., 1810. Insecta Suecica. Classis I. Coleoptera sive Eleuterata. Tom. I, Pars II. - Scaris, 1-660.
- 1813. Insecta Suecica. Classis I. Coleoptera sive Eleuterata. Tom. I, Pars III. - Scaris, 1-734.
- 1827. Insecta Suecica. Classis I. Coleoptera sive Eleuterata. Tom. I. Pars IV. - Lipsiae, 1-762.
Haliday, A. H., 1841. Genus *Amphyginus* of Haliday. - Entomologist (E. Newman), London, 11: 175.
Hansen, V., 1940. *Harpalus tardoides* n.sp., *Phyllodrepa melis* n.sp. and *Bledius larseni* n.sp. Three new beetles from Denmark. - Ent. Meddr. 20: 577-584.
- 1944. *Badister striatulus* n.sp., a new species from Denmark, and *Rhamphus oxyacanthae* Marsh., a separate species. - Ent. Meddr. 24: 93-96.
Hayward, R., 1900. A study of the species of *Tachys* of Boreal America. - Trans. Am. Ent. Soc. 26: 191-238.
Heer, O., 1837. Die Kaefer der Schweiz. Teil 2. - Neuchatel, 1-55.
- 1841. Fauna Coleopterorum Helvetica. Pars I. - Turici, 1-652.
Hellén, W., 1934. Koleopterologische Mitteilungen aus Finnland XI. - Notul. Ent. 14: 52-59.
- 1935. Eine neue *Europhilus*-Art aus Finnland (Col.) - Notul. Ent. 15: 87-89.
Helliesen, T. A., 1892. Bidrag til kundskaben om Norges coleopterfauna. III. - Stavanger Mus. Aarsh. 1892: 30-57.

Herbst, J. F. W., 1784. Kritisches Verzeichniss meiner Insecten-Sammlung. (Coleoptera). - Fuessly Arch. Insectengesch. 4-5: 1-151.
- 1786. Erste Mantisse zum Verzeichniss der ersten Klasse meiner Insecten-Sammlung. - Fuessly Arch. Insectengesch. 6: 153-182.
Heydemann, B., 1962. Die biozönotische Entwicklung vom Vorland zum Koog. Vergleichend-ökologische Untersuchungen an der Nordseekuste. II. Teil: Käfer (Coleoptera). Abh. Math. - Naturw. Kl. Akad. Wiss. Mainz 11: 765-964.
- 1968. Das Freiland- und Laborexperiment zur Ökologie der Grenze Land-Meer. - Verh. Deut. Zool. Ges. Heidelberg 1967: 256-309.
Heyden, L. v., 1870. Entomologische Reise nach dem südlichen Spanien, der Sierra Guadarrama und Sierra Morena, Portugal und den Cantabrischen Gebirgen. - Berlin, 1-218.
Hieke, F., 1976. Gattung *Amara* Bonelli 1809. - *In* Freude, H., Harde, K. W. & Lohse, G. A., Die Käfer Mitteleuropas 2: 225-249.
Hoffmann, J. J., 1803. Entomologische Hefte II. - Frankfurt am Main, 1-119.
Hope, F. W., 1838. The Coleopterists Manual., Vol. II. - London, 1-168.
Horion, A., 1941. Faunistik der deutschen Käfer. I. - Wien, 1-463.
Houston, W. W. K., 1981. The life cycles and age of *Carabus glabratus* Paykull and *C. problematicus* Herbst (Col.: Carabidae) on moorland in northern England. - Ecol. Ent. 6: 263-271.
Hummel, A. D., 1827. Insectes de 1826. - Essais Ent., St. Pétersbourg, 6: 1-20.
Hůrka, K., 1973. Fortpflanzung und Entwicklung der mitteleuropäischen *Carabus*- und *Procerus*-Arten. - Stud. Čsl. Akad. Věd. 9: 1-78.
Illiger, K., 1798. Verzeichniss der Käfer Preussens. - Halle, 1-510.
- 1801. Nachtrag und Berichtigungen zum Verzeichniss der Käfer Preussens. - Mag. Insektenk. 1: 1-94.
- 1802. Vermischte Nachrichten und Bemerkungen. V. Einige Bemerkungen und Berichtigungen zu den ersten beiden Heften. - Mag. Insektenk. 1: 489-491.
Jacquelin du Val, C., 1857. Genera des Coléoptères d'Europe. Tome ler. - Paris, 1-140.
Jarmer, G., 1973. Ein Vergleich der Carabidenfauna an eutrophen und dystrophen Gewässern in der Umgebung der Station Grietherbusch am Niederrhein. - Staatsexamensarbeit, Köln.
Jeannel, R., 1941-42. Coléoptères Carabiques. - Faune de France 39-40: 1-1173.
Jedlička, A., 1936. Nová *Amara* z okolí Pražského. - Čas. Čsl. Spol. Ent. 33: 4-5.
Kangas, E., 1978. Kolme Suomen faunalle uutta maakiitäjäislajia (Carabidae). - Notul. Ent. 58: 175.
- 1980. Merkmale und Verbreitung einer neuen und zweier bekannter Unterarten des Laufkäfers *Bembidion petrosum* (Coleoptera: Carabidae) - Ent. Gener. 6: 363-365.
Kirby, W., 1837. Insects. - *In* Richardson, J., Fauna Boreali-Americana, 4: 1-325.
Koch, D., 1984. *Pterostichus nigrita,* ein Komplex von Zwillingsarten. - Ent. Blätter 79: 141-152.
Koch, D. & Thiele, H. U., 1980. Zur ökologisch-physiologischen Differenzierung und Speziation der Laufkäfer-Art *Pterostichus nigrita* (Coleoptera: Carabidae). - Ent. Gener. 6: 135-150.
Kolenati, F. A., 1845. Insecta Caucasi cum distributione geographica. Coleopterorum Pentamera Carnivora. - Melet. Ent., Petropoli, 1: 1-88.
Krehan, I., 1970. Die Steuerung von Jahresrhytmik und Diapause bei Larval- und Imagoüberwinterern der Gattung *Pterostichus* (Col., Carab.). - Oecologia (Berl.) 6: 58-105.
Krogerus, R., 1932. Über die Ökologie und Verbreitung der Arthropoden der Triebsandgebiete an den Küsten Finnlands. - Acta Zool. Fenn. 12: 1-308.
- 1960. Ökologische Studien über nordische Moorarthropoden. - Comment. biol. 21: 1-238.
Kryzhanovskij, O. L., 1976. An attempt of revised classification of the family Carabidae (Coleoptera). - Ent. Obozr. 55: 80-91. (in Russian).
Kult, K., 1949. *Amara Pulpani* sp.n. a nové subspecie druhů rodu *Pterostichus z* ČSR. - Ent. Listy, Brno, 12: 77-88.

Kůrka, A., 1972. Bionomy of the Czechoslovak species of the genus *Calathus* Bon., with notes on their rearing (Col., Carabidae). - Věst. Čs. spol. zool. 36: 101-114.
Kvamme, T., 1978. *Stenolophus mixtus* Hbst., an expanding Carabid beetle new to Norway. - Norw. J. Ent. 25: 227-228.
Küster, H. C., 1847. Die Käfer Europa's. IX. Heft. - Nürnberg, 1-100.
Laczó, J., 1912. Új bojarak Trencsén-vármegböl. - Rovart. Lapok 19: 3-5.
Laporte de Castelnau, F. L., 1835. Études Entomologiques. - Paris, 1-159.
Larochelle, A., 1974. A world list of prey of *Chlaenius* (Coleoptera: Carabidae). - Great Lakes Entomologist 7: 137-142.
Larsen, E. B., 1936. Biologische Studien über die tunnelgrabenden Käfer auf Skallingen. - Vidensk. Medd. dansk naturh.Foren. 100: 1-231.
Larsson, S. G., 1939. Entwicklungstypen und Entwicklungszeiten der dänischen Carabiden. - Ent. Meddr. 20: 277-560.
Latreille, P. A., 1802. Histoire Naturelle, générale et particulière des Crustacés et des Insectes. Tome 3me. - Paris, 1-467.
- 1829. Tome IV. Crustacés, Arachnides et partie des Insectes. - *In* Cuvier, Le Règne Animal. (ed. 2.) Paris, 1-584.
Latreille, P. A. & Dejean, P., 1822. Histoire Naturelle et Iconographie des Insectes Coléoptères d'Europe. 1re Livr. - Paris, 1-198.
Lauterbach, A. W., 1964. Verbreitungs- und Aktivitätsbestimmende Faktoren bei Carabiden in sauerländischen Wäldern. - Abh. Landesmus. Naturk. Münster 26: 1-103.
LeConte, J. L., 1848. A descriptive catalogue of the geodephagous Coleoptera inhabiting the United States east of the Rocky Mountains. - Ann. Lyc. Nat. Hist. New York 4: 173-474.
- 1853. Notes on the classification of the Carabidae of the United States. - Trans. Amer. Phil. Soc. 10: 363-403.
- 1854. Synopsis of the species of *Platynus* and allied genera, inhabiting the United States. - Proc. Acad. Nat. Sci. Philadelphia 7: 35-59.
Leske, N. G., 1785. Reise durch Sachsen in Rücksicht der Naturgeschichte und Ökonomie. - Leipzig, 1-548.
Letzner, K., 1852. Systematische Beschreibung der Laufkäfer Schlesiens. - Zschr. Ent., Breslau, 6: 187-292.
Lindroth, C. H., 1939a. *Bradycellus ponderosus* n.sp. aus Finnland (Col.). - Notul. Ent. 18: 117-119.
- 1939b. Zur Systematik fennoskandischer Carabiden. 2-3. - Ent. Tidskr. 60: 54-68.
- 1942. Skalbaggar, Coleoptera, Sandjägare och jordlöpare. Fam. Carabidae. - Svensk Insektfauna 35: 1-258.
- 1943. Zur Systematik fennoskandischer Carabiden. 13-33. - Ent. Tidskr. 63: 1-68.
- 1945a. Die Fennoskandischen Carabidae. I. Spezieller Teil. - Göteborgs Kgl. Vet. Vitterh. Samh. Handl., Ser. B 4 (1): 1-709.
- 1945b. Die Fennoskandischen Carabidae. II. Die Karten. - Göteborgs Kgl. Vet. Vitterh. Samh. Handl., Ser. B 4 (2): 1-277.
- 1946. Inheritance of wing dimorphism in *Pterostichus anthracinus* Ill. - Hereditas 32: 27-40.
- 1949. Die Fennoskandischen Carabidae. III. Allgemeiner Teil. - Göteborgs Kgl. Vet. Vitterh. Samh. Handl., Ser. B 4: 1-911.
- 1954. Die Larve von *Lebia chlorocephala* Hoffm. (Coleoptera, Carabidae). - Opusc. Entomol. 19: 29-33.
- 1957. The Linnaean species of Carabid beetles. - J. Linn. Soc. London (Zool.) 43: 325-341.
- 1961. Skalbaggar. Coleoptera. Sandjägare och jordlöpare. Fam. Carabidae. (2nd ed.) - Svensk Insektfauna 35: 1-209.

- 1961-69. The ground-beetles (Carabidae excl. Cicindelinae) of Canada and Alaska. I-VI. - Opuscula Ent., Suppl. 20, 24, 29, 33, 34, 35: 1-1192.
- 1972. Changes in the Fennoscandian Groundbeetle fauna (Coleoptera, Carabidae) during the twentieth century. - Ann. Zool. Fenn. 9: 49-64.
- 1974a. Coleoptera Carabidae. - Handb. Ident. Brit. Ins. IV (2): 1-148.
- 1974b. On the elytral microsculpture of Carabid beetles. - Ent. Scand. 5: 251-264.
Linnaeus, C., 1758. Systema Naturae, ed. 10. - Holmiae, 1-824.
- 1761. Fauna Suecica. Ed. 2. - Stockholmiae, 1-578.
- 1767. Systema Naturae, ed. 12. Tom. I. Pars II. - Holmiae, 533-1327.
Lohse, G. A., 1983. Die *Asaphidion*-Arten aus der Verwandschaft des *A. flavipes* L. - Entomol. Bl. 79: 33-36.
Lundberg, S., 1973. En lokal för *Trachypachus zetterstedti* Gyll. vid Messaure i Lule lappmark (Col. Carabidae). - Ent. Tidskr. 94: 34-36.
- 1980. Fynd av för Sverige nya skalbaggsarter rapporterade under åren 1978-79. - Ent. Tidskr. 101: 91-93.
- 1981. Återfynd av jordlöparna *Chlaenius costulatus* och *Harpalus nigritarsis* i Sverige - Ent. Tidskr. 102: 13-15.
- 1984.Den brända skogens skalbaggsfauna i Sverige. - Ent. Tidskr. 105: 129-141.
Lutshnik, V., 1915. Sous-genre *Morphnosoma* mihi (= *Omaseus* auct.) du genre *Platysma* (Bon.) Tschitsch. et ses especes (Coleoptera, Carabidae). - Ent. Obozr. 14: 424-426.
Löser, S., 1970. Brutfürsorge und Brutpflege bei Laufkäfern der Gattung *Abax*. - Zool. Anz., Suppl. 33: 322-326.
- 1972. Art und Ursachen der Verbreitung einiger Carabidenarten (Coleoptera) im Grenzraum Ebene-Mittelgebirge. - Zool. Jb. Syst. 99: 213-262.
Makolski, J., 1952. Revue of Central-European species from the *Badister bipustulatus* Fabr. group with description of a new species. - Ann. Mus. Zool. Polon. 15: 7-23.
Mandl, K., 1962. Die fennoskandischen Formen des *Carabus violaceus* L. (Carabidae, Col.). - Opusc. Ent. 27: 193-209.
Manley, G. V. 1971. A see-cacheing carabid, *Synuchus impunctatus* Say (Coleoptera, Carabidae). - Ann. ent. Soc. Am. 64: 1474-75.
Mannerheim, C. G., 1823. Novae species. - *In* Hummel, Essais entomol., St. Petersb., 3: 43-44.
- 1825. Novae coleopterorum species Imperii Rossici incolae. - *In* Hummel, Essais entomol., St. Petersb., 4: 19-41.
- 1853. Dritter Nachtrag zur Käfer-Fauna der Nord-Americanischen Länder des Russischen Reiches. - Bull. Soc. Nat. Moscou 26 (2): 95-273.
Marsham, Th., 1802. Entomologia Britannica. Tom. I. Coleoptera. - London, 1-547.
Martin, O., 1984. Nyt fund af guldløberen (*Carabus auratus* Linnaeus, 1758) i Danmark (Coleoptera, Carabidae) - Ent. Meddr 51: 102.
Mitchell, B., 1963a. Ecology of two carabid beetles, *Bembidion lampros* (Herbst) and *Trechus quadristriatus* (Schrank). I. Life cycles and feeding behaviour. - J. Anim. Ecol. 32: 289-299.
- 1963b. Ecology of two carabid beetles, *Bembidion lampros* (Herbst) and *Trechus quadristriatus* (Schrank). II. Studies on populations of adults in the field, with special reference to the technique of pit-fall trapping. - J. Anim. Ecol. 32: 377-392.
Mjöberg, E., 1915. *Nebria Klinckowströmi* n.sp. - Ent. Tidskr. 36: 285.
Morawitz, A., 1862. Vorläufige Diagnosen neuer Coleopteren aus Südost-Sibirien. - Mél. Biol. Bull. Acad. Sci. St. Pétersb. 4: 180-228.
Mossakowski, D., 1970a. Das Hochmoor-Ökoareal von *Agonum ericeti* (Panz.) (Coleoptera, Carabidae) und die Frage der Hochmoorbildung. - Faun.-ökol. Mitt. 3: 378-392.
- 1970b. Ökologische Untersuchungen an epigäischen Coleopteren atlantischer Moor- und Hei-

destandorte. - Z. wiss. Zool. 181: 233-316.
Motschulsky, V. de, 1838-39. Coléoptères du Caucase et des provinces Transcaucasiennes. - Bull. Soc. Nat. Moscou 11: 175-188, 12: 68-93.
- 1844. Insectes de la Sibérie rapportés d'un Voyage fait en 1839 et 1840. - Mém. Acad. Sci. St. Pétersb. 5: 1-274.
- 1845. Remarque sur la collection de Coléoptères Russes de M. Article I. - Bull. Soc. Nat. Moscou 18 (1): 1-127.
- 1847. Antwort an Dr Gebler auf einige seiner Bemerkungen in der No II dieses Bulletins (1847). - Bull. Soc. Nat. Moscou 20 (2): 218-228.
- 1850. Die Käfer Russlands. - Moskva, 1-91.
- 1859. Catalogue des insectes rapportés des environs du fl. Amour, depuis la Schilka jusqu' a Nikolaevsk, examinés et enumerés. Bull. Soc. Nat. Moscou 32 (2): 407-507.
- 1862. Etudes Entomologiques 11. - Dresden, 1-55.
- 1864. Énumeration des nouvelles especes de Coléoptères rapportés de ses voyages. 4-eme article. - Bull. Soc. Nat. Moscou 37 (2): 171-240, 297-355.
Munster, T., 1923. *Dyschirius* Bonelli (Col.). De norske arter. - Norsk ent. Tidskr. 1: 244-250.
- 1924. Nova etc. ex Norvegia - Norsk ent. Tidskr. 1: 288-294.
- 1927. To bidrag till Norges koleopterfauna. - Nyt Mag. Naturv. 65: 275-306.
- 1930. Tillaeg og bemaerkninger til Norges koleopterfauna. - Norsk ent. Tidskr. 2: 353-357.
- 1932. Bembidiini I. - Norsk ent. Tidskr. 3: 80-82.
Murdoch, W. W., 1966. Aspects of the population dynamics of some marsh Carabidae. - J. Anim. Ecol. 35: 127-156.
Mäklin, F. W., 1877. Diagnoser öfver några nya siberiska insektarter. - Öfvers. Finska Vet. Soc. Förh. 19: 15-32.
Müller, G., 1930. Carabiden-Studien. - Coleopt. Centralbl. 5: 1-19.
Müller, J., 1921. Über neue und bekannte Carabiden. - Wien. ent. Zeit. 38: 133-141.
- 1922. Bestimmungstabelle der *Dyschirius*-Arten Europas. - Kol. Rdsch. 10: 33-120.
Müller, O. F., 1764. Fauna insectorum Fridrichsdalina. - Hafniae et Lipsiae, 1-96.
- 1776. Zoologiae Danicae prodromus. - Havniae, 1-274.
Netolitzky, F., 1910. Bemerkungen zur Systematik in der Gattung *Bembidion* Latr. - Wien. ent. Zeit. 29: 209-228.
- 1911. *Bembidion*-Studien. - Wien. ent. Zeit. 30: 179-194.
- 1914a. Die Bembidiini in Winklers Catalogus. - Ent. Bl. 10: 50-55.
- 1914b. Die Bembidiini in Winklers Catalogus. Zweite Mitteilung. - Ent. Bl. 10: 164-176.
- 1918. Neue Bembidiini Europas (Carabidae). - Kol. Rdsch. 6: 19-25.
- 1920. Zwei neue Bembidien-Untergattungen und eine neue Art. - Kol. Rdsch. 8: 96.
- 1942-43. Bestimmungstabelle der *Bembidion*-Arten des paläarktischen Gebietes. - Kol. Rdsch. 28: 29-124, 29: 1-70.
Nicolai, A., 1822. Dissertatio Coleopterorum species Agri Halensis. - Halae, 1-48.
Olivier, A. G., 1795. Entomologie, ou Histoire Naturelle des Insectes. Tome III. - Paris, 1-557.
Paarmann, W., 1966. Vergleichende Untersuchungen über die Bindung zweier Carabidenarten (*P. angustatus* Dft. und *P. oblongopunctatus* F.) an ihre verschiedenen Lebensräume. - Z. wiss. Zool. 174: 83-176.
Palm, T., 1981. Skalbaggsstudier vid ett återbesök i Abisko. - Ent. Tidskr. 102: 65-70.
- 1982. Förändringar i den svenska skalbaggsfaunan. - Ent. Tidskr. 103: 25-32.
Panzer, G. W. F., 1793-1813. Faunae Insectorum Germaniae initia 1-109. - Nürnberg.
Paykull, G., 1790. Monographia Caraborum Sueciae. - Upsaliae, 1-138.
- 1792. Monographia Curculionum Sueciae. - Upsaliae, 1-151.
- 1798. Fauna Suecica. Insecta. Tom. I. - Upsaliae, 1-360.

Perrault, G. G., 1981. Etudes sur la tribu des Bembidiini (Coleoptera, Carabidae). I. Notes sur la classification supraspécifique. - Nouv. Rev. Ent. 11: 237-250.
Perris, E., 1866. Descriptions de quelques insectes nouveaux (Coleoptera). - Ann. Soc. Ent. France (4) 6: 181-196.
Piller, M. & Mitterpacher, L., 1783. Iter per Poseganam Sclavoniae provinciam mensibus Junio et Julio Anno MDCCLXXXII susceptum. - Budae, 1-147.
Pollard, E., 1968. Hedges. IV. A comparison between the Carabidae of a hedge and field site and those of a woodland glade. - J. Appl. Ecol. 5: 649-657.
Pontoppidan, E., 1763. Den Danske Atlas. Tom. I. - Kjøbenhavn, 1-724.
Poppius, B., 1906. Beiträge zur Kenntnis der Coleopteren-Fauna des Lena-Thalen in Ostsibirien. II. Cicindelidae und Carabidae. - Öfvers. Finska Vet. Soc. Förh. 48 (3): 1-65.
Puel, L., 1925. Notes sur les Carabiques (cont.). - Misc. Ent. 28: 1-48.
- 1937. Notes sur les Carabiques (cont.). - Misc. Ent. 38: 89-135.
Putzeys, J., 1846a. Prémices entomologiques, II. - Mém. Soc. R. Sci. Liege 2: 365-417.
- 1846b. Monographie des *Clivina* et genres voisins, précédée d'un tableau synoptique des genres de la tribu des Scaritides. - Mém. Soc. R. Sci. Liege 2: 521-663.
Ragusa, E., 1884. Catalogo ragionato dei Coleotteri di Sicilia. - Nat. Sicil. 4: 1-6.
Reitter, E., 1894. Uebersicht der mir bekannten *Trichocellus*-Arten. - Deutsche ent. Zschr. 1894: 36-39.
- 1897. Die Arten der Coleopteren-Gattung *Notiophilus* Duméril aus Europa und den angrenzenden Ländern. - Ent. Nachr. 23: 361-364.
- 1900. Bestimmungs-Tabelle der europäischen Coleopteren. XLI. Carabidae: Harpalini und Licinini. - Verh. naturf. Ver. Brünn 38: 33-155.
- 1905. Zur systematischen Gruppeneinteilung des Coleopteren-Genus *Dromius* Bonelli und Übersicht der mir bekannten Arten. - Wien. ent. Zeit. 24: 229-239.
- 1908. Fauna Germanica. Die Käfer des Deutschen Reiches. I. - Stuttgart, 1-248.
Richoux, P., 1972. Ecologie et Ethologie de la faune des fissures intertidales de la Région malouine (1). - Bull. Lab. Marit. Dinard 1: 145-206.
Rossi, P., 1790. Fauna Etrusca. Tom. I. Liburni, 1-272.
Roth, C. D. E., 1898. Bidrag till en bild av Skånes insektfauna (forts.). - Ent. Tidskr. 18: 127-138.
Sahlberg, C. R., 1817-34. Insecta Fennica Pars I. - Aboae et Helsingforsiae, 1-519.
Sahlberg, J., 1870. Anteckningar till Lapplands Coleopterfauna. - Notis. Sällsk. Fauna Flora Fenn. Förh. 11: 385-440.
- 1875. Enumeratio Coleopterorum Carnivorum Fenniae. - Notis. Sällsk. Fauna Flora Fenn. Förh. 14: 41-200.
- 1880. Bidrag till nordvestra Sibiriens Insektfauna. Coleoptera insamlade under expeditionerna till Obi och Jenessej 1876 och 1877. I. Cicindelidae, Carabidae, Hydrophilidae, Gyrinidae, Dryopidae, Georyssidae, Limnichidae, Heteroceridae, Staphylinidae och Micropeplidae. - Kongl. Vet. Akad. Handl. 17 (4): 1-115.
- 1900. Coleoptera nova vel minus cognita Faunae fennicae. - Acta Soc. Fauna Flora Fenn. 19 (3): 1-23.
Sahlberg, R. F., 1834. Novae Coleopterorum fennicorum species. - Bull. Soc. Nat. Moscou 7: 267-280.
- 1844. In Faunam Insectorum Rossicam Symbola novas ad Ochotsk lectas Carabicorum Species continens. - Disp., Helsingfors, 1-69.
Samouelle, G., 1819. The Entomologist's Useful Compendium. - London, 1-496.
Say, T., 1823. Descriptions of Insects of the families Carabici and Hydrocanthari of Latreille, inhabiting North America. - Trans. Amer. Philos. Soc. 2: 1-109.
Schaller, J. G., 1783. Neue Insekten. - Abh. Naturf. Ges. Halle 1: 217-332.

Schauberger, E., 1923. Beiträge zur Kenntnis der paläarktischen *Harpalus*-Arten. - Ent. Anz. 3: 115-118.
- 1926a. Beitrag zur Kenntnis der paläarktischen Harpalinen. - Col. Centralbl. 1: 24-51.
- 1926b. Beitrag zur Kenntnis der paläarktischen Harpalinen, II. - Col. Centralbl. 1: 153-182.
- 1933. Zur Kenntnis der paläarktischen Harpalinen. (13. Beitrag). - Kol. Rdsch. 19: 123-133.
Schaum, H., 1843. Beitrag zur Kenntnis der norddeutschen Salzkäfer - Zschr. Ent. (Germar) 4: 172-193.
- 1860. Naturgeschichte der Insecten Deutschlands. I. Abteilung. Coleoptera. Band 1. - Berlin, 1-791.
Schildknecht, H. & Holoubek, K., 1961. Die Bombardierkäfer und ihre Explosionschemie. - Angew. Chem. 73: 1-7.
Schildknecht, H. & al., 1968. Die Explosionschemie der Bombardierkäfer (Coleoptera, Carabidae). III. Mitt.: Isoliering und Charakterisierung der Explosionskatalysatoren. - Z. Naturforsch. 23 b: 1213-1218.
- 1970. Die Explosionschemie der Bombardierkäfer: Struktur und Eigenschaften der Brennkammerenzyme. - J. Insect Physiol. 16: 749-789.
Schilsky, J., 1888. Beitrag zur Kenntniss der deutschen Käferfauna. - Deutsche ent. Zschr. 32: 177-190.
Schiødte, J., 1837. Forsøg til en monographisk Fremstilling af de i Danmark hidtil opdagede Arter af Insect-Slægten *Amara* Bonelli. - Naturh. Tidsskr. 1: 38-65, 138-171, 242-252.
- 1841. Genera og species af Danmarks Eleutherata. I. Bind. - Kjøbenhavn, 1-612.
- 1861. Danmarks Harpaliner. - Naturh. Tidsskr. (3) 1: 149-192.
Schjøtz-Christensen, B., 1957. The beetle fauna of the Corynephoretum in the ground of the Mols Laboratory. - Natura Jutlandica 6-7: 13-119.
- 1965. Biology and population studies of Carabidae of the Corynephoretum. - Natura Jutlandica 11: 1-173.
- 1966a. Biology of some ground beetles (*Harpalus* Latr.) of the Corynephoretum. - Natura Jutlandica 12: 225-229.
- 1966b. Some notes on the biology of *Bradycellus collaris* Payk. and *B. similis* Dej. (Col., Carabidae). - Natura Jutlandica 12: 230-234.
- 1968. Some notes on the biology and ecology of *Carabus hortensis* L. (Col., Carabidae). - Natura Jutlandica 14: 127-151.
Schmidt-Goebel, H. M., 1846. Faunula Coleopterorum Birmaniae, adjectis nonnullis Bengaliae indigenis. - Prag, 1-94.
Schrank, F., 1781. Enumeratio Insectorum Austriae indigenorum. - Augustae Vindelicorum, 1-548.
- 1798. Fauna Boica. Vol. 1. - Nürnberg, 1-720.
Schönherr, C. J., 1806. Synonymia Insectorum. Band I. Eleutherata oder Käfer. Teil 1. - Stockholm, 1-294.
Seidlitz, G., 1887-91. Fauna Baltica. Die Kaefer (Coleoptera) der deutschen Ostseeprovinzen Russlands. (2nd ed.). - Königsberg, 1-818.
Semenow, A., 1895. Coleoptera nova Rossiae europaeae Caucasique. II. - Hor. Soc. Ent. Ross. 29: 303-327.
Silfverberg, H., 1977. Nomenclatoric notes on Coleoptera Adephaga. - Notul. Ent. 57: 41-44.
Smit, H., 1957. Onderzoek naar het voedsel von *Calathus erratus* Sahlb. en *Calathus ambiguus* Payk. aan de hand van hun magen inhouden. - Ent. Ber. Amsterd. 17: 199-209.
Spaeth, F., 1899. Uebersicht der paläarktischen Arten des Genus *Notiophilus* Duméril. - Verh. zool. -bot. Ges. Wien 49: 510-523.
Sparre-Schneider, J., 1910. Maalselvens insektfauna. I. - Tromsø Mus. Aarsh. 30: 37-216.

Sperk, F., 1835. Beschreibung einiger Coleopteren des südlichen Russland's. - Bull. Soc. Nat. Moscou 8: 151-159.
Stephens, J. F., 1827-28. Illustrations of British Entomology. Mandibulata. Vol. I. - London, 1-188.
- 1828-29. Illustrations of British Entomology. Mandibulata. Vol. II. - London, 1-200.
- 1835. Illustrations of British Entomology. Mandibulata. Vol. V. - London, 1-447.
- 1839. A Manual of British Coleoptera. - London, 1-443.
Steven, C., 1829. Museum Historiae Naturalis Universitatis Caesareae Mosquensis. Pars II. Insecta. - Mosquae, 1-143.
Strand, A., 1935. De fennoskandiske former av *Carabus problematicus* Hbst. (Col.). - Norsk ent. Tidsskr. 4: 56-74.
Ström, H., 1768. Beskrivelse over Norske Insecter. Andet stykke. - Norske Vidensk. Selsk. Skr. 4: 313-371.
- 1788. Nogle Insect-Larver med deres Forvandlinger. - Nye Saml. Norske Vidensk. Selsk. Skr. 2: 375-400.
Sturani, M., 1962. Osservazioni e ricerche biologiche sul genere *Carabus* Linnaeus (sensu lato) (Coleoptera, Carabidae). - Mem. Soc. Ent. Ital. 41: 85-202.
Sturm, J, 1815. Deutschlands Fauna. V. Insecten. Band 3. - Nürnberg, 1-192.
- 1818. Deutschlands Fauna. V. Insecten. Band 4. - Nürnberg, 1-179.
- 1824. Deutschlands Fauna. V. Insecten. Band 5. - Nürnberg, 1-220.
- 1825. Deutschlands Fauna. V. Insecten. Band 6. - Nürnberg, 1-188.
- 1827. Deutschlands Fauna. V. Insecten. Band 7. - Nürnberg, 1-186.
Sømme, L., 1974. The overwintering of *Pelophila borealis* Payk. III. Freezing tolerance. - Norsk ent. Tidsskr. 21: 131-134.
Thiele, H. U., 1964. Experimentelle Untersuchungen über die Ursachen der Biotopbindung bei Carabiden. - Z. Morphol. Ökol. Tiere 53: 387-452.
- 1967. Ein Beitrag zur experimentellen Analyse von Euryökie und Stenökie bei Carabiden. - Z. Morphol. Ökol. Tiere 58: 355-372.
- 1969. The control of larval hibernation and of adult aestivation in the Carabid beetles *Nebria brevicollis* F. and *Patrobus atrorufus* Stroem. - Oecologia (Berl.) 2: 347-361.
- 1977. Carabid Beetles in Their Environments. - Zoophysiol. & Ecol. 10: 1-369.
Thomson, C. G., 1857. Skandinaviens Coleoptera. Häft. I: Carabici. - Lund, 1-64.
- 1859. Skandinaviens Coleoptera. Tom. I. - Lund, 1-290.
- 1867. Skandinaviens Coleoptera. Tom. IX. - Lund, 1-407.
- 1870. Bidrag till Sveriges Insect-fauna. (1). - Opuscula Entomologica 3: 322-340.
- 1872. Bidrag till Sveriges insect-fauna. (2). - Opuscula Entomologica 4: 361-452.
- 1873. Några för Sveriges fauna nya Coleoptera. - Opuscula Entomologica 5: 527-530.
- 1884. Petites notices entomologiques. Deuxième suite. - Ann. Soc. Ent. France (6) 3: cxx-cxxi.
Thunberg, C. P., 1784. Novae Insectorum species descriptae. - Nova Acta R. Soc. Sci. Upsala 4: 1-28.
- 1787. Museum naturalium Academiae Upsaliensis. - Diss., Upsaliae, 43-58.
Treherne, J. E. & Foster, W. A., 1977. Diel activity of an intertidal beetle, *Dicheirotrichus gustavi* Crotch. - J. Anim. Ecol. 46: 127-138.
Tschitschérine, T., 1899a. Carabiques nouveaux ou peu connus. - l'Abeille 29: 45-75, 93-114, 269-283.
- 1899b. Mémoire sur le genre *Trichocellus* (Ganglb.). - Hor. Soc. Ent. Ross. 32: 444-477.
- 1902. Platysmatini (Coleoptera, Carabidae) nouveaux ou peu connus de l'Asie orientale. - Hor. Soc. Ent. Ross. 35: 494-501.
Turin, H., Haeck, J. & Hengeveld, R., 1977. Atlas of the carabid beetles of The Netherlands. -

Verh. Afdel. Natuurk. (2) 68. Amsterdam.
Villa, A. & Villa, J. B., 1833. Coleoptera Europae dupleta in collectione Villa quae pro mutua commutatione offerri possunt. - Mediolani, 32-36.
Villers, C., 1789. Caroli Linnaei Entomologia, Faunae Suecicae descriptionibus aucta. Tom. I. - Lugduni, 1-766.
Vorbringer, G., 1898. *Dromius cordicollis,* nov. spec. - Ent. Nachr. 24: 286-287.
Wagner, H., 1915. Beiträge zur Coleopterenfauna der Mark Brandenburg. - Ent. Mitt. 4: 240-245.
- 1927. Beschreibungen neuer Coleopteren der europäischen Fauna, nebst kritischen Bemerkungen zu bekannten Arten. - Col. Centralbl. 2: 85-97.
Wasner, U., 1979. Zur Ökologie und Biologie sympatrischer *Agonum (Europhilus)*-Arten (Carabidae, Coleoptera). I. Individualentwicklung und Gonadenreifung, Generationsaufbau, Eiproduktion und Fruchtbarkeit. - Zool. Jb. Syst. 106: 105-123.
Waterhouse, G. R., 1833. Monographia Notiophilon Angliae. - Ent. Mag. 1: 202-211.
Weber, F., 1801. Observationes entomologicae. - Kiliae, 1-117.
Wesmael, C., 1835. Revue des Coléoptères de la famille des Carnassiers de Belgique. - Bull. Acad. R. Sci. Bruxelles 2: 47-48.
Wissmann, O. L., 1846. Entomologische Notizen. - Ent. Zeit., Stettin 7: 24-26.
Witzke, G., 1976. Beiträge zur Kenntnis der Biologie und Ökologie des Laufkäfers *Pterostichus (Platysma) niger* Schaller 1783 (Col., Carabidae). - Z. angew. Zool. 63: 145-162.
Zetterstedt, J. W., 1828. Fauna Insectorum Lapponica. - Hammone, 1-563.
- 1838-40. Insecta Lapponica. - Lipsiae, 1-1140.
Zimmermann, Ch., 1832. Ueber die bisherige Gattung *Amara.* - Faunus 1: 5-40.
Østbye, E. & Sømme, L., 1972. The overwintering of *Pelophila borealis* Payk. I. Survival rates and cold- hardiness. - Norsk ent. Tidsskr. 19: 165-168.

Index

Reference is given to the keys and the main treatment. Synonyms are in italics.

Appendix

by H. Silfverberg

FAMILY RHYSODIDAE

Elongate, black or dark brown, generally glabrous beetles, which are easily recognized on the moniliform antennae. Head and pronotum in most species with deep longitudinal grooves. Elytra with at most 7 striae, scutellar stria absent. Antennae comparatively short and broad, strongly moniliform. Legs short, fore tibiae at apex with a pair of curved processes, but without true spurs, middle and hind tibiae with short spurs; male hind tibiae toothed near apex. Abdomen with 6 visible segments, the 1st segment with a visible portion between hind coxae; 2nd and 3rd segments fused.

Larva broad and soft-bodied; urogomphi absent. Most segments dorsally with a pair of tubercles carrying a row of denticles.

Rhysodidae live both as larvae and imagines in decomposing wood. They are known from all zoogeographical regions, showing the greatest diversity in tropical and subtropical areas. The species are generally rare, many are known only from one or a few specimens.

Rhysodidae has formerly been considered a separate family of Adephaga, but lately it has generally been included in Carabidae as a specialized branch, adapted to the life in decaying wood (cf. pt. 1, p. 23). It is listed here as a family from purely practical considerations, not intended as an opinion on its systematic position. For additional information see Bell & Bell (1978-79) and Burakowski (1975).

Genus ***Rhysodes*** Germar, 1822

Rhysodes Germar, 1822, Fauna Ins. Eur. 6: 1.

Type-species: *Rhysodes europaeus* Germar, 1822 (= *sulcatus* Fabricius).

Head with an elongate median lobe, and on each side with an indented temporal lobe. Pronotum with a deep median groove, and on each side with a somewhat shallower discal groove, resulting in four distinct longitudinal ridges. Elytra with 7 well developed striae, not tuberculate at humerus.

Rhysodes sulcatus (Fabricius, 1787)

Cucuius sulcatus Fabricius, 1787, Mant. Ins. 1: 165.
Rhysodes europaeus Germar, 1822, Fauna Ins. Eur. 6: 1.
Rhysodes exaratus Dalman, 1823, Analect. Entomol.: 93.

6.8-8.3 mm. Shining black or dark broxh. Pronotum almost as broad as long, discal groove ending well before fore margin; hind angles acute. Elytral intervals rather flat, marginal stria near apex with 5-7 setae, also near apex with a short side branch bearing 3-4 setae.

Distribution. In our area only found in Sweden, where several specimens were reported in the early 19th century from Värnanäs in Blekinge (in fact Småland); there are also a few contemporary specimens, labelled Bl. and Smol. (Palm 1955). It is not possible to decide whether all specimens came from one locality, but anyway no more recent finds have been reported, and the species must be considered extinct. Otherwise it has been found in France (the Pyrenees), Italy, Poland, Rumania, Ukraine, the Caucasus and Turkey; it was also reported in the 19th century from W. Germany. It is everywhere local and apparently confined to relict areas.

Biology. Living in decaying wood of both conifers and deciduous trees, mainly in fallen trunks in moist places. The biology was thoroughly studied by Burakowski (1975).

References

Bell, R.T. & Bell, J.R. 1978: Rhysodini of the World. Part I. A New Classification of the Tribe, and a synopsis of *Omoglymnius* Subgenus *Nitiglymnius,* New Subgenus (Coleoptera: Carabidae or Rhysodidae). - Quaest. Entomol. 14: 43-88.

\- 1979: Rhysodini of the World. Part II. Revisions of the smaller genera (Coleoptera: Carabidae or Rhysodidae). - Quaest. Entomol. 15: 377-446.

Burakowski, B. 1975: Descriptions of larva and pupa of *Rhysodes sulcatus* (F). (Coleoptera, Rhysodidae) and notes on the bionomy of this species. - Ann. Zool. (Warszawa) 32: 271-287.

Dalman, J.W. 1823: Analecta Entomologica. - Holmiae, 1-108.

Fabricius, J.C. 1787: Mantissa Insectorum. I. - Hafniae, 1-384.

Germar, E.F. 1822: Fauna Insectorum Europae. Fasc. VI. - Halae, 25 tt.

Palm, T. 1955: Coleoptera med isolerad nordeuropeisk förekomst i Sverige. - Opusc. entomol. 20: 105-131.